Introduction to
LOGISTICS ENGINEERING

Introduction to
LOGISTICS ENGINEERING

Edited by
G. Don Taylor

CRC Press
Taylor & Francis Group
Boca Raton London New York

CRC Press is an imprint of the
Taylor & Francis Group, an **informa** business

This material was previously published in *Logistics Engineering Handbook* © Taylor & Francis, 2008.

CRC Press
Taylor & Francis Group
6000 Broken Sound Parkway NW, Suite 300
Boca Raton, FL 33487-2742

First issued in paperback 2019

© 2009 by Taylor & Francis Group, LLC
CRC Press is an imprint of Taylor & Francis Group, an Informa business

No claim to original U.S. Government works

ISBN-13: 978-1-4200-8851-9 (hbk)
ISBN-13: 978-0-367-38626-9 (pbk)

Library of Congress Cataloging-in-Publication Data

Taylor, G. Don.
    Introduction to logistics engineering / editor, G. Don Taylor.
      p. cm.
    " A CRC title."
    Includes bibliographical references and index.
    ISBN 978-1-4200-8851-9 (hardcover : alk. paper)
    1. Systems engineering. 2. Logistics. I. Title.

TA168.T39 2009
658.5--dc22
                                             2008039016

Visit the Taylor & Francis Web site at
http://www.taylorandfrancis.com

and the CRC Press Web site at
http://www.crcpress.com

This handbook is dedicated to my children, Alex and Caroline.

Alex always makes me laugh and he is the best pal I've ever had.
We think so much alike it seems that we are almost the same guy!
My time with him is treasured.

Caroline is the sweetest little person I've ever known.
She has stolen my heart forever and has made the word "Daddy"
my most cherished title.

Contents

Preface ix

About the Editor xi

Contributors xiii

1   Logistics from a Historical Perspective 1-1
    Joel L. Sutherland

2   Economic Impact of Logistics 2-1
    Rosalyn A. Wilson

3   Logistics Engineering Tool Chest 3-1
    Dušan Teodorović and Katarina Vukadinović

4   Logistics Metrics 4-1
    Thomas L. Landers, Alejandro Mendoza, and John R. English

5   Facilities Location and Layout Design 5-1
    Benoit Montreuil

6   Inventory Control Theory: Deterministic and Stochastic Models 6-1
    Lap Mui Ann Chan and Mustafa Karakul

7   Material Handling System 7-1
    Sunderesh S. Heragu

8   Warehousing 8-1
    Gunter P. Sharp

9   Distribution System Design 9-1
    Marc Goetschalckx

10 Transportation Systems Overview 10-1
    Joseph Geunes and Kevin Taaffe

11 Logistics in Service Industries 11-1
    Manuel D. Rossetti

12 Logistics as an Integrating System's Function 12-1
    Benjamin S. Blanchard

Index I-1

Preface

Logistics activities are critical integrating functions in any type of business. Annual expenditures on logistics in the United States alone are equivalent to approximately 10% of the U.S. gross domestic product. Logistics expenditures represent an even larger percentage of the world economy. Thus, achieving state-of-the-art excellence in logistics functions, and attaining the inherent cost reductions associated with outstanding logistics efforts, are very important in terms of competitiveness and profitability. As logistics tools evolve in comprehensiveness and complexity and as the use of such tools becomes more pervasive in industry, it is increasingly difficult to maintain a position of leadership in logistics functions. In spite of the importance of the topic, logistics education often lags industry requirements, especially in terms of engineering-based needs. This book seeks to fill this void by providing a brief but comprehensive volume that could be effectively used as an engineering textbook or as a professional reference.

This book is based on the CRC Press *Logistics Engineering Handbook*. It is designed to be a brief version of that book that covers only the most basic material in the field of logistics engineering. The original handbook utilizes 30 chapters divided into 5 major sections to cover introductory materials, logistics activities, enabling technologies, and emerging trends. It also has a full section on topics in transportation management. This book utilizes only 12 of those chapters and covers only introductory materials and major logistics activities. It is a more suitable book for those persons who wish to have only a good overview of the topic. The chapters selected for inclusion in this book are written by leading experts in their fields, and they represent the core elements of the original book.

Although this book is not organized into sections, the reader should find that its layout and order make sense. The book begins by discussing logistics from a historical perspective and by introducing the reader to the importance of logistics from an economic viewpoint. Next, the basic tools required for the study and practice of logistics are discussed, along with the metrics that can be used to evaluate progress toward logistics goals. Following these introductory chapters, the book delves into activities that commonly fill the workdays of logisticians. These activities include facilities location and layout design, inventory control theory, material handling systems, warehousing, distribution center design, and transportation system design. Finally, the book discusses logistics activities in service industries and ends with an excellent chapter on logistics as an integrating systems function.

In spite of the growing importance of logistics as a necessary condition for business success, no comprehensive engineering-oriented handbook existed to support educational and reference needs for this topic prior to the publication of the *Logistics Engineering Handbook*. Although colleges and universities are starting to pay greater attention to logistics, business schools seem to be well ahead of engineering schools in terms of the development of educational materials, degree programs, and continuing education for logisticians. While business schools produce very capable logisticians, there is certainly also a great need for more technical logisticians, whether they come from industrial, systems, or even civil engineering or related programs. The comprehensive *Logistics Engineering Handbook* supports education and reference needs for the more technically oriented logisticians, but it is perhaps more information than required for those persons seeking a more focused education and reference volume based solely on the basics of the topic. This briefer book is designed to support that constituency. Thus, in addition to the engineering/technical orientation, this book offers a more concise coverage of the critical topics introduced in the original handbook.

As with the *Logistics Engineering Handbook*, a final distinguishing factor for this book is that each chapter includes either a brief 'case study' overview of an industrially motivated problem or a tutorial using fabricated data designed to highlight important issues. In most cases, this is a discussion that focuses on applications of one or more topics discussed in the chapter, in the form of either a separate section or as a "breakout" at the end of the chapter. In some cases, the case study environment is embedded within the chapter so that key points can be illustrated with actual case data throughout the chapter. This feature helps to ensure that the topics are relevant and timely in terms of industry needs. It also enables the reader to see direct application of the techniques presented in the chapters. Furthermore, having a required case study in every chapter served as a reminder to the contributing authors that this book has been designed to be a useful teaching and reference tool, not a forum for theoretical work.

The book should be equally useful as either a textbook or as part of a professional reference library. Beginning with the initial chapters, the book can be used as either a course introduction or as a professional refresher. The comprehensive coverage of logistics activities and topics presented subsequently is likewise useful in either a classroom or business setting. Hopefully, the reader will agree that the chapters in this book have been written by the world's leading experts in their fields and that the book provides a "one-stop shopping" location for the most basic logistics engineering reference materials.

About the Editor

G. Don Taylor, Jr. is the Charles O. Gordon Professor and Department Head of the Grado Department of Industrial and Systems Engineering at Virginia Polytechnic Institute and State University in Blacksburg, Virginia. In addition to leading this distinguished department, he has broad-based research interests in several aspects of logistics systems. He has particular interest in seeking state-of-the-art solutions to large-scale, applied logistics problems using simulation and optimization techniques. His recent work has been primarily in the truckload trucking and barge transportation industries.

Prior to joining Virginia Polytechnic Institute and State University, Professor Taylor held the Mary Lee and George F. Duthie Endowed Chair in Engineering Logistics at the University of Louisville, where he was co-founder of a multi-university center, the Center for Engineering Logistics and Distribution. He has also held the rank of full professor at the University of Arkansas, where he was also the Arkansas Director of The Logistics Institute. He has held a visiting position at Rensselaer Polytechnic Institute and industrial positions at Texas Instruments and Digital Equipment Corporation.

Professor Taylor has a PhD in Industrial Engineering and Operations Research from the University of Massachusetts and MSIE and BSIE degrees from the University of Texas at Arlington. He has served as Principal Investigator (PI) or Co-PI on approximately 70 funded projects and has written more than 200 technical papers. This handbook is his ninth edited book or proceedings. He is a registered professional engineer in Arkansas and an active leader in the field of industrial and systems engineering.

Contributors

Benjamin S. Blanchard
Virginia Tech
Blacksburg, Virginia

Lap Mui Ann Chan
Virginia Tech
Blacksburg, Virginia

John R. English
Kansas State University
Manhattan, Kansas

Joseph Geunes
University of Florida
Gainesville, Florida

Marc Goetschalckx
Georgia Tech
Atlanta, Georgia

Sunderesh S. Heragu
University of Louisville
Louisville, Kentucky

Mustafa Karakul
York University, Ontario
Toronto, Canada

Thomas L. Landers
Oklahoma University
Norman, Oklahoma

Alejandro Mendoza
University of Arkansas
Fayetteville, Arkansas

Benoit Montreuil
University of Laval, Quebec
Quebec City, Canada

Manuel D. Rossetti
University of Arkansas
Fayetteville, Arkansas

Gunter P. Sharp
Georgia Tech
Atlanta, Georgia

Joel L. Sutherland
Lehigh University
Bethlehem, Pennsylvania

Kevin Taaffe
Clemson University
Clemson, South Carolina

Dušan Teodorović
University of Belgrade
Belgrade, Serbia
and
Virginia Tech
Blacksburg, Virginia

Katarina Vukadinović
University of Belgrade
Belgrade, Serbia

Rosalyn A. Wilson
R. Wilson, Inc.
Sterling, Virginia

1

Logistics from a Historical Perspective

1.1   Defining Logistics 1-1
      Definition of Logistics Management • Definition of
      Supply Chain Management
1.2   Business Logistics and Engineering Logistics 1-2
1.3   Historical Examples of Military Logistics 1-3
      Alexander the Great • The Romans • Napoleon in
      Russia • World War I • World War II • The Korean
      War • Vietnam • Today
1.4   Emergence of Logistics as a Science 1-8
1.5   Case Study: The Gulf War 1-9
      Background • Lessons Learned from the Gulf War
      • Applying Lessons Learned from the Gulf War

Joel L. Sutherland
Lehigh University

1.1   Defining Logistics

Logistics is a word that seems to be little understood, if at all, by nearly anyone not directly associated with this professional and very important discipline. Many, when hearing someone say they work in the logistics field, associate it with some quantitative, technological, or mathematical practice. Some even confuse *logistics* with the study of language (i.e., *linguistics*). The fact is, logistics is a very old discipline that has been, currently is, and always will be, critical to our everyday lives.

The origin of the term logistics comes from the French word "logistique," which is derived from "loger" meaning quarters (as in quartering troops). It entered the English language in the nineteenth century.

The practice of logistics in the military sector has been in existence for as long as there have been organized armed forces and the term describes a very old practice: the supply, movement, and maintenance of an armed force both in peacetime and in battle conditions. Logistics considerations are generally built into battle plans at an early stage, for it is logistics that determine the forces that can be delivered to the theater of operations, what forces can be supported once there, and what will then be the tempo of operations. Logistics is not only about the supply of materiel to an army in times of war, it also includes the ability of the national infrastructure and manufacturing base to equip, support and supply the armed forces, the national transportation system to move the forces to be deployed, and its ability to resupply that force once they are deployed.

The practice of logistics in the business sector, starting in the later half of the twentieth century, has been increasingly recognized as a critical discipline. The first professional association of logisticians was formed in 1963, when a group of practitioners and academicians formed the National Council of Physical Distribution Management, which in 1985 became the Council of Logistics Management, and then in 2004 the Council of Supply Chain Management Professionals ("The Council"). Today, this organization has thousands of members around the world. A sister organization, The International Society of Logistics (or SOLE), was founded in 1966 as the Society of Logistics Engineers. Today, there are numerous professional associations throughout the world with essentially the same objectives: to conduct research, provide education, and disseminate knowledge for the advancement of the logistics discipline worldwide.

The Council, early on, recognized that there was confusion in the industry regarding the meaning of the term logistics. Over the years, they have provided, and adjusted to changing needs, a definition of logistics that is the most widely accepted definition worldwide. Just as important, they recognized that the relationship between logistics and supply chain management was not clearly understood by those who used these terms—often interchangeably. The Council struggled with the development of a broader definition of logistics and its relationship to supply chain management that would be widely accepted by practitioners around the world. In 2003, the Council published the following definitions, and boundaries and relationships, for logistics and supply chain management:

1.1.1 Definition of Logistics Management

Logistics management is that part of supply chain management that plans, implements, and controls the efficient, effective forward and reverse flow and storage of goods, services, and related information between the point of origin and the point of consumption in order to meet customers' requirements.

1.1.1.1 Logistics Management—Boundaries and Relationships

Logistics management activities typically include inbound and outbound transportation management, fleet management, warehousing, materials handling, order fulfillment, logistics network design, inventory management, supply–demand planning, and management of third-party logistics services providers. To varying degrees, the logistics function also includes sourcing and procurement, production planning and scheduling, packaging and assembly, and customer service. It is involved in all levels of planning and execution—strategic, operational, and tactical. Logistics management is an integrating function, which coordinates and optimizes all logistics activities, as well as integrates logistics activities with other functions including marketing, sales manufacturing, finance, and information technology.

1.1.2 Definition of Supply Chain Management

Supply chain management encompasses the planning and management of all activities involved in sourcing and procurement, conversion, and all logistics management activities. Importantly, it also includes coordination and collaboration with channel partners, which can be suppliers, intermediaries, third-party service providers, and customers. In essence, supply chain management integrates supply and demand management within and across companies.

1.1.2.1 Supply Chain Management—Boundaries and Relationships

Supply chain management is an integrating function with primary responsibility for linking major business functions and business processes within and across companies into a cohesive and high-performing business model. It includes all of the logistics management activities stated earlier, as well as manufacturing operations, and it drives coordination of processes and activities with and across marketing, sales, product design, finance, and information technology.

1.2 Business Logistics and Engineering Logistics

Before moving on, it is probably helpful to understand the differences that exist between business logistics and engineering logistics. The fact is, there are few, if any, significant differences between the two except that logistics engineers are often charged with handling the more "mathematical" or "scientific" applications in logistics. For example, the business logistician might be concerned with building information systems to support supply chain management, whereas the logistics engineer might be looking for an optimal solution to a vehicle routing problem within defined time windows. This is important to understand as examples are provided throughout the remainder of this chapter.

1.3 Historical Examples of Military Logistics

Without supplies, no army is brave—Frederick II of Prussia, in his *Instruction for his Generals,* 1747

Business logistics is essentially an offshoot of military logistics. So it behooves us to look at the military side of the logistical coin first. For war is not just about tactics and strategy. War is very often about logistics.

Looking at most wars throughout history, a point can be identified at which the victory of one side could no longer be prevented except by a miracle—a point after which the pendulum was tipped heavily to one side and spending less and less time on the other. Logistics is absolutely the main factor that tends to tip the pendulum. The following examples illustrate the importance of logistics in military campaigns of the past.

1.3.1 Alexander the Great

Alexander the Great and his father Philip recognized the importance and improved upon the art of logistics in their time. Philip realized that the vast baggage train that traditionally followed an army limited the mobility of his forces. In order to compensate he made the troops carry their own weapons, armor, and some provisions while marching, minimizing the need for a transportation infrastructure. Oxen and oxcarts were not used as they were in many other campaigns during earlier "ancient" times. Oxen could achieve a speed of only 2 miles per hour, their hooves were unsuitable for carrying goods for long distances, and they could not keep up with the army's daily marches, which averaged 15 miles per day. The army did not use carts or servants to carry supplies, as was the practice of contemporary Greek and Roman armies; horses, camels, and donkeys were used in Alexander's baggage train because of their speed and endurance. As necessary, road builders preceded the army on its march to keep the planned route passable.

Alexander also made extensive use of shipping, with a reasonable sized merchant ship able to carry around 400 tons, while a horse could carry 200 lbs (but needed to eat 20 lbs of fodder a day, thus consuming its own load every 10 days). He never spent a winter or more than a few weeks with his army on campaign away from a sea port or navigable river. He even used his enemy's logistics weaknesses against them, as many ships were mainly configured for fighting but not for endurance, and so Alexander would blockade the ports and rivers the Persian ships would use for supplies, thus forcing them back to base. He planned to use his merchant fleet to support his campaign in India, with the fleet keeping pace with the army, while the army would provide the fleet with fresh water. However, the monsoons were heavier than usual, and prevented the fleet from sailing. Alexander lost two-thirds of his force, but managed to get to a nearby port where he reprovisioned. The importance of logistics was central to Alexander's plans, indeed his mastery of it allowed him to conduct the longest military campaign in history. At the farthest point reached by his army, the river Beas in India, his soldiers had marched 11,250 miles in eight years. Their success depended on his army's ability to move fast by depending on comparatively few animals, by using the sea wherever possible, and on good logistic intelligence.

1.3.2 The Romans

The Roman legions used techniques broadly similar to the old methods (large supply trains, etc.), however, some did use those techniques pioneered by Philip and Alexander, most notably the Roman consul Marius. The Romans' logistics were helped, of course, by the superb infrastructure, including the roads they built as they expanded their empire. However, with the decline in the Western Roman Empire in AD fifth century, the art of warfare degenerated, and with it, logistics was reduced to the level of pillage and plunder. It was with the coming of Charlemagne in AD eighth century, that provided the basis for feudalism, and his use of large supply trains and fortified supply posts called "burgs," enabled him to campaign up to 1000 miles away, for extended periods.

The Eastern Roman (Byzantine) Empire did not suffer from the same decay as its western counterpart. It adopted a defensive strategy that, in many ways, simplified their logistics operations. They had interior lines of communication, and could shift base far easier in response to an attack, than if they were in conquered territory—an important consideration due to their fear of a two-front war. They used shipping and considered it vital to keep control of the Dardanelles, Bosphorous, and Sea of Marmara; and on campaign made extensive use of permanent magazines (i.e., warehouses) to supply troops. Hence, supply was still an important consideration, and thus logistics were fundamentally tied up with the feudal system—the granting of patronage over an area of land, in exchange for military service. A peacetime army could be maintained at minimal cost by essentially living off the land, useful for princes with little hard currency, and allowed the man-at-arms to feed himself, his family, and retainers from what he grew on his own land and given to him by the peasants.

1.3.3 Napoleon in Russia

As the centuries passed, the problems facing an army remained the same: sustaining itself while campaigning, despite the advent of new tactics, of gunpowder and the railway. Any large army would be accompanied by a large number of horses, and dry fodder could only really be carried by ship in large amounts. So campaigning would either wait while the grass had grown again, or pause every so often. Napoleon was able to take advantage of the better road system of the early 19th century, and the increasing population density, but ultimately still relied upon a combination of magazines and foraging. While many Napoleonic armies abandoned tents to increase speed and lighten the logistics load, the numbers of cavalry and artillery pieces (pulled by horses) grew as well, thus defeating the objective. The lack of tents actually increased the instance of illness and disease, putting greater pressure on the medical system, and thereby increasing pressure on the logistics system because of the larger medical facilities required and the need to expand the reinforcement system.

There were a number of reasons that contributed to Napoleon's failed attempt to conquer Russia in 1812. Faulty logistics is considered a primary one. Napoleon's method of warfare was based on rapid concentration of his forces at a key place to destroy his enemy. This boiled down to moving his men as fast as possible to the place they were needed the most. To do this, Napoleon would advance his army along several routes, merging them only when necessary. The slowest part of any army at the time was the supply trains. While a soldier could march 15–20 miles a day, a supply wagon was generally limited to about 10–12 miles a day. To avoid being slowed down by the supply trains, Napoleon insisted that his troops live as much as possible off the land. The success of Napoleon time after time in Central Europe against the Prussians and the Austrians proved that his method of warfare worked. However, for it to work, the terrain had to cooperate. There had to be a good road network for his army to advance along several axes and an agricultural base capable of supporting the foraging soldiers.

When Napoleon crossed the Nieman River into Russia in June 1812, he had with him about 600,000 men and over 50,000 horses. His plan was to bring the war to a conclusion within 20 days by forcing the Russians to fight a major battle. Just in case his plans were off, he had his supply wagons carry 30 days of food. Reality was a bit different. Napoleon found that Russia had a very poor road network. Thus he was forced to advance along a very narrow front. Even though he allowed for a larger supply train than usual, food was to be supplemented by whatever the soldiers could forage along the way. But this was a faulty plan. In addition to poor roads, the agricultural base was extremely poor and could not support the numbers of soldiers that would be living off the land. Since these 600,000 men were basically using the same roads, the first troops to pass by got the best food that could easily be foraged. The second troops to go by got less, and so forth. If you were at the rear, of course there would be little available. The Russians made the problem worse by adopting a scorched earth policy of destroying everything possible as they retreated before the French. As time went by, soldiers began to straggle, due to having to forage further away from the roads for food and weakness from lack of food.

The situation was just as bad for the horses. Grazing along the road or in a meadow was not adequate to maintain a healthy horse. Their food had to be supplemented with fodder. The further the army went into Russia, the less fodder was available. Even the grass began to be thinned out, for like food the first horses had the best grazing, and those bringing up the rear had it the worse. By the end of the first month, over 10,000 horses had died!

Poor logistics, leading to inadequate food supplies and increasingly sick soldiers, decimated Napoleon's army. By the time Napoleon had reached Moscow in September, over 200,000 of his soldiers were dead and when the army crossed into Poland in early December, less than 100,000 exhausted, tattered soldiers remained of the 600,000 proud soldiers who had crossed into Russia only five months before.

1.3.4 World War I

World War I was unlike anything that had happened before. Not only did the armies initially outstrip their logistics systems with the amount of men, equipments, and horses moving at a fast pace, but they totally underestimated the ammunition requirements, particularly for artillery. On an average, ammunition was consumed at ten times the prewar estimates, and the shortage of ammunition posed a serious issue, forcing governments to vastly increase ammunition production. But rather than the government of the day being to blame, it was faulty prewar planning, for a campaign on the mainland of Europe, for which the British were logistically unprepared. Once the war became trench bound, supplies were needed to build fortifications that stretched across the whole of the Western Front. Given the scale of the casualties involved, the difficulty in building up for an attack (husbanding supplies), and then sustaining the attack once it had started (if any progress was made, supplies had to be carried over the morass of "no-man's land"), it was no wonder that the war in the west was conducted at a snail's pace, given the logistical problems.

It was not until 1918, that the British, learning the lessons of the previous four years, finally showed how an offensive should be carried out, with tanks and motorized gun sleds helping to maintain the pace of the advance, and maintain supply well away from the railheads and ports. World War I was a milestone for military logistics. It was no longer true to say that supply was easier when armies kept on the move due to the fact that when they stopped they consumed the food, fuel, and fodder needed by the army. From 1914, the reverse applied, because of the huge expenditure of ammunition, and the consequent expansion of transport to lift it forward to the consumers. It was now far more difficult to resupply an army on the move. While the industrial nations could produce huge amounts of war materiel, the difficulty was in keeping the supplies moving forward to the consumer.

1.3.5 World War II

World War II was global in size and scale. Not only did combatants have to supply forces at ever greater distances from the home base, but these forces tended to be fast moving and voracious in their consumption of fuel, food, water, and ammunition. Railways proved indispensable, and sealift and airlift made ever greater contributions as the war dragged on (especially with the use of amphibious and airborne forces, as well as underway replenishment for naval task forces). The large-scale use of motorized transport for tactical resupply helped maintain the momentum of offensive operations, and most armies became more motorized as the war progressed. After the fighting had ceased, the operations staffs could relax to some extent, whereas the logisticians had to supply not only the occupation forces, but also relocate those forces that were demobilizing, repatriate Prisoners of War, and feed civil populations of often decimated countries.

World War II was, logistically, as in every other sense, the most testing war in history. The cost of technology had not yet become an inhibiting factor, and only a country's industrial potential and access to raw materials limited the amount of equipment, spares, and consumables a nation could produce. In this regard, the United States outstripped all others. Consumption of war material was never a problem for the United States and its allies. Neither was the fighting power of the Germans diminished by their huge expenditure of war material, nor the strategic bomber offensives of the Allies. They conducted a stubborn, often brilliant defensive strategy for two-and-a-half years, and even at the end, industrial production was still rising. The principal logistic legacy of World War II was the expertise in supplying far-off operations and a sound lesson in what is, and what is not, administratively possible.

During World War II, America won control of the Atlantic and Pacific oceans from the German and Japanese navies, and used its vast wartime manufacturing base to produce, in 1944, about 50 ships, 10 tanks, and 5 trained soldiers for every one ship, tank, and soldier the Axis powers put out. German soldiers captured by Americans in North Africa expressed surprise at the enormous stockpiles of food, clothing, arms, tools, and medicine their captors had managed to bring over an ocean to Africa in just a few months. Their own army, though much closer to Germany than the American army was to America, had chronic shortages of all vital military inventory, and often relied on captured materiel.

Across the world, America's wartime ally, the Soviet Union, was also outproducing Germany every single year. Access to petroleum was important—while America, Britain, and the Soviet Union had safe and ready access to sources of petroleum, Germany and Japan obtained their own from territories they had conquered or pressed into alliance, and this greatly hurt the Axis powers when these territories were attacked by the Allies later in the war. The 1941 Soviet decision to physically move their manufacturing capacity east of the Ural mountains and far from the battlefront took the heart of their logistical support out of the reach of German aircraft and tanks, while the Germans struggled all through the war with having to convert Soviet railroads to a gauge their own trains could roll on, and with protecting the vital converted railroads, which carried the bulk of the supplies German soldiers in Russia needed, from Soviet irregulars and bombing attacks.

1.3.6   The Korean War

The Korean War fought between the U.S.-led coalition forces against the Communists offered several lessons on the importance of logistics. When the North Korean Army invaded South Korea on June 25, 1950, South Korea, including the United States, was caught by surprise. Although there were signs of an impending North Korean military move, these were discounted as the prevailing belief was that North Korea would continue to employ guerrilla warfare rather than military forces.

Compared to the seven well-trained and well-equipped North Korea divisions, the Republic of Korea (ROK) armed forces were not in a good state to repel the invasion. The U.S. 8th Army, stationed as occupation troops in Japan, was subsequently given permission to be deployed in South Korea together with the naval and air forces already there, covering the evacuation of Americans from Seoul and Inchon. The U.S. troops were later joined by the UN troops and the forces put under U.S. command.

In the initial phase of the war, the four divisions forming the U.S. 8th Army were not in a state of full combat readiness. Logistics was also in a bad shape: for example, out of the 226 recoilless rifles in the U.S. 8th Army establishment, only 21 were available. Of the 18,000 jeeps and 4×4 trucks, 55% were unserviceable. In addition, only 32% of the 13,800 6×6 trucks available were functional.

In the area of supplies, the stock at hand was only sufficient to sustain troops in peacetime activities for about 60 days. Although materiel support from deactivated units was available, they were mostly unserviceable. The lack of preparedness of the American troops was due to the assumptions made by the military planners that after 1945 the next war would be a repeat of World War II. However, thanks to the availability of immense air and sea transport resources to move large quantities of supplies, they recovered quickly.

As the war stretched on and the lines of communication extended, the ability to supply the frontline troops became more crucial. By August 4, 1950, the U.S. 8th Army and the ROK Army were behind the Nakton River, having established the Pusan perimeter. While there were several attempts by the North Koreans to break through the defense line, the line held. Stopping the North Koreans was a major milestone in the war. By holding on to the Pusan perimeter, the U.S. Army was able to recuperate, consolidate, and grow stronger.

This was achieved with ample logistics supplies received by the U.S. Army through the port at Pusan. The successful logistics operation played a key role in allowing the U.S. Army to consolidate, grow, and carry on with the subsequent counteroffensive. Between July 2, 1950 and July 13, 1950, a daily average of 10,666 tons of supplies and equipment were shipped and unloaded at Pusan.

The Korean War highlights the need to maintain a high level of logistics readiness at all times. Although the U.S. 8th Army was able to recover swiftly thanks to the availability of vast U.S. resources, the same cannot be said for other smaller armies. On hindsight, if the U.S. 8th Army had been properly trained and logistically supported, they would have been able to hold and even defeat the invading North Koreans in the opening phase of the war. The war also indicates the power and flexibility of having good logistics support as well as the pitfalls and constraints due to their shortage.

1.3.7 Vietnam

In the world of logistics, there are few brand names to match that of the Ho Chi Minh Trail, the secret, shifting, piecemeal network of jungle roadways that helped the North win the Vietnam War.

Without this well-thought-out and powerful logistics network, regular North Vietnamese forces would have been almost eliminated from South Vietnam by the American Army within one or two years of American intervention. The Ho Chi Minh Trail enabled Communist troops to travel from North Vietnam to areas close to Saigon. It has been estimated that the North Vietnamese troops received 60 tons of aid per day from this route. Most of this was carried by porters. Occasionally bicycles and horses would also be used.

In the early days of the war it took six months to travel from North Vietnam to Saigon on the Ho Chi Minh Trail. But the more people who traveled along the route the easier it became. By 1970, fit and experienced soldiers could make the journey in six weeks. At regular intervals along the route, the North Vietnamese troops built base camps. As well as providing a place for them to rest, the base camps provided medical treatment for those who had been injured or had fallen ill on the journey.

From the air the Ho Chi Minh Trail was impossible to be identified and although the United States Air Force tried to destroy this vital supply line by heavy bombing, they were unable to stop the constant flow of men and logistical supplies.

The North Vietnamese also used the Ho Chi Minh Trail to send soldiers to the south. At times, as many as 20,000 soldiers a month came from Hanoi through this way. In an attempt to stop this traffic, it was suggested that a barrier of barbed wire and minefields called the McNamara Line should be built. This plan was abandoned in 1967 after repeated attacks by the North Vietnamese on those involved in constructing this barrier.

The miracle of the Ho Chi Minh Trail "logistics highway" was that it enabled the "impossible" to be accomplished. A military victory is not determined by how many nuclear weapons can be built, but by how much necessary materiel can be manufactured and delivered to the battlefront. The Ho Chi Minh Trail enabled the steady, and almost uninterrupted, flow of logistics supplies to be moved to where it was needed to ultimately defeat the enemy.

1.3.8 Today

Immediately after World War II, the United States provided considerable assistance to Japan. In the event, the Japanese have become world leaders in management philosophies that have brought about the greatest efficiency in production and service. From organizations such as Toyota came the then revolutionary philosophies of Just in Time (JIT) and Total Quality Management (TQM). From these philosophies have arisen and developed the competitive strategies that world class organizations now practice. Aspects of these that are now considered normal approaches to management include kaizen (or continuous improvement), improved customer–supplier relationships, supplier management, vendor managed inventory, collaborative relationships between multiple trading partners, and above all recognition that there is a supply chain along which all efforts can be optimized to enable effective delivery of the required goods and services. This means a move away from emphasizing functional performance and a consideration of the whole supply chain as a total process. It means a move away from the silo mentality to thinking and managing outside the functional box. In both commercial and academic senses the recognition that supply chain management is an enabler of competitive advantage is increasingly accepted. This has resulted in key elements being seen as best practice in their own right, and includes value for money, partnering, strategic procurement policies, integrated supply chain/network management, total cost of ownership, business process reengineering, and outsourcing.

The total process view of the supply chain necessary to support commercial business is now being adopted by, and adapted within, the military environment. Hence, initiatives such as "Lean Logistics" and "Focused Logistics" as developed by the U.S. Department of Defense recognize the importance of logistics within a "cradle-to-grave" perspective. This means relying less on the total integral stockholding and transportation systems, and increasing the extent to which logistics support to military operations is outsourced to civilian contractors—as it was in the 18th century. From ancient days to modern times, tactics and strategies have received the most attention from amateurs, but wars have been won by logistics.

1.4 Emergence of Logistics as a Science

In 1954, Paul Converse, a leading business and educational authority, pointed out the need for academicians and practitioners to examine the physical distribution side of marketing. In 1962, Peter Drucker indicated that distribution was the "last frontier" and was akin to the "dark continent" (i.e., it was an area that was virtually unexplored and, hence, unknown). These and other individuals were early advocates of logistics being recognized as a science. For the purpose of this section we define the science of logistics as, the study of the physical movement of product and services through the supply chain, supported by a body of observed facts and demonstrated measurements systematically documented and reported in recognized academic journals and publications.

In the years following the comments of Converse and Drucker, those involved in logistics worked hard to enlighten the world regarding the importance of this field. At the end of the twentieth century, the science of logistics was firmly in place. Works by Porter and others were major contributors in elevating the value of logistics in strategic planning and strategic management. Other well-known writers, such as Heskett, Shapiro, and Sharman, also helped elevate the importance of logistics through their writings in the most widely read and respected business publications. Because these pioneers were, for the most part, outsiders (i.e., not logistics practitioners) they were better able to view logistics from a strategic and unbiased perspective.

The emergence of logistics as a science has been steady and at times even spectacular. Before the advent of transportation deregulation in the 1980s, particularly in the 1960s and 1970s, "traffic managers" and then "distribution managers" had the primary responsibility for moving finished goods from warehouses to customers on behalf of their companies. Little, if any, attention was given to managing the inbound flows. Though many of these managers no doubt had the capacity to add significant value to their organization, their contribution was constrained by the strict regulatory environment in which they operated. That environment only served to intensify a silo mentality that prevailed within many traffic, and other logistics related, departments.

The advent of transport deregulation in the 1980s complemented, and in many cases accelerated, a parallel trend taking place—the emergence of logistics as a recognized science. The rationale behind this was that transportation and distribution could no longer work in isolation of those other functional areas involved in the flow of goods to market. They needed to work more closely with other departments such as purchasing, production planning, materials management, and customer service as well as supporting functions such as information systems and logistics engineering. The goal of logistics management, a goal that to this day still eludes many organizations, was to integrate these related activities in a way that would add value to the customer and profit to the bottom line.

In the 1990s, many leading companies sought to extend this integration end-to-end within the organization—that is, from the acquisition of raw materials to delivery to the end customer. Technology would be a great enabler in this effort, particularly the enterprise resource planning (ERP) systems and supply chain planning and execution systems that connect the internal supply chain processes. The more ambitious of the leaders sought to extend the connectivity outward to their trading partners both upstream and downstream. They began to leverage Internet-enabled solutions that allowed them to extend connectivity and provide comprehensive visibility over product flow.

As we turned the corner into the 21st century, the rapid evolution of business practices has changed the nature and scope of the job. Logistics professionals today are interacting and collaborating in new ways within their functional area, with other parts of the organization, and with extended partners. As the traditional roles and responsibilities change, the science of logistics is also changing. Logistics contributions in the future will be measured within the context of the broader supply chain.

1.5 Case Study: The Gulf War

1.5.1 Background

The Gulf War was undoubtedly one of the largest military campaigns seen in recent history. The unprecedented scale and complexity of the war presented logisticians with a formidable logistics challenge.

On July 17, 1990, Saddam Hussein accused Kuwait and the United Arab Emirates of overproduction of oil, thereby flooding the world market and decreasing its income from its sole export. Talks between Iraq and Kuwait collapsed on August 1, 1990. On August 2, Iraq, with a population of 21 million, invaded its little neighbor Kuwait, which had a population of less than two million. A few days later, Iraqi troops massed along the Saudi Arabian border in position for attack. Saudi Arabia asked the United States for help. In response, severe economic sanctions were implemented, countless United Nations resolutions passed, and numerous diplomatic measures initiated. In spite of these efforts Iraq refused to withdraw from Kuwait. On January 16, 1991, the day after the United Nations deadline for Iraqi withdrawal from Kuwait expired, the air campaign against Iraq was launched. The combat phase of the Gulf War had started.

There were three phases in the Gulf War worthy of discussion: deployment (Operation Desert Shield); combat (Operation Desert Storm); and redeployment (Operation Desert Farewell). Logistics played a significant role throughout all three phases.

1.5.1.1 Operation Desert Shield

The Coalition's challenge was to quickly rush enough troops and equipment into the theater to deter and resist the anticipated Iraqi attack against Saudi Arabia. The logistical system was straining to quickly receive and settle the forces pouring in at an hourly rate. This build-up phase, Operation Desert Shield, lasted six months. Why the six-month delay? A large part of the answer is supply.

Every general knows that tactics and logistics are intertwined in planning a military campaign. Hannibal used elephants to carry his supplies across the Alps during his invasion of the Roman Empire. George Washington's colonial militias had only nine rounds of gunpowder per man at the start of the

Revolution, but American privateers brought in two million pounds of gunpowder and saltpeter in just one year. Dwight Eisenhower's plans for the June 1944 invasion of Normandy hinged on a massive buildup of war materiel in England. The most brilliant tactics are doomed without the ability to get the necessary manpower and supplies in the right place at the right time.

During the six-month buildup to the Gulf War, the United States moved more tonnage of supplies—including 1.8 million tons of cargo, 126,000 vehicles, and 350,000 tons of ordnance—over a greater distance than during the two-year buildup to the Normandy invasions in World War II.

Besides the massive amount of supplies and military hardware, the logistics personnel also had to deal with basic issues such as sanitation, transport, and accommodation. A number of these requirements were resolved by local outsourcing. For example, Bedouin tents were bought and put up by contracted locals to house the troops; and refrigerated trucks were hired to provide cold drinks to the troops.

Despite the short timeframe given for preparation, the resourceful logistics team was up to the given tasks. The effective logistics support demonstrated in Operation Desert Shield allowed the quick deployment of the troops in the initial phase of the operation. It also provided the troops a positive start before the commencement of the offensive operation.

1.5.1.2 Operation Desert Storm

It began on January 16, 1991, when the U.S. planes bombed targets in Kuwait and Iraq. The month of intensive bombing that followed badly crippled the Iraqi command and control systems. Coalition forces took full advantage of this and on February 24, 1991, the ground campaign was kicked off with a thrust into the heart of the Iraqi forces in central Kuwait. The plan involved a wide flanking maneuver around the right side of the Iraqi line of battle while more mobile units encircled the enemy on the left, effectively cutting lines of supply and avenues of retreat. These initial attacks quickly rolled over Iraqi positions and on February 25, 1991, were followed up with support from various infantry and armored divisions.

To the logisticians, this maneuver posed another huge challenge. To support such a maneuver, two Army Corps' worth of personnel and equipment had to be transported westward and northward to their respective jumping off points for the assault. Nearly 4,000 heavy vehicles were used. The amount of coordination, transport means, and hence the movement control required within the theater, was enormous.

One reason Iraq's army was routed in just 100 hours, with few U.S. casualties, was that American forces had the supplies they needed, where they needed them, when they needed them, and in the necessary quantities.

1.5.1.3 Operation Desert Farewell

It was recognized that the logistical requirements to support the initial buildup phase and the subsequent air and land offensive operations were difficult tasks to achieve. However, the sheer scope of the overall redeployment task at the end of the war was beyond easy comprehension. To illustrate, the King Khalid Military City (KKMC) main depot was probably the largest collection of military equipment ever assembled in one place. A Blackhawk helicopter flying around the perimeter of the depot would take over an hour. While the fighting troops were heading home, the logisticians, who were among the first to arrive at the start of the war, were again entrusted with a less glamorous but important "clean up job." Despite the massive amount of supplies and hardware to be shipped back, the logisticians who remained behind completed the redeployment almost six months ahead of schedule.

Throughout the war, the Commanding General, Norman Schwarzkopf, had accorded great importance to logistics. Major General William G. (Gus) Pagonis was appointed as the Deputy Commanding General for logistics and subsequently given a promotion to a three-star general during the war. This promotion symbolized the importance of a single and authoritative logistical point of contact in the Gulf War. Under the able leadership of General Pagonis, the efficient and effective logistical support system set up in the Gulf War, from deployment phase to the pull-out phase, enabled the U.S.-led coalition forces to achieve a swift and decisive victory over the Iraqi.

Both at his famous press conferences as well as later in his memoirs, Stormin' Norman called Desert Storm a "logistician's war," handing much of the credit for the Coalition's lightning-swift victory to his chief logistician, General Gus Pagonis. Pagonis, Schwarzkopf declared, was an "Einstein who could make anything happen," and, in the Gulf War, did. Likewise, media pundits from NBC's John Chancellor on down also attributed the successful result of the war to logistics.

1.5.2   Lessons Learned from the Gulf War

1.5.2.1   "Precision-Guided" Logistics

In early attempts inside and outside of the Pentagon to assess the lessons learned from the Gulf War, attention has turned to such areas as the demonstrated quality of the joint operations, the extraordinary caliber of the fighting men and women, the incredible efficacy of heavy armor, the impact of Special Forces as part of joint operations on the battlefield, and the success of precision-guided weapons of all kinds. Predictably lost in the buzz over celebrating such successes was the emergence and near-seamless execution of what some have termed "precision-guided" logistics.

Perhaps, this is as it should be. Logistics in war, when truly working, should be transparent to those fighting. Logistics is not glamorous, but it is critical to military success. Logisticians and commanders need to know "what is where" as well as what is on the way and when they will have it. Such visibility, across the military services, should be given in military operations.

1.5.2.2   "Brute Force" Logistics

In 1991, the United States did not have the tools or the procedures to make it efficient. The Gulf War was really the epitome of "brute force" logistics. The notion of having asset visibility—in transit, from factory to foxhole—was a dream. During the Gulf War, the United States did not have reliable information on almost anything. Materiel would enter the logistics pipeline based on fuzzy requirements, and then it could not be readily tracked in the system.

There were situations where supply sergeants up front were really working without a logistics plan to back up the war plan. They lacked the necessary priority flows to understand where and when things were moving. It was all done on the fly, on a daily basis, and the U.S. Central Command would decide, given the lift they had, what the priorities were. Although progress was eventually made, often whatever got into the aircraft first was what was loaded and shipped to the theater. It truly was brute force.

Even when air shipments were prioritized there was still no visibility. Although it is difficult to grasp today, consider a load being shipped and then a floppy disk mailed to the receiving unit in the theater. Whether that floppy disk got where it was going before the ship got there was in question. Ships were arriving without the recipients in the theater knowing what was on them.

Generally speaking, if front-line commanders were not sure of what they had or when it would get there, they ordered more. There were not enough people to handle this flow, and, in the end, far more materiel was sent to the theater than was needed. This was definitely an example of "just-in-case" logistics. When the war ended, the logistics pipeline was so highly spiked that there were still 101 munitions ships on the high seas. Again, it was brute-force logistics.

The result was the oft-referenced "iron mountains" of shipping containers. There was too much, and, worse yet, little, if any, knowledge of what was where. This led, inevitably, to being forced to open something like two-thirds of all of the containers simply to see what was inside. Imagine the difficulty in finding things if you shipped your household goods to your new house using identical unmarked boxes. Since there were a great number of individual users, imagine that the household goods of all of your neighbors also were arriving at your new address, and in the same identical boxes.

That there was this brute force dilemma in the Gulf War was no secret. There just wasn't any other way around it. The technology used was the best available. Desert Storm was conducted using 286-processor technology with very slow transfer rates, without the Internet, without the Web, and

without encrypted satellite information. Telexes and faxes represented the available communication technology.

1.5.2.3 "Flying Blind" Logistics

This was an era of green computer screens, when it took 18 keystrokes just to get to the main screen. When the right screen was brought up, the data were missing or highly suspect (i.e., "not actionable"). In contrast to today, there were no data coming in from networked databases, and there was no software to reconcile things. There were also no radio frequency identification tags. In effect, this was like "flying blind."

In fact, nothing shipped was tagged. Every shipment basically had a government bill of lading attached to it, or there were five or six different items that together had one bill of lading. When those items inevitably got separated, the materiel was essentially lost from the system. Faced with this logistics nightmare, and knowing that there was often a critical need to get particular things to a particular place at a particular time, workarounds were developed.

As a result of our experience in the Gulf War, the Department of Defense (DOD) has subsequently been refining its technologies and testing them through military joint exercises and deployments and contingencies in such places as Bosnia, Kosovo, and Rwanda. Specifically, the DOD has focused on the issue of logistics management and tracking and on how technology can enable improvements in this mission critical area. The DOD has improved its logistics management and tracking through policy directives and by engaging with innovative technology companies in the development and leveraging of technical solutions.

The DOD now has clear knowledge of when things are actually moving—the planes, the ships, what is going to be on them, and what needs to be moved. Communication is now digital and that represents a quantum leap in capability and efficiency from the first war in Iraq. Operators now get accurate information, instantaneously, and where needed. The technology exists to absorb, manage, and precisely guide materiel.

1.5.3 Applying Lessons Learned from the Gulf War

1.5.3.1 Operation Enduring Freedom

While troops raced toward Baghdad in the spring of 2003, digital maps hanging from a wall inside the Joint Mobility Operations Center at Scott Air Force Base, Illinois, blinked updates every four minutes to show the path cargo planes and ships were taking to the Middle East. During the height of the war in Iraq, every one of the military's 450 daily cargo flights and more than 120 cargo ships at sea were tracked on the screen, as was everything stowed aboard them—from Joint Direct Attack Munitions to meals for soldiers.

In rows of cubicles beneath the digital displays, dozens of military and civilian workers from the U.S. Transportation Command (TRANSCOM) looked at the same maps on their computer screens. The maps, along with an extensive database with details on more than five million items and troops in transit, came in handy as telephone calls and e-mail queries poured in from logisticians at ports and airfields in the Persian Gulf: How soon would a spare part arrive? When would the next shipment of meals arrive? When was the next batch of troops due? With just a few mouse clicks, TRANSCOM workers not only could report where a ship or plane was and when it was due to arrive, but also could determine which pallet or shipping container carried what. In many cases, logisticians in the field also could go online, pull up the map and data and answer their own questions.

Vice Admiral Keith Lippert, director of the Defense Logistics Agency (DLA) says the war in Iraq validated a new business model that moves away from "stuffing items in warehouses" to relying on technology and contractors to provide inventory as needed. The agency, which operates separately from TRANSCOM, is responsible for ordering, stocking, and shipping supplies shared across the services. In addition, the Army, Navy, Air Force, and Marines have their own supply operations to ship items unique to each service. The DLA supplied several billion dollars worth of spare parts, pharmaceuticals, clothing and 72 million ready-to-eat meals to troops during the war.

Military logisticians have won high marks for quickly assembling the forces and supplies needed in Iraq. Advances in logistics tracking technology, investments in a new fleet of cargo airplanes and larger ships, and the prepositioning of military equipment in the region allowed troops to move halfway around the world with unprecedented speed. Troops were not digging through containers looking for supplies they had ordered weeks earlier, nor were they placing double and triple orders in hopes that one of their requests would be acted upon, as they did during the Gulf War in 1991. While the military transportation and distribution system may never be as fast or efficient as FedEx or UPS, its reliability has increased over the past decade.

Nonetheless, challenges remain. Several changes to the way troops and supplies are sent to war are under consideration, including:

* Further improvement of logistics information technology systems
* Development of a faster way to plan troop deployments
* Consolidated management of the Defense supply chain

While TRANSCOM has gotten positive reviews for moving troops and supplies to the Middle East, concerns have been raised about how the services moved supplies after they arrived in the field.

Perhaps the most valuable logistics investment during the war was not in expensive cargo aircraft or advanced tracking systems, but in thousands of plastic radio frequency identification labels that cost $150 apiece. The tags, which measure eight inches long by about two inches wide, contain memory chips full of information about when a shipment departed, when it is scheduled to arrive and what it contains. They are equipped with small radio transponders that broadcast information about the cargo's status as it moves around the world. The tags enable the Global Transportation Network to almost immediately update logistics planners on the location of items in the supply chain.

These tags were a key factor in avoiding the equipment pileups in warehouses and at desert outposts that came to symbolize logistics failings during the first Gulf War. The tags also saved hundreds of millions of dollars in shipping costs, logisticians say. For example, British soldiers spent almost a full day of the war searching cargo containers for $3 million in gear needed to repair vehicles. Just as they were about to place a second order for the gear, a U.S. logistician tapped into a logistics tracking system and was able to locate the supplies in the American supply network.

Rapid response to shifting requirements is clearly the fundamental challenge facing all logisticians, as relevant in the commercial sector as it is in the military environment. The commercial logistician requires the same thing that the combatant commander requires: situational awareness. We all need an in-depth, real-time knowledge of the location and disposition of assets.

Indeed, Wal-Mart, arguably the channel master for the world's largest, most globally integrated commercial supply chain, has embarked on a passive RFID initiative that is very similar to the Department of Defense's plans. The retailer mandated that suppliers tag inbound materiel with passive RFID tags beginning at the case and pallet level. Wal-Mart established a self-imposed January 2005 deadline to RFID-enable its North Texas operation, along with 100 of its suppliers. The first full-scale operational test began on April 30, 2005. Based on the success of this initial test Wal-Mart expanded its supplier scope and deployment plan for RFID and by early 2007 reported that some 600 suppliers were RFID-enabled.

While there have been some solid successes early on, there are now many suppliers (in particular the smaller ones) that are dragging their feet on RFID adoption due to an elusive return on investment (ROI). Current generation RFID tags cost about 15 cents, while bar codes cost a fraction of a cent. Suppliers have also had to absorb the cost of buying hardware—readers, transponders, antennas—and software to track and analyze the data. The tags also have increased labor. Bar codes are printed on cases at the factory, but because most manufacturers have yet to adopt RFID, tags have to be put on by hand at the warehouse. The retail giant also experienced difficulties rolling out RFID in their distribution network. Wal-Mart had hoped to have up to 12 of its roughly 137 distribution centers using RFID technology by the end of 2006, but had installed the technology at just five. Now Wal-Mart has shifted gears from their distribution centers to their stores where they believe they will be better able to drive sales for their suppliers and to get product on the shelf, where it needs to be for their customers to buy.

Regardless of where Wal-Mart places their priorities, with this retail giant leading the charge, and driving industry compliance, it is expected that this initiative will have a greater, and more far-reaching, impact on just the retail supply chain. Virtually every industry, in every corner of the planet, will be fundamentally impacted sometime in the not-too-distant future. Clearly the lessons learned in military logistics are being applied to business logistics and as a result engineering logistics.

2

Economic Impact of Logistics

Rosalyn A. Wilson

R. Wilson, Inc.

2.1 Expenditures in the United States and Worldwide **2**-1
2.2 Breakdown of Expenditures by Category **2**-2
    Carrying Costs • Transportation Costs • Administrative
    Costs
2.3 Logistics Productivity over the Past 25 Years **2**-7

2.1 Expenditures in the United States and Worldwide

As the world continues to develop into a homogenized global marketplace the growth in world merchandise trade has outpaced the growth in both global production and the worldwide economy. In 2006, world merchandise trade increased 8%, while the global economy rose only 3.7%.* Globalization has dramatically shifted where logistics dollars are spent as developing countries now account for over one-third of world merchandise exports. Increased world trade means higher demand for logistics services to deliver the goods. Expenditures for logistics worldwide are estimated at well over $4 trillion in 2006 and now account for about 15% to 20% of finished goods cost.† Growth in world merchandise trade, measured as export volume, has exceeded the growth in the worldwide economy, as measured by Gross Domestic Product (GDP), for close to two decades. Although the worldwide economy slowed to some extent in late 2006 and early 2007, trade volumes are predicted to continue to rise well into the next decade.

This phenomenal growth in world trade has profound implications for logistics. In the past five years the demand for shipping has outstripped the capacity in many markets, altering the supply demand equilibrium and pushing up prices. It now costs from 15% to 20% more to move products than it did in 2002. Shifts in global manufacturing as the United States continues to move manufacturing facilities to other global markets with lower labor costs, such as China, India, and South Korea, are redrawing the landscape for transportation strategies. The growth was led by Asia and the so-called transition economies (Central and Eastern Europe and the Russian Trade Federation). In real terms these regions experienced 10–12% growth rates in merchandise exports and imports. China, for instance, has seen the most dramatic trade growth, with a 27% jump in 2006. The World Trade Organization (WTO) recently

---

* World Trade Organization Press Release, "World Trade 2006, Prospects for 2007," April 12, 2007.

† Estimated from a 2003 figure for global logistics of $3.43 trillion. Report from the Ad Hoc Expert Meeting on Logistics Services by the United Nations Conference on Trade and Development's (UNCTAD) Trade and Development Board, Commission on Trade in Goods and Services, and Commodities, Geneva, July 13, 2006.

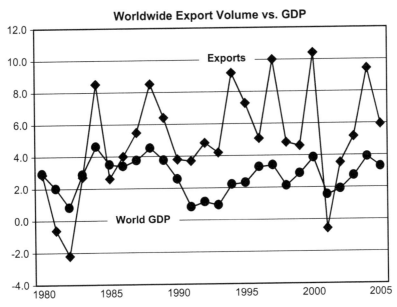

FIGURE 2.1 Worldwide export volume vs. GDP. (From World Trade Organization, International Trade Statistics, 2006.)

reported that China's merchandise exports actually exceeded those of the United States, the market leader, for the second half of 2006. Worldwide export volumes as a percentage of world GDP appear in Figure 2.1.

Studies have shown that total expenditures as a percentage of GDP are generally lower in more efficient industrialized countries, usually 10% or less. Conversely less-developed countries expend a much greater portion of their GDP, 10–20%, on logistics. Where a country falls on the spectrum depends on factors such as the size and disbursement of the population, the level of import and export activity, and the type and amount of infrastructure development. The relative weights for the components of total logistic costs vary significantly by country, with carrying costs accounting for 15–30%, transportation expenditures for another 60–80%, and administrative costs for the remaining 5–10%. Logistics costs in the United States have been holding steady at just under 10% of GDP. The breakout for the components of U.S. logistics costs are 33% for carrying costs, 62% for transportation costs, and about 4% for administrative costs. Additional detail is provided in Figure 2.2.

During 2005, the cost of the U.S. business logistics system increased to $1.18 trillion, or the equivalent of 9.5% of nominal GDP. Logistics costs have gone up over 50% during the last decade. The year 2005 was a year of record highs for many of the components of the model, especially transportation costs, mostly trucking. Transportation costs jumped 14.1% over 2004 levels, and 77.1% during the past decade. Yet, total logistics costs remained below 10% of GDP.

2.2 Breakdown of Expenditures by Category

The cost to move goods encompasses a vast array of activities including supply and demand planning, materials handling, order fulfillment, management of transportation and third-party logistics (3PLs) providers, fleet management, and inventory warehouse management. To simplify, logistics can be defined as the management of inventory in motion or at rest. Transportation costs are those incurred when the inventory is in motion, and inventory carrying costs are those from inventory at rest awaiting the production process or in storage awaiting consumption. The third broad category of logistics cost is administrative costs, which encompass the other costs of carrying out business logistics that is not

2005 U.S. Business Logistics System Cost

		$ Billions
Carrying Costs - $1.763 Trillion All Business Inventory		
Interest		58
Taxes, Obsolescence, Depreciation, Insurance		245
Warehousing		90
	Subtotal	393
Transportation Costs		
Motor Carriers:		
Truck - Intercity		394
Truck - Local		189
	Subtotal	583
Other Carriers:		
Railroads		48
Water I 29 D 5		34
Oil Pipelines		9
Air I 15 D 25		40
Forwarders		22
	Subtotal	153
Shipper Related Costs		8
Logistics Administration		46
TOTAL LOGISTICS COST		1183

FIGURE 2.2 Breakdown of U.S. business logistics system costs. (From 17th Annual State of Logistics Report, Rosalyn Wilson, CSCMP, 2006.)

directly attributable to the first two categories. The cost of the U.S. business logistics system as measured by these three categories was $1,183 billion in 2005.*

2.2.1 Carrying Costs

Carrying costs are the expenses associated with holding goods in storage, whether that be in a warehouse or, as is increasingly done today, in a shipping container, trailer, or railcar. There are three subcomponents that comprise carrying cost. The first is interest and that represents the opportunity cost of money invested in holding inventory. This expense will vary greatly depending on the level of inventory held and the interest rate used. The second subcomponent covers inventory risk costs and inventory service costs and comprises about 62% of carrying cost expense. These are measured by using expenses for obsolescence, depreciation, taxes, and insurance. Obsolescence includes damages to inventory and shrinkage or pilferage, as well as losses from inventory which cannot be sold at value because it was not moved through the system fast enough. In today's fast paced economy with quick inventory turns, obsolescence represents a significant cost to inventory managers. The taxes are the *ad valorem* taxes collected on inventory and will vary with inventory levels. Insurance costs are the premiums paid to protect inventory and mitigate losses. The final subcomponent is warehousing. Warehousing is the cost of storing goods and has traditionally included both public and private warehouses, including those in manufacturing plants. The market today includes a wide variety of storage possibilities from large megadistribution centers, to smaller leased facilities, to container and trailer-storage yards.

---

* Logistics expenditures for the United States have been measured consistently and continuously for the "Annual State of Logistics Report" developed by Robert V. Delaney of Cass Logistics in the mid-1980s and continued today by Rosalyn Wilson. The methodology used by Mr. Delaney was based on a model developed by Nicholas A. Glaskowsky, Jr., James L. Heskett, Robert M. Ivie in *Business Logistics*, 2nd edition, New York, Ronald Press, 1973. The Council for Supply Chain Management Professionals (CSCMP) has sponsored the report since 2004.

In 2005, inventory carrying costs rose 17%—the highest level since 1971. The increase was due to both significantly higher interest rates than in 2004 and a rise in inventories. The average investment in all business inventories was $1.74 trillion, which surpassed 2004's record high by $101 billion. Both the inventory-to-sales ratio and the inventory-to-factory shipments ratio have been rising steadily in recent years. Inventories have been slowly creeping up since 2000, reversing the trend to leaner inventories from the previous decade. The globalization of production has driven the economy away from the lean just-in-time inventory management model of the 1990s. Stocks are increasingly maintained at a higher level in response to longer and sometimes unpredictable delivery times, as well as changes in distribution patterns. Manufacturers and retailers have struggled to achieve optimum inventory levels as they refine their supply chains to mitigate uncertain delivery times, add new sources of supply, and become more adept at shifting existing inventories to where they are most advantageous. On an annualized basis, the value of all business inventory has risen every year since 2001, as depicted graphically in Figure 2.3.

2.2.2 Transportation Costs

Transportation costs are the expenditures to move goods in various states of production. This could include the movement of raw materials to manufacturing facilities, movement of components to be included in the final product, to the movement of final goods to market. Transportation costs are measured by carriers' revenues collected for providing freight services. All modes of transportation are included: trucking, intercity and local; freight rail; water, international and domestic; oil pipeline; both international and domestic airfreight transport; and freight forwarding costs, not included in carrier revenue. Transportation includes movement of goods by both public and private, or company-owned, carriers. The freight forwarder expenditures are for other value-added services provided by outside providers exclusive of actual transportation revenue which is included in the modal numbers. Transportation costs are the single largest contributor to total logistics costs, with trucking being the most significant subcomponent. Figure 2.4 shows recent values for these costs.

Trucking costs account for roughly 50% of total logistics expenditures and 80% of the transportation component. Truck revenues are up 21% since 2000, but that does not tell the whole story. In 2002,

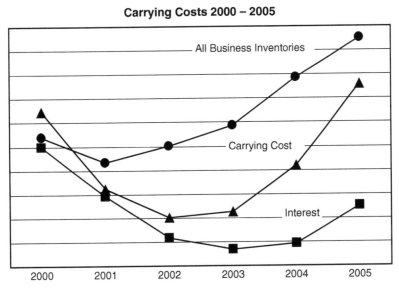

FIGURE 2.3 Costs associated with inventories. (From 17th Annual State of Logistics Report, Rosalyn Wilson, CSCMP, 2006.)

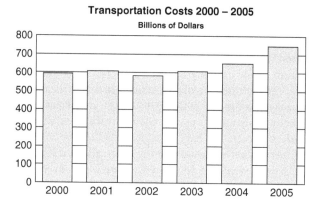

FIGURE 2.4 Transportation costs. (From 17th Annual State of Logistics Report, Rosalyn Wilson, CSCMP, 2006.)

trucking revenues declined for the first time since the 1974–1975 recession. During this period demand was soft and rates were dropping, fuel prices were soaring, and insurance rates were skyrocketing. The trucking industry was forced to undergo a dramatic reconfiguration. About 10,000 motor carriers went bankrupt between 2000 and 2002, and many more were shedding their terminal and other real estate and non-core business units to survive.* While the major impact was the elimination of many smaller companies with revenues in the $5–$20 million range, there were some notable large carriers including Consolidated Freightways. Increased demand and tight capacity enabled trucking to rebound in 2003 and it has risen steadily since.

Trucking revenues in 2005 increased by $74 billion over 2004, but carrier expenses rose faster than rates, eroding some of the gain. The hours-of-service rules for drivers have had a slightly negative impact by reducing the "capacity" of an individual driver, and at the same time a critical driver shortage is further straining capacity. The American Trucking Association (ATA) has estimated that the driver shortage will grow to 111,000 by 2014. Fuel ranks as a top priority at trucking firms as substantially higher fuel prices have cut margins. However, for many the focus has shifted from the higher price level to the volatility of prices. The U.S. trucking industry consumes more than 650 million gallons of diesel per week, making it the second largest expense after labor. The trucking industry spent $87.7 billion for diesel in 2005, a big jump over the $65.9 billion spent in 2004.

Rail transportation has enjoyed a resurgence as it successfully put capacity and service issues behind. Freight ton-mile volumes have reached record levels for nine years in a row. Despite a growth of 33% since 2000, rail freight revenue accounts for only 6.5% of total transportation cost. Intermodal shipping has given new life to the rail industry, with rail intermodal shipments more than tripling since 1980, up from 3.1 to 9.3 million trailers and containers. Sustained higher fuel prices have made shipping by rail a more cost-effective mode than an all truck move. High demand kept the railroad industry operating at near capacity throughout 2005, bumping revenue 14.3%. The expansion of rail capacity has become a paramount issue. The Association of American Railroads (AAR) has reported that railroads will spend record amounts of private capital to add new rail lines to double and triple track existing corridors where needed. In addition, freight railroads are expected to hire 80,000 new workers by 2012.

Water transportation is comprised of two major segments—domestic and international or oceangoing. The international segment has been the fastest growing segment leaping over 60% since 2000, from $18 billion to $29 billion. This tracks with the dramatic growth in global trade. Domestic water traffic, by comparison, has actually declined 30% since 2000, falling from $8 billion to $5 billion in 2005. The United States continues to struggle with port capacity problems, both in terms of available berths for unloading and throughput constraints which slow down delivery.

---

* Donald Broughton tracks bankruptcies in a proprietary database for A.G. Edwards and Sons.

Water transportation faces many obstacles to its continued health. Given the expected growth in international trade U.S. ports are rapidly becoming inadequate. Many ports are over fifty years old and are showing signs of neglect and obsolescence and many have narrow navigation channels and shallow harbors that do not permit access by deep draft vessels which are becoming predominant in the world-wide fleet. The U.S. ports system is close to reaching the saturation point. The World Shipping Council estimates that over 800 ocean freight vessels make over 22,000 calls at U.S. ports every year, or over 60 vessels a day at the nation's 145 ports. Even worse, while the U.S. has done little more than maintain our ports, ports throughout Asia and Europe have become more modern and efficient, giving them an edge in the global economy. As global trading partners build port facilities to handle the larger ships the U.S. places itself at an even greater competitive disadvantage.

The domestic waterway system, the inland waterways, and Great Lakes, has also been the victim of underinvestment. For too many years there has been a lack of resources aimed at maintaining and improving this segment of our transportation network and it is beginning to have dramatic impacts on the capacity of the system. Dredging has fallen behind and the silt built up is hampering navigation and the nation's lock systems are aged and crumbling, with 50% of them obsolete today. Revitalizing this important transportation segment and increasing its use could have a significant impact on reducing congestion and meeting demand for capacity. Although it is not very prevalent now, waterways could even handle containers. A single barge can move the same amount of cargo as 58 semi-trucks at one-tenth the cost.

The air cargo industry has both a domestic and an international side. It is primarily composed of time-sensitive shipments for which customers are willing to pay a premium. Both markets are strong with international revenue up almost 88% since 2000 and domestic revenues up 32% during the same period. Although the air cargo market is thriving and growing, it is still a relatively small share of the whole, representing only about 5% of transport costs. Airfreight revenues increased by $6 billion during 2005, which was an increase of 17.6% over 2004. Along with the growth in revenue came skyrocketing expenses, especially for fuel. In 2003, fuel represented about 14% of operating expenses and in 2005 the percentage had grown to 22%.

The next segment, oil pipeline transportation, accounts for slightly over 1% of total transportation costs. It includes the revenue for the movement of crude and refined oil. We have not added much capacity in the last decade and costs have remained stable, so revenues have been largely constant since 2000.

The final segment, forwarders, has increased over two and half times since 2000, rising from $6 billion to $22 billion. It is important to note that this segment does not include actual transportation expenses; those are picked up in the figures for each mode. Freight forwarders provide an ever increasing array of services as they adapt to meet the changing needs of shippers who choose to outsource their freight needs. The most basic function of a forwarder is to procure carrier resources and facilitate the freight movement. Globalization was a boon to such third-party providers as they specialized in the processes and documentation necessary to engage in international trade. Today forwarders offer such services as preparation of export and import documentation, consolidation and inspection services, and supply chain optimization consulting.

2.2.3 Administrative Costs

The final component of logistics cost is administrative costs and it has two subcomponents: shipper-related costs and logistics administration costs. Shipper-related costs are expenses for logistics-related functions performed by the shipper that are in addition to the actual transportation charges, such as the loading and unloading of equipment, and the operation of traffic departments. Shipper costs actually amount to less than 1% of total logistics costs.

Logistics administration costs represent about 4% of total logistics costs. It includes corporate management and support staff who provide logistics support, such as supply chain planning and analysis staff and physical distribution staff. Computer software and hardware costs attributable to logistics are included in this category if they cannot be amortized directly elsewhere.

2.3 Logistics Productivity over the Past 25 Years

There has been a dramatic improvement in the U.S. business logistics system in the past 20 years. Inventory carrying costs as a percentage of GDP has declined about 40%. Transportation costs as a percentage of GDP has dropped by 8% and total logistics costs declined by 23%. Logistics costs as a percentage of nominal GDP has been below 10% since 2000, despite a 25% increase in the last two years. Imports into the United States, as measured by TEUs, has jumped from under 50 million units to over 400 million in the past 26 years, despite the fact that the capacity growth rate of the nation's transportation infrastructure has been static.

Logistics costs in the United States, and to some extent Europe, have dropped significantly since the deregulation of the transportation modes in the 1980s. Much of the gain was due to reductions in inventory costs. The improved performance of the U.S. logistics sector can be traced to the regulatory reforms in the 1980s. All modes were substantially deregulated, including trucking, rail and air, and after a period of six to eight years of adjustment the economy began to reap the benefits of enhanced productivity, rationalized rail lines, and expanded use of rate contracts. Investments in public infrastructure, particularly the interstate highway system and airports, initially contributed to improved performance in the industry. For the last decade the United States has seriously lagged behind in the necessary investment to sustain the growth however. Much of the gain has come from private innovations and companies agile enough to change rapidly with the times. Examples are the appearance and then explosive growth of the express shipping market, just in time and lean inventory practices which are now being replaced with carefully managed inventories that can be redirected instantaneously, mega retail stores like Wal-Mart and Target with clout to influence logistics practices, and logistics outsourcing.

Over the last 15 years, there has not been a dramatic shift in the relative weights for each of the components that make up total logistics costs. Carrying costs represented 39% of total logistics costs in 1989 and account for 32% today, while transportation costs have climbed from a 56% share to a 62% share of the total. With the exception of carrying costs, each of the other components has risen over 60% since 1989, with both transportation and shipper-related costs jumping 75%. (See Fig. 2.5 for a graphical depiction of trends.)

The nation's railroads move over 50% of all international cargo entering the United States for some portion of the move. International freight is expected to double its current level by 2025. Although the railroads have made heavy investments in recent years in equipment and additional labor, average train speed is falling. Truck vehicle-miles traveled on U.S. highways have nearly doubled in the last 25 years. According to the Federal Highway Administration (FHWA), the volume of freight traffic on the U.S. road system will increase 70% by 2020. Also by 2020, the highway system will have to carry an additional 6.6 billion tons of freight—an increase of 62%. Slower trains mean higher costs and more congestion. Statistics published by the AAR show that average train speed for the entire United States declined from 23 miles per hour in 2000 to less than 22 miles per hour in 2005. The rail freight network was rationalized shortly after the passage of the Staggers Act in 1980 and is now about one-half the size it was, prior to 1980. The leaner system is more productive however, and carries almost double the number of ton-miles the old system carried. Yet, shippers are pushing for even more efficiencies in this area. Will the old strategies applied so successfully in the past work in the rapidly changing global environment? Perhaps the evidence will show that to maintain the gains we have made and to improve the U.S. world competitiveness will require innovation and a re-engineering of supply chain management. Leading the pack in this arena is the contract logistics market.

Market location has become one of the most important drivers of logistics cost. The push by the United States to locate manufacturing facilities offshore to take advantage of less expensive labor and abundant resources has caused a shift in trade patterns. Logistics services that were traditionally

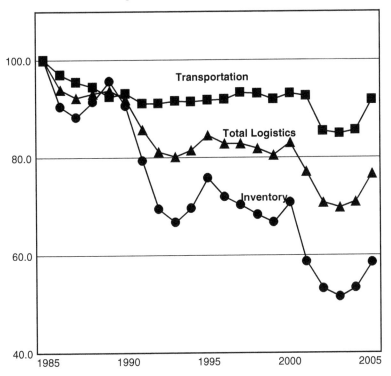

FIGURE 2.5 Logistics series as a percentage of GDP. (From 17th Annual State of Logistics Report, Rosalyn Wilson, CSCMP, 2006.)

performed largely by developed nations are now increasingly being carried out by emerging economies. Now developing countries move finished goods, in addition to raw materials.

The growth and market clout of mega-retailers like Wal-Mart increased the pressure to reduce costs and increase efficiency, forcing many companies to outsource pieces of their supply chain, often to offshore resources. However, global manufacturing is driving many companies to devise innovative strategies for ensuring reliable sources of goods. The ongoing shift of manufacturing to Asia has added stress to an already congested and overburdened domestic transportation system, particularly on shipping in the Pacific. The region has already been operating at full capacity.

Another interesting demographic is the number of small companies now participating in global trade, which had been the purview of large multinational companies until the late 1990s. Over 80% of corporations surveyed in 2002, ranging from small businesses to global giants, indicated that they operated on a global scale. Most operate distribution, sales or marketing centers outside of their home markets.

The globalization of trade and logistics operations has led to the development of international operators based in the regional hubs of developing regions, including Hong Kong, Singapore, United Arab Emirates, and the Philippines. These entities have refined their processes and often employed state-of-the-art equipment to enhance their productivity. The infrastructure has often been built from the ground up with today's global climate in mind. These companies now account for over 30% of global terminal operations.

Many U.S. shippers are contracting their logistics out to non-U.S.-based providers. The estimated value globally for contract logistics services has exceeded $325 billion, with the U.S. portion estimated to be about $150 billion. Shippers are now outsourcing one or more of their supply chain management activities to 3PLs service suppliers. These providers specialize in providing integrated logistics services that meet the needs of today's highly containerized freight system. These companies have proven to be particularly adaptable to the changing global environment including the use of larger and faster ships, containerization of freight, increased security requirements, new technologies to track and monitor shipments, and the rise in air transport for time-sensitive shipments. The global marketplace seemed to emerge overnight and most companies were not prepared or agile enough to respond to the changes. A new knowledge base needed to be acquired and the rules were constantly changing. Third-party providers provided the answers to these problems. These companies filled the niche and became experts, enabling even the smallest firms to operate multinationally. The most successful of these companies control a major share of the market and they play a key role in our ability to expand our supply chains into international markets.

3

Logistics Engineering Tool Chest

3.1 Introduction **3**-1
3.2 Operations Research: Basic Concepts **3**-2
Problem Solving Steps
3.3 Mathematical Programming **3**-4
Linear Programming • Integer Programming
3.4 Heuristic Algorithms **3**-7
"Classical" Heuristic Algorithms • Heuristic Algorithm
Based on Random Choice • "Greedy" Heuristic
Algorithms • Exchange Heuristic Algorithms
• Decomposition Based Heuristic Algorithms
3.5 Algorithm's Complexity **3**-13
3.6 Randomized Optimization Techniques **3**-15
Simulated Annealing Technique • Genetic Algorithms
3.7 Fuzzy Logic Approach to Dispatching
in Truckload Trucking **3**-18
Basic Elements of Fuzzy Sets and Systems • Trucks
Dispatching by Fuzzy Logic
Bibliography **3**-28

Dušan Teodorović
*University of Belgrade and
Virginia Polytechnic Institute
and State University*

Katarina Vukadinović
University of Belgrade

3.1 Introduction

Logistic systems are systems of big dimensions that are geographically dispersed in space. Their complexity is caused by many factors, including interactions between decision-makers, drivers, workers and clients; vehicles, transportation and warehousing processes; communication systems and modern computer technologies, which are very complex. Logistics has been defined by the Council of Logistics Management as "... the process of planning, implementing, and controlling the efficient, effective flow and storage of goods, services, and related information from point of origin to point of consumption for the purpose of conforming to customer requirements." This definition includes inbound, outbound, internal, and external movements, and return of materials for environmental purposes.

Many aspects of logistic systems are stochastic, dynamic, and nonlinear causing logistic systems to be highly sensitive even to small perturbations. Management and control of modern logistic systems are based on many distributed, hierarchically organized levels. Decision-makers, dispatchers, drivers, workers, and clients have different interests and goals, different educational levels, and diverse work experience. They perceive situations in different ways, and make a lot of decisions based on subjective perceptions and subjectively evaluated parameters.

Management and control of modern logistic systems are based on Management Science (MS), Operations Research (OR), and Artificial Intelligence (AI) techniques. Implementation of specific control actions is possible because of a variety of classical and modern electronic, communication, and information technologies that are vital parts of logistic infrastructure. These technologies significantly contribute to the efficient distribution, lower travel times and traffic congestion, lower production and transportation costs, and higher level of service.

Observation, analysis, prediction of future development, control of complex systems, and optimization of these systems represent some of the main research tasks within OR. Analysis of system behavior assumes development of specific theoretical models capable of accurately describing various system processes. The developed mathematical models are used to predict system behavior in the future, to plan future system development, and to define various control strategies and actions. Logistic systems characterized by complex and expensive infrastructure and equipment, great number of various users, and uncertain value of many parameters, have been one of the most important and most challenging OR areas.

Artificial Intelligence is the study and research in computer programs with the ability to display "intelligent" behavior. (AI is defined as a branch of computer science that studies how to endow computers with capabilities of human intelligence.) In essence, AI tries to mimic human intelligent behavior. AI techniques represent convenient tools that can reasonably describe behavior and decision-making of various decision-makers in production, transportation, and warehousing. Distributed AI and multi-agent systems are especially convenient tools for the analysis of various logistic phenomena.

During the last decade, significant progress has been made in merging various OR and AI techniques.

3.2 Operations Research: Basic Concepts

The basic OR concepts can be better described with the help of an example. Let us consider the problem of milk distribution in one city. Different participants in milk distribution are facing various decision problems. We assume that the distributor has a fleet composed of a few vehicles. These vehicles should deliver milk and dairy products to 50 different stores. The whole distribution process could be organized in many different ways. There are a number of feasible vehicle routes. The dispatcher in charge of distribution will always try to discover vehicle routes that facilitate lowest transportation costs.

Store managers are constantly facing the problem of calculating the proper quantity of milk and dairy products that should be ordered from the distributor. Unsold milk and other products significantly increase the costs. On the other hand, potential revenue could be lost in a case of shortage of products.

Both decision problems (faced by distributor dispatcher and store managers) are characterized by limited resources (the number of vehicles that can participate in the milk distribution, the amount of money that could be invested in milk products), and by the necessity to discover the optimum course of action (the best set of vehicle routes, the optimal quantities of milk and dairy products to be ordered).

Operations Research could be defined as a set of scientific techniques searching for the best course of action under limited resources. The beginning of OR is related to the British Air Ministry activities in 1936, and the name Operations Research (Operational Research) has its roots in research of military operations. The real OR boom started after World War II when OR courses were established at many American universities, together with extensive use of OR methods in the industry and public sectors. The development of modern computers further contributed to the success of OR techniques.

Formulation of the problem (in words) represents the first step in the usual problem solving scheme. In the next step, verbal description of the problem should be replaced by corresponding mathematical formulation. Mathematical formulation describes the problem mathematically. Variables, objective function, and constraints are the main components of the mathematical model. To build a mathematical model, analysts try to establish various logical and mathematical relationships between specific variables. The analysts define the objective function, as well as the set of constraints that must be satisfied. Depending on the problem context, the constraints could be by their nature physical, institutional, or financial resources. The generated feasible solutions are evaluated by corresponding objective function values. The set of feasible solutions is composed of all problem solutions that satisfy a given set of constraints. It is very difficult (and in the majority of cases impossible) to produce a mathematical model that will capture all different aspects of the problem considered. Consequently, mathematical models represent simplified description of the real problem. Practically, all mathematical models represent the compromise between the wish to accurately describe the real-life problem and the capability to solve the mathematical model.

3.2.1 Problem Solving Steps

Many real-life logistic and transportation problems can be relatively easily formulated in words (Fig. 3.1). After such formulation of the problem, in the next step, engineers usually translate a problem's verbal description into a mathematical description.

Main components of the mathematical description of the problem are variables, constraints, and the objective. Variables are sometimes called *unknowns*. While some of the variables are under the control of the analyst, some are not. Constraints could be physical resources, caused by some engineering rules, laws, guidelines, or due to various financial reasons. One cannot accept more than 100 passengers for the planned flight, if the capacity of the aircraft equals 100 seats. This is a typical example of physical constraint. Financial constraints are usually related to various investment decisions. For example, one cannot invest more than $10,000,000 in road improvement if the available budget equals $10,000,000. Solutions could be feasible or infeasible. Solutions are feasible when they satisfy all the defined constraints. An objective represents the end result that the decision-maker wants to accomplish by selecting a specific program or action. Revenue maximization, cost minimization, or profit maximization are typical objectives of profit-oriented organizations. Providing the highest level of service to the customers represents the usual objective of a nonprofit organization.

Mathematical description of a real-world problem is called a mathematical model of the real-world problem. An algorithm represents some quantitative method used by an analyst to solve the defined mathematical model. Algorithms are composed of a set of instructions, which are usually followed in a

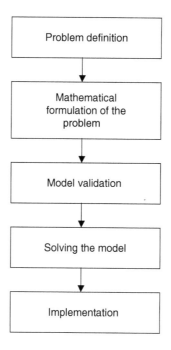

FIGURE 3.1 Problem solving steps.

defined step-by-step procedure. An algorithm produces a feasible solution to a defined model with the goal to find an optimal solution. The optimal solution to the defined problem is the best possible solution among all feasible solutions. Depending on a defined objective function, the optimal solution corresponds to maximum revenue, minimum cost, maximum profit, and so on.

3.3 Mathematical Programming

In the past three decades, linear, nonlinear, dynamic, integer, and multiobjective programming have been successfully used to solve various engineering, management, and control problems. Mathematical programming techniques have been used to address problems dealing with the most efficient allocation of limited resources (supplies, capital, labor, etc.) to meet the defined objectives. Typical problems include market share maximization, production scheduling, personnel scheduling and rostering, vehicle routing and scheduling, locating facilities in a network, planning fleet development, etc. Their solutions can be found using one of the mathematical programming methods.

3.3.1 Linear Programming

Let us consider a rent-a-car company's operations. The total number of vehicles that the company owns equals 100. The potential clients are offered 2 tariff classes at \$150 per week and \$100 per week. The potential client pays \$100 per week if he or she makes the reservation at least 3 days in advance. We assume that we are able to predict exactly the total number of requests in both client-tariff classes. We expect 70 client requests in the first class and 80 client requests in the second class during the considered time period. We decide to keep at least 10 vehicles for the clients paying higher tariffs. We have to determine the total numbers of vehicles rented in different client tariff classes to reach the maximum company revenue.

Solution:

As we wish to determine the total numbers of vehicles rented in different client tariff classes, the variables of the model can be defined as:

x_1—the total number of vehicles planned to be rented in the first client-tariff class
x_2—the total number of vehicles planned to be rented in the second client-tariff class

Because each vehicle from the first class rents for \$150, the total revenue from renting x_1 vehicles is $150x_1$. In the same way, the total company revenue from renting the x_2 vehicles equals $100x_2$. The total company revenue equals the sum of the two revenues, $150x_1 + 100x_2$.

From the problem formulation we conclude that there are specific restrictions on vehicle renting and demand. The vehicle renting restrictions may be expressed verbally in the following way:

- Total number of vehicles rented in both classes together must be less than or equal to the total number of vehicles.
- Total number of vehicles rented in any class must be less than or equal to the total number of client requests.
- Total number of vehicles rented in the first class must be at least 10.
- Total number of vehicles rented in the second class cannot be less than zero (non-negativity restriction).

The following is the mathematical model for the rent-a-car revenue management problem:
Maximize

$$F(X) = 150x_1 + 100x_2$$

subject to:

$$x_1 + x_2 \leq 100$$
$$x_1 \leq 70$$
$$x_2 \leq 80$$
$$x_1 \geq 10$$
$$x_2 \geq 0$$

In our problem, we allow variables to take the fractional values (we can always round the fractional value to the closest feasible integer value). In other words, all our variables are continuous variables. We also have only *one* objective function. We try to maximize the total company's revenue. Our objective function and all our constraints are linear, meaning that any term is either a constant or a constant multiplied by a variable. Any mathematical model that has one objective function, all continuous variables, a linear objective function and all linear constraints is called a linear program (LP). It has been seen through many years that many real-life problems can be formulated as linear programs. Linear programs are usually solved using a widely spread Simplex algorithm (there is also an alternative algorithm called the Interior Point Method).

As we have only two variables, we can also solve our problem graphically. The graphical method is impractical for mathematical models with more than two variables. To solve the earlier-stated problem graphically, we plot the feasible solutions (solution space) that satisfy all constraints simultaneously. Figure 3.2 shows our solution space.

All feasible values of the variables are located in the first quadrant. This is caused by the following constraints: $x_1 \geq 10$, and $x_2 \geq 0$. The straight-line equations $x_1 = 10, x_1 = 70, x_2 = 80, x_2 = 0$, and $x_1 + x_2 = 100$ are obtained by substituting "≤" by "=" for each constraint. Then, each straight line is plotted. The region in which each constraint is satisfied when the inequality is put in power is indicated by the direction of the arrow on the corresponding straight line. The resulting solution space of the rent-a-car problem is shown in Figure 3.3. Feasible points for the problem considered are all points within the boundary or on the boundary of the solution space. The optimal solution is discovered by studying the direction in which the objective function $F = 150 x_1 + 100 x_2$ rises. The optimal solution is shown in Figure 3.3.

The parallel lines in Figure 3.3 represent the objective function $F = 150 x_1 + 100 x_2$. They are plotted by arbitrarily assigning increasing values to F. In this way, it is possible to make conclusions about the slope and the direction in which the total company revenue increases.

To discover the optimal solution, we move the revenue line in the direction indicated in Figure 3.3 to the point "*O*" where any further increase in company revenue would create an infeasible solution. The optimal solution happens at the intersection of the following lines:

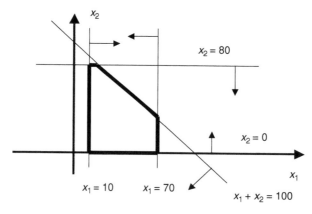

FIGURE 3.2   Solution space of the rent-a-car revenue management problem.

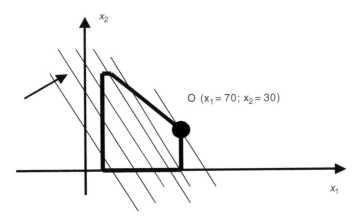

FIGURE 3.3   The optimal solution of the rent-a-car problem.

$$x_1 + x_2 = 100$$
$$x_1 = 70$$

After solving the system of equations we get:

$$x_1 = 70$$
$$x_2 = 30$$

The corresponding rent-a-car company revenue equals:

$$F = 150\,x_1 + 100\,x_2 = 150(70) + 100(30) = 13{,}500$$

The problem considered is a typical resource allocation problem. Linear Programming helps us to discover the best allocation of limited resources. The following is a Linear Programming Model:

Maximize

$$F(X) = c_1 x_1 + c_2 x_2 + c_3 x_3 + \cdots + c_n x_n$$

subject to:

$$a_{11}x_1 + a_{12}x_2 + a_{13}x_3 + \cdots + a_{1n}x_n \leq b_1 \qquad (3.1)$$
$$a_{21}x_1 + a_{22}x_2 + a_{23}x_3 + \cdots + a_{2n}x_n \leq b_2$$

$$a_{m1}x_1 + a_{m2}x_2 + a_{m3}x_3 + \cdots + a_{mn}x_n \leq b_m$$
$$x_1, x_2, \ldots, x_n \geq 0$$

The variables $x_1, x_2, \ldots, x_n$ describe the level of various economic activities (number of cars rented to the first class of clients, number of items to be kept in the stock, number of trips per day on specific route, number of vehicles assigned to a particular route, etc.).

3.3.2   Integer Programming

Analysts frequently realize that some or all of the variables in the formulated linear program must be integers. This means that some variables or all take exclusively integer values. To make the formulated problem easier, analysts often allow these variables to take fractional values. For example, analysts know that the number of first class clients must be in the range between 30 and 40. A linear program could produce the "optimal solution" that tells us that the number of first class clients equals 37.8. In this case, we can neglect the fractional part, and we can decide to protect 37 (or 38) cars for the first class clients. In this way, we are making a small numerical error, but we are capable of easily solving the problem.

In some other situations, it is not possible for analysts to behave in this way. Imagine that we have to decide about a new warehouse layout. You must choose one out of numerous generated alternatives. This is kind of "yes/no" ("1/0") decision: "Yes" if the alternative is chosen, "No," otherwise. In other words, we can introduce binary variables into the analysis. The variable has value 1 if the i-th alternative is chosen and value 0 otherwise. The value 0.7 of the variable means nothing to us. We are not able to decide about the best warehouse layout if the variables take fractional values. When we solve problems similar to the warehouse layout problem we work exclusively with integer variables. These kinds of problems are known as integer programs, and the corresponding area is known as Integer Programming. Integer programs usually describe the problems in which one, or more, alternatives must be selected from a finite set of generated alternatives. Problems of determining the best schedule of activities, finding the optimal set of vehicle routes, or discovering the shortest path in a transportation network are typical problems that are formulated as integer programs. There are also problems in which some variables can take only integer values, while some other variables can take fractional values. These problems are known as mixed-integer programs. It is much harder to solve Integer Programming problems than Linear Programming problems.

The following is the Integer Programming Model formulation:

Maximize

$$F(X) = \sum_{j=1}^{n} c_j x$$

subject to:

$$\sum_{j=1}^{n} a_{ij} x_j \leq b_i \qquad \text{for } i = 1, 2, ..., m$$

$$0 \leq x_j \leq u_j \qquad \text{integer for } j = 1, 2, ..., n$$

(3.2)

There are numerous software systems that solve linear, integer, and mixed-integer linear programs (CPLEX, Excel and Quattro Pro Solvers, FortMP, LAMPS, LINDO, LINGO, MILP88, MINTO, MIPIII, MPSIII, OML, OSL).

A combinatorial explosion of possible solutions characterizes many of the Integer Programming problems. In cases when the number of integer variables in a considered problem is very large, finding the optimal solution becomes very difficult, if not impossible. In such cases, various heuristic algorithms are used to discover "good" solutions. These algorithms do not guarantee the optimal solution discovery.

3.4 Heuristic Algorithms

Many logistic problems are combinatorial by nature. Combinatorial optimization problems could be solved by exact or by heuristic algorithms.

The exact algorithms always find the optimal solution(s). The wide usage of the exact algorithms is limited by the computer time needed to discover the optimal solution(s). In some cases, this computer time is enormously large.

The word "heuristic" has its roots in the Greek word "ευρισκω" that means "to discover," or "to find." A heuristic algorithm could be described as a combination of science, invention, and problem solving skills. In essence, a heuristic algorithm represents a procedure invented and used by the analyst(s) in order to "travel" (search) through the space of feasible solutions. A good heuristics algorithm should generate quality solutions in an acceptable computer time. Complex logistic problems of big dimensions are usually solved with the help of various heuristic algorithms. Good heuristic algorithms are capable of discovering optimal solutions for some problem instances, but heuristic algorithms do not guarantee optimal solution discovery.

There are a few reasons why heuristic algorithms are widely used. Heuristic algorithms are used to solve the problems in situations in which an exact algorithm would require a solution time that increases exponentially with the size of a problem. For example, in the case of a problem that is characterized by 3,000 binary variables (that can take values 0 or 1), the number of potential solutions is equal to 2^{3000}.

In some cases, the costs of using the exact algorithm are much higher than the potential benefits of discovering the optimal solution. Consequently, in such situations analysts usually use various heuristic algorithms.

It could frequently happen that the problem considered is not well "structured." This means that all relevant information is not known by the analyst, and that the objective function(s) and constraints are not precisely defined. An attempt to find the "optimal" solution for the ill-defined problem could generate the "optimal" solution that is in reality a poor solution to the real problem.

The decision-makers are frequently interested in discovering a "satisfying" solution of real-life problems. Obtaining adequate information about considered alternatives is usually very costly. At the same time, the consequences of many possible decisions are not known precisely causing decision-makers to come across with a course of action that is acceptable, sufficient, and logical. In other words, a "satisfying" solution represents the solution that is satisfactory to the decision-makers. A satisfactory solution could be generated by various heuristic algorithms, after a limited search of the solution space.

A great number of real-life logistic problems could be solved only by heuristic algorithms. A large number of heuristic algorithms are based on relatively simple ideas, and many of them have been developed without previous mathematical formulation of the problem.

3.4.1 "Classical" Heuristic Algorithms

The greedy and interchange heuristics are the widely used heuristic algorithms. Let us clarify the basic principles of these algorithms by analyzing the traveling salesman problem (TSP). The TSP is one of the most well-known problems in OR and computer science. This problem can be defined as follows: Find the shortest itinerary which starts in a specific node, goes through all other nodes exactly once, and finishes in the starting node. In different traffic, transportation, and logistic problems, the traveling salesman can represent airplanes, boats, trucks, buses, crews, etc. Vehicles visiting nodes can deliver or pick up goods, or simultaneously perform pickup and delivery.

A typical solution process of the TSP is stepwise as in the following: (a) First an initial tour is constructed; (b) Any remaining unvisited nodes are inserted; (c) The created tour is improved. There are many developed algorithms for each step.

Before discussing various heuristic algorithms, let us define the "scenario" of the TSP. A traveling salesman starting and finishing a tour at one fixed point must visit $(n - 1)$ points. The transportation network connecting these n points is completely connected. This means that it is possible to reach any node from any other node, directly, without going through the other nodes (an air transportation network is a typical example of this type of network). The shortest distance between any two nodes equals the length of the branches between these nodes.

From this, it is certain that the following inequality is satisfied:

$$d(a,b) < d(a,c) + d(c,b) \qquad (3.3)$$

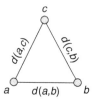

FIGURE 3.4 "Triangular inequality."

for any three nodes a, b, and c.

We also assume that the matrix of shortest distances between the nodes is symmetrical. The nodes a, b, and c are shown in Figure 3.4.

3.4.2 Heuristic Algorithm Based on Random Choice

The TSP could be easily solved by the following simple heuristic algorithm:

Step 1: Arbitrarily choose starting node.
Step 2: Randomly choose the next node to be included in the traveling salesman tour.
Step 3: Repeat Step 2 until all nodes are chosen. Connect the first and the last node of the tour.

This algorithm is based on the idea of random choice. The next node to be included in the partial traveling salesman tour is chosen at random. In other words, the sequence of nodes to be visited is generated at random. It is intuitively clear that one cannot expect that this algorithm would give very good results, as it does not use any relevant information when choosing the next node that is to be included in the tour. On the other hand, generating sequences of nodes at random can be repeated two, three, …, or ten thousand times. The repetition of generating various solutions represents the main power of this kind of an algorithm. Obviously, the decision-maker can choose the best solution among all solutions generated at random. The greater the number of solutions generated, the higher the probability that one can discover a "good" solution.

3.4.3 "Greedy" Heuristic Algorithms

"Greedy" heuristic algorithms build the solution of the studied problem in a step-by-step procedure. In every step of the procedure the value is assigned to one of the variables in order to maximally improve the objective function value. In every step, the greedy algorithm is looking for the best current solution with no look upon future cost or consequences. Greedy algorithms use local information available in every step. The fundamental concept of greedy algorithms is similar to the "Hill-climbing" technique. In the case of the "Hill-climbing" technique the current solution is continuously replaced by the new solution until it is not possible to produce further improvements in the objective function value. "Greedy" algorithms and the "Hill-climbing" technique are similar to the hiker who is trying to come to the mountaintop by never going downwards (Fig. 3.5).

As it can be seen from Figure 3.5, the hiker's wish to never move down while climbing, can trap him or her at some of the local peaks (local maximums), and prevent him or her from reaching the mountaintop (global maximum). "Greedy" algorithms and the "Hill-climbing" technique consider only local improvements.

The Nearest Neighbor (NN) heuristic algorithm is a typical representative of "Greedy" algorithms. This algorithm, which is used to generate the traveling salesman tour, is composed of the following algorithmic steps:

Step 1: Arbitrarily (or randomly) choose a starting node in the traveling salesman tour.

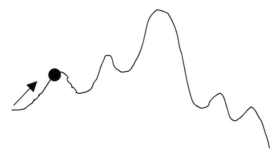

FIGURE 3.5   Hiker who is trying to come to the mountaintop by going up exclusively.

Step 2:   Find the nearest neighbor of the last node that was included in the tour. Include this nearest neighbor in the tour.

Step 3:   Repeat Step 2 until all nodes are not included in the traveling salesman tour. Connect the first and the last node of the tour.

The NN algorithm finds better solutions than the algorithm based on random choice, as it uses the information related to the distances between nodes.

Let us find the traveling salesman tour starting and finishing in node 1, using the NN heuristic algorithm (Fig. 3.6). The distances between all pairs of nodes are given in Table 3.1.

The route must start in node 1. The node 2 is the NN of node 1. We include this NN in the tour. The current tour reads: (1, 2). Node 3 is the NN of node 2. We include this NN in the tour. The updated tour reads: (1, 2, 3). Continuing in this way, we obtain the final tour that reads: (1, 2, 3, 4, 5, 6, 7, 1). The final tour is shown in Figure 3.7.

Both algorithms shown ("random choice" and "greedy") repeat the specific procedure a certain number of times unless a solution has been generated. Many of the heuristic algorithms are based on a specific procedure that is repeated until a solution is generated.

When applying a "greedy" approach, the analyst is forced, after a certain number of steps, to start to connect the nodes (in the case of the TSP) quite away from each other. Connecting the nodes distant from each other is forced by previous connections that significantly decrease the number of possible connections left.

3.4.4   Exchange Heuristic Algorithms

Exchange heuristic algorithms are based on the idea of interchange and they are widely used. The idea of interchange is the idea to start with the existing solution and check if this solution could be improved. The exchange heuristic algorithm first creates or selects an initial feasible solution in some arbitrary way (randomly or using any other heuristic algorithm), and then tries to improve the current solution by specific exchanges within the solution.

The good illustration of this concept is two-optimal tour (2-OPT) heuristic algorithms for the TSP [3-OPT and k-optimal tour (k-OPT) algorithms are based on the same idea]. Within the first

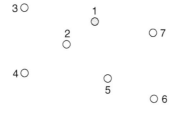

FIGURE 3.6   Network in which a traveling salesman tour should be created using NN heuristic algorithm.

TABLE 3.1 The Distances between All Pairs of Nodes

	1	2	3	4	5	6	7
1	0	75	135	165	135	180	90
2	75	0	90	105	135	210	150
3	135	90	0	150	210	300	210
4	165	105	150	0	135	210	210
5	135	135	210	135	0	90	105
6	180	210	300	210	90	0	120
7	90	150	210	210	105	120	0

step of the 2-OPT algorithm, an initial tour is created in some arbitrary way (randomly or using any other heuristic algorithm). The two links are then broken (Fig. 3.8). The paths that are left are joined so as to form a new tour. The length of the new tour is compared with the length of the old tour. If the new tour length is less than the old tour length, the new tour is retained. In a systematic way, two links are broken at a time, paths are joined, and a comparison is made. Eventually, a tour is found whose total length cannot be decreased by the interchange of any two links. Such a tour is known as a two-optimal tour (2-OPT).

After breaking links (a, j) and (d, e), the node a has to be connected with node e. The node d should be connected with node j. The connection between node a and node d, as well as the connection between node j and node e would prevent creating the traveling salesman tour. In the case of the 3-OPT algorithm in a systematic way three links are broken, a new tour is created, tour lengths are compared, and so on.

The 2-OPT algorithm is composed of the following algorithmic steps:

Step 1: Create an initial traveling salesman tour.
Step 2: The initial tour is the following tour: $(a_1, a_2, ..., a_n, a_1)$. The total length of this tour is equal to D. Set $i = 1$.
Step 3: $j = i + 2$.
Step 4: Break the links (a_i, a_{i+1}) and (a_j, a_{j+1}) and create the new traveling salesman tour. This tour is the following tour: $(a_1, a_2, ..., a_i, a_j, ..., a_{i+1}, a_{j+1}, a_{j+2}, ..., a_1)$. If the length of the new tour is less than D, then keep this tour and return to Step 2. Otherwise go to Step 5.

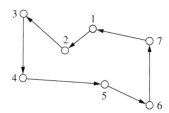

FIGURE 3.7 Traveling salesman tour obtained by the *NN* heuristic algorithm.

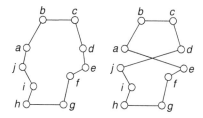

FIGURE 3.8 Interchange of two links during 2-OPT algorithm.

Step 5:   Set $j = j + 1$. If $j \leq n$ go to Step 4. In the opposite case, increase i by 1 ($i = i + 1$). If $i \leq n - 2$ go to Step 3. Otherwise, finish with the algorithm.

By using the 2-OPT algorithm, we will try to create the traveling salesman tour for the network shown in Figure 3.6. The distances between nodes are given in Table 3.1. The traveling salesman should start his trip from node 1. The initial tour shown in Figure 3.7 is generated by the NN algorithm. It was not possible to decrease the total length of the initial tour by interchanging of any two links (Table 3.2). Our initial tour is 2-OPT.

The *k*-opt algorithm for the TSP assumes breaking *k* links in a systematic way, joining the paths, and performing the comparison. Eventually a tour is found whose total length cannot be decreased by the interchange of any *k* links. Such a tour is known as k-OPT.

3.4.5   Decomposition Based Heuristic Algorithms

In some cases it is desirable to decompose the problem considered into smaller problems (subproblems). In the following step every subproblem is solved separately. Final solution of the original problem is then obtained by "assembling" the subproblem solutions. We illustrate this solution approach in the case of the standard vehicle routing problem (VRP).

There are *n* nodes to be served by a homogeneous fleet (every vehicle has identical capacity equal to *V*). Let us denote by v_i ($i = 1, 2, ..., n$) demand at node *i*. We also denote by D the vehicle depot (all vehicles start their trip from D, serve certain number of nodes and finish route in node D).

Vehicle capacity *V* is greater than or equal to demand at any node. In other words, every node could be served by one vehicle, that is, vehicle routes are composed of one or more nodes.

The problem to be solved could be described in the following way: Create a set of vehicle routes in such a way as to minimize the total distance traveled by all vehicles.

A real-life VRP could be very complex. One or more of the following characteristics could appear when solving a real-life VRP: (a) Some nodes must be served within prescribed time intervals (time windows); (b) Service is performed by a heterogeneous fleet of vehicles (vehicles have different capacities); (c) Demand at nodes is not known in advance; (d) There are few depots in the network.

The Sweep algorithm is one of the classical heuristic algorithms for the VRP. This algorithm is applied to polar coordinates, and the depot is considered to be the origin of the coordinate system. Then the depot is joined with an arbitrarily chosen point that is called the *seed point*. All other points are joined to the depot and then aligned by increasing angles that are formed by the segments that connect the

TABLE 3.2   Steps in the 2-OPT Algorithm

Broken Links	New Traveling Salesman Tour	Tour Length
(1, 2), (3, 4)	(1, 3, 2, 4, 5, 6, 7, 1)	765
(1, 2), (4, 5)	(1, 4, 3, 2, 5, 6, 7, 1)	840
(1, 2), (5, 6)	(1, 5, 3, 4, 2, 6, 7, 1)	1020
(1, 2), (6, 7)	(1, 6, 3, 4, 5, 2, 7, 1)	1140
(1, 2), (7, 1)	(1, 7, 3, 4, 5, 6, 2, 1)	960
(2, 3), (4, 5)	(1, 2, 4, 3, 5, 6, 7, 1)	840
(2, 3), (5, 6)	(1, 2, 5, 4, 3, 6, 7, 1)	1005
(2, 3), (6, 7)	(1, 2, 6, 4, 5, 3, 7, 1)	1140
(2, 3), (7, 1)	(1, 2, 7, 4, 5, 6, 3, 1)	1095
(3, 4), (5, 6)	(1, 2, 3, 5, 4, 6, 7, 1)	930
(3, 4), (6, 7)	(1, 2, 3, 6, 5, 4, 7, 1)	990
(3, 4), (7, 1)	(1, 2, 3, 7, 5, 6, 4, 1)	945
(4, 5), (6, 7)	(1, 2, 3, 4, 6, 5, 7, 1)	810
(4, 5), (7, 1)	(1, 2, 3, 4, 7, 6, 5, 1)	870
(5, 6), (7, 1)	(1, 2, 3, 4, 5, 7, 6, 1)	855

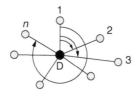

FIGURE 3.9 Sweep algorithm.

points to the depot and the segment that connects the depot to the seed point. The route starts with the seed point, and then the points aligned by increasing angles are included, respecting given constraints. When a point cannot be included in the route as this would violate a certain constraint, this point becomes the seed point of a new route, and so on. The process is completed when all points are included in the routes (Fig. 3.9).

In the case when a large number of nodes need to be served, the Sweep algorithm should be used within the "clustering-routing" approach. In this case, considering a clockwise direction, the ratio of cumulative demand and vehicle capacity should be checked (including all other constraints). The node that cannot be included because of the violation of vehicle capacity or other constraints becomes the first node in another cluster. In this way, the whole region is divided into clusters (zones). In the following step, the VRP is solved within each cluster separately. Clustering is completed when all nodes are assigned to clusters (Fig. 3.10). It is certain that one vehicle can serve all nodes within one cluster. In this way, the VRP is transformed into a few TSPs.

The final solution depends on a choice of the seed point. By changing locations of the seed point it is possible to generate various sets of vehicle routes. For the final solution the set of routes with minimal total length should be chosen.

3.5 Algorithm's Complexity

Various heuristic algorithms could be used to solve a specific problem. Decision-makers prefer to use algorithms that have relatively short CPU time (execution time) and provide reasonably good solutions. One might ask, which one of the developed algorithms is better for solving the TSP? The execution time highly depends on the CPU time, programming language, speed of a computer, etc. To objectively compare various algorithms, a measurement of algorithm's complexity has been proposed that is independent of all computer types and programming languages. The "goodness" of the algorithm is highly influenced by the algorithm's complexity. The complexity of the algorithm is usually measured through the total number of elementary operations (additions, subtractions, comparisons, etc.) that the algorithm requires to solve the problem under the worst case conditions.

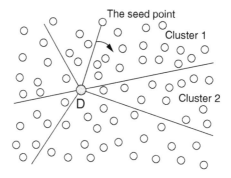

FIGURE 3.10 Clustering by Sweep algorithm.

Let us assume that we have to solve the TSP. We denote by n the total number of nodes. We also denote by E the total number of elementary operations. Let us assume that E equals:

$$E = 4n^4 + 5n^3 + 2n + 7 \tag{3.4}$$

As n increases, the E value is largely determined by the term n^4. We can describe this fact by using the "O-notation." The "O-notation" is used to describe the algorithm's complexity. In the considered example, we write that the algorithm's complexity is $O(n^4)$, or that solution time is of the order $O(n^4)$. The "O-notation" neglects smaller terms, as well as proportional factors. It could happen that for small input sizes an inefficient algorithm may be faster than an efficient algorithm. Practically, the comparison of the algorithms based on "O-notation" is practical only for large input sizes. For example, the algorithm whose complexity is $O(n^2)$ is better than the algorithm whose complexity is $O(n^3)$.

Many real-life problems can be solved by the algorithms whose solution time grows as a polynomial function of the problem size. We call such algorithms polynomial algorithms. The problems that can be solved by polynomial algorithms are considered as *easy problems*. Large instances of easy problems can be solved in "reasonable" computer times using an adequate algorithm and a "fast" computer.

All optimization problems can be classified into two sets. By P we denote the set of problems that can be solved by polynomial algorithms. All other problems, whose solution is difficult or impossible, belong to the set that is called *NP-Complete*. No polynomial time algorithms have been created for the problems that belong to the set *NP-Complete*.

Polynomial algorithms are "good" algorithms [e.g., the algorithms whose complexity is $O(n^2)$, $O(n^5)$, or $O(n^6)$]. The algorithm whose complexity is $O(n \log n)$ also belongs to the class of polynomial algorithms, as ($n \log n$) is bounded by (n^2). Developing an appropriate polynomial algorithm could be in some cases difficult, time consuming, or costly.

Non-polynomial algorithms [e.g., the algorithms whose complexity is $O(3^n)$ or $O(n!)$] are not "good" algorithms. When the algorithm's complexity is, for example, $O(3^n)$, we see that the function in the parentheses is *exponential* in n. One might ask, "Could a faster computer help us to successfully solve 'difficult' problems?" The development of faster computers in the future will enable us to solve larger sizes of these problems; however, there is no indication that we will be able to find optimal solutions in these cases. Every specific problem should be carefully studied. In some cases, it is not an easy task to recognize an "easy" problem and to make the decision regarding the solution approach (optimization vs. heuristic). All heuristic algorithms are evaluated according to the quality of the solutions generated, as well as computer time needed to reach the solution. In other words, a good heuristics algorithm should generate quality solutions in an acceptable computer time. Simplicity and easiness to implement these algorithms are the additional criteria that should be taken into account when evaluating a specific heuristic algorithm.

Heuristic algorithms do not guarantee the optimal solution discovery. The closer the solution produced is to the optimal solution, the better the algorithm. It is an usual practice to perform "Worst Case Analysis," as well as "Average Case Analysis" for every considered heuristic algorithm. Worst Case Analysis assumes generating special numerical examples (that appear rarely in real life) that can show the worst results generated by the proposed heuristic algorithm. For example, we can conclude that the worst solution generated by the proposed heuristic algorithm is 5% far from the optimal solution. Within the Average Case Analysis, a great number of typical examples are usually generated and analyzed. By performing statistical analysis related to the solutions generated, the conclusions are derived about the quality of the solutions generated in the "average case." The more real-life examples are tested, the easier it is to evaluate a specific heuristic algorithm.

3.6   Randomized Optimization Techniques

Many heuristic techniques that have been developed are capable of solving only a specific problem, whereas metaheuristics can be defined as general combinatorial optimization techniques. These techniques are designed to solve many different combinatorial optimization problems. The developed metaheuristics are based on local search techniques, or on population search techniques. Local search-based metaheuristics (Simulated Annealing, Tabu Search, etc.) are characterized by an investigation of the solution space in the neighborhood of the current solution. Each step in these metaheuristics represents a move from the current solution to another potentially good solution in the current solution's neighborhood. In the case of a population search, as opposed to traditional search techniques, the search is run in parallel from a population of solutions. These solutions are combined and the new generation of solutions is generated. Each new generation of solutions is expected to be "better" than the previous one.

3.6.1   Simulated Annealing Technique

The simulated annealing technique is one of the methods frequently used in solving complex combinatorial problems. This method is based on the analogy with certain problems in the field of statistical mechanics. The term, simulated annealing, comes from the analogy with physical processes. The process of annealing consists of decreasing the temperature of a material, which in the beginning of the process is in the molten state, until the lowest state of energy is attained. At certain points during the process the so-called thermal equilibrium is reached. In the case of physical systems we seek to establish the order of particles that has the lowest state of energy. This process requires that the temperatures at which the material remains for a while are previously specified.

The basic idea of simulated annealing consists in performing small perturbations (small alterations in the positions of particles) in a random fashion and computing the energy changes between the new and the old configurations of particles, ΔE. In the case when $\Delta E < 0$, it can be concluded that the new configuration of particles has lower energy. The new configuration then becomes a new initial configuration for performing small perturbations. The case when $\Delta E > 0$ means that the new configuration has higher energy. However, in this case the new configuration should not be automatically excluded from the possibility of becoming a new initial configuration. In physical systems, "jumps" from lower to higher energy levels are possible. The system has higher probability to "jump" to a higher energy state when the temperature is higher. As the temperature decreases, the probability that such a "jump" will occur diminishes. Probability P that at temperature T the energy will increase by ΔE equals:

$$P = e^{-\frac{\Delta E}{T}} \qquad (3.5)$$

The decision of whether a new configuration of particles for which $\Delta E > 0$ should be accepted as a new initial configuration is made upon the generation of a random number r from the interval [0, 1]. The generated random number is uniformly distributed. If $r < P$, the new configuration is accepted as a new initial configuration. In the opposite case, the generated configuration of particles is excluded from consideration.

In this manner, a successful simulation of attaining thermal equilibrium at a particular temperature is accomplished. Thermal equilibrium is considered to be attained when, after a number of random perturbations, a significant decrease in energy is not possible. Once thermal equilibrium has been attained, the temperature is decreased, and the described process is repeated at a new temperature.

The described procedure can also be used in solving combinatorial optimization problems. A particular configuration of particles can be interpreted as one feasible solution. Likewise, the energy of a physical system can be interpreted as the objective function value, while temperature assumes the role of a control parameter. The following is a pseudo-code for a simulated annealing algorithm:

Select an initial state $i \in S$;
Select an initial temperature $T > 0$;
Set temperature change counter $t := 0$;
Repeat
    Set repetition counter $n := 0$;
    Repeat
        Generate state j, a neighbor of i;
        Calculate $\Delta E := f(j) - f(i)$
        if $\Delta E < 0$ then $i := j$
        else if random $(0, 1) < \exp(-\Delta E/T)$ then $i := j$;
        Inc(n);
    Until $n = N(t)$;
    Inc(t);
    $T := T(t)$;
Until stopping criterion true.
where:
S—finite solution set,
i—previous solution,
j—next solution,
$f(x)$—criteria value for solution x, and
$N(t)$—number of perturbations at the same temperature.

It has been a usual practice that during the execution of the simulated annealing algorithm, the best solution obtained thus far is always remembered. The simulated annealing algorithm differs from general local search techniques as it allows the acceptance of improving as well as nonimproving moves. The benefit of accepting nonimproving moves is that the search does not prematurely converge to a local optimum and it can explore different regions of the feasible space.

3.6.2 Genetic Algorithms

Genetic algorithms represent search techniques based on the mechanics of nature selection used in solving complex combinatorial optimization problems. These algorithms were developed by analogy with Darwin's theory of evolution and the basic principle of the "survival of the fittest." In the case of genetic algorithms, as opposed to traditional search techniques, the search is run in parallel from a population of solutions. In the first step, various solutions to the considered maximization (or minimization) problem are generated. In the following step, the evaluation of these solutions, that is, the estimation of the objective (cost) function is made. Some of the "good" solutions yielding a better "fitness" (objective function value) are further considered. The remaining solutions are eliminated from consideration. The chosen solutions undergo the phases of *reproduction, crossover,* and *mutation.* After that, a new generation of solutions is produced to be followed by a new one, and so on. Each new generation is expected to be "better" than the previous one. The production of new generations is stopped when a prespecified stopping condition is satisfied. The final solution of the considered problem is the best solution generated during the search. In the case of genetic algorithms an encoded parameter set is used. Most frequently, binary coding is used. The set of decision variables for a given problem is encoded into a bit string (chromosome, individual).

Let us explain the concept of encoding in the case of finding the maximum value of function $f(x) = x^3$ in the domain interval of x ranging from 0 to 15. By means of binary coding, the observed values of variable x can be presented in strings of length 4 (as $2^4 = 16$). Table 3.3 shows 16 strings with corresponding decoded values.

TABLE 3.3 Encoded Values of Variable x

String	Value of Variable x	String	Value of Variable x
0000	$0 = 0*2^3 + 0*2^2 + 0*2^1 + 0*2^0$	1000	$8 = 1*2^3 + 0*2^2 + 0*2^1 + 0*2^0$
0001	$1 = 0*2^3 + 0*2^2 + 0*2^1 + 1*2^0$	1001	$9 = 1*2^3 + 0*2^2 + 0*2^1 + 1*2^0$
0010	$2 = 0*2^3 + 0*2^2 + 1*2^1 + 0*2^0$	1010	$10 = 1*2^3 + 0*2^2 + 1*2^1 + 0*2^0$
0011	$3 = 0*2^3 + 0*2^2 + 1*2^1 + 1*2^0$	1011	$11 = 1*2^3 + 0*2^2 + 1*2^1 + 1*2^0$
0100	$4 = 0*2^3 + 1*2^2 + 0*2^1 + 0*2^0$	1100	$12 = 1*2^3 + 1*2^2 + 0*2^1 + 0*2^0$
0101	$5 = 0*2^3 + 1*2^2 + 0*2^1 + 1*2^0$	1101	$13 = 1*2^3 + 1*2^2 + 0*2^1 + 1*2^0$
0110	$6 = 0*2^3 + 1*2^2 + 1*2^1 + 0*2^0$	1110	$14 = 1*2^3 + 1*2^2 + 1*2^1 + 0*2^0$
0111	$7 = 0*2^3 + 1*2^2 + 1*2^1 + 1*2^0$	1111	$15 = 1*2^3 + 1*2^2 + 1*2^1 + 1*2^0$

We assume that in the first step the following four strings were randomly generated: 0011, 0110, 1010, and 1100. These four strings form the initial population P(0). In order to make an estimation of the generated strings, it is necessary to decode them. After decoding, we actually obtain the following four values of variable x: 3, 6, 10, and 12. The corresponding values of function $f(x) = x^3$ are equal to $f(3) = 27$, $f(6) = 216$, $f(10) = 1000$ and $f(12) = 1728$. As can be seen, string 1100 has the best fitness value.

Genetic algorithms is a procedure where the strings with better fitness values are more likely to be selected for mating. Let us denote by f_i the value of the objective function (fitness) of string i. The probability p_i for string i to be selected for mating is equal to the ratio of f_i to the sum of all strings' objective function values in the population:

$$p_i = \frac{f_i}{\sum_j f_j} \tag{3.6}$$

This type of reproduction, that is, selection for mating represents a proportional selection known as the "roulette wheel selection." (The selections of roulette are in proportion to probabilities p_i.) In addition to the "roulette wheel selection," several other ways of selection for mating have been suggested in the literature.

In order to generate the next population P(1), we proceed to apply the other two genetic operators to the strings selected for mating. The crossover operator is used to combine the genetic material. At the beginning, pairs of strings (parents) are randomly chosen from a set of previously selected strings. Later, for each selected pair the location for crossover is randomly chosen. Each pair of parents creates two offspring (Fig. 3.11).

After completing crossover, the genetic operator mutation is used. In the case of binary coding, mutation of a certain number of genes refers to the change in value from 1 to 0 or vice versa. It should be noted that the probability of mutation is very small (of order of magnitude 1/1000). The purpose of mutation is to prevent an irretrievable loss of the genetic material at some point along the string. For example, in the overall population a particularly significant bit of information might be missing (e.g., none of the strings have 0 at the seventh location), which can considerably influence the determination of the optimal or near-optimal solution. Without mutation, none of the strings in all future populations could have 0 at the seventh location. Nor could the other two genetic operators help to overcome the given problem. Having generated population P(1) [which has the same number of members as popula-

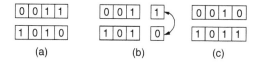

FIGURE 3.11 A single-point crossover operator: (a) two parents; (b) randomly chosen location is before the last bit; (c) two offspring.

tion P(0)], we proceed to use the operators reproduction, crossover, and mutation to generate a sequence of populations P(2), P(3), and so on.

In spite of modifications that may occur in some genetic algorithms (regarding the manner in which the strings for reproduction are selected, the manner of doing crossover, the size of population that depends on the problem being optimized, and so on), the following steps can be defined within any genetic algorithm:

Step 1: Encode the problem and set the values of parameters (decision variables).

Step 2: Form the initial population P(0) consisting of n strings. (The value of n depends on the problem being optimized.) Make an evaluation of the fitness of each string.

Step 3: Considering the fact that the selection probability is proportional to the fitness, select n parents from the current population.

Step 4: Randomly select a pair of parents for mating. Create two offspring by exchanging strings with the one-point crossover. To each of the created offspring, apply mutation. Apply crossover and mutation operators until n offspring (new population) are created.

Step 5: Substitute the old population of strings with the new population. Evaluate the fitness of all members in the new population.

Step 6: If the number of generations (populations) is smaller than the maximal prespecified number of generations, go back to Step 3. Otherwise, stop the algorithm. For the final solution choose the best string discovered during the search.

3.7 Fuzzy Logic Approach to Dispatching in Truckload Trucking

3.7.1 Basic Elements of Fuzzy Sets and Systems

In the classic theory of sets, very precise bounds separate the elements that belong to a certain set from the elements outside the set. For example, if we denote by A the set of signalized intersections in a city, we conclude that every intersection under observation belongs to set A if it has a signal. Element x's membership in set A is described in the classic theory of sets by the membership function $\mu_A(x)$, as follows:

$$\mu_A(x) = \begin{cases} 1, \text{if and only if } x \text{ is member of A} \\ 0, \text{if and only if } x \text{ is not member of A} \end{cases} \tag{3.7}$$

Many sets encountered in reality do not have precisely defined bounds that separate the elements in the set from those outside the set. Thus, it might be said that the waiting time of a vessel at a certain port is "long." If we denote by A the set of "long waiting time at a port," the question logically arises as to the bounds of such a defined set. In other words, we must establish which element belongs to this set. Does a waiting time of 25 hours belong to this set? What about 15 hours or 90 hours?

The membership function of a fuzzy set can take any value from the closed interval [0, 1]. Fuzzy set **A** is defined as the set of ordered pairs **A** = {$x, \mu_A(x)$}, where $\mu_A(x)$ is the grade of membership of element x in set **A**. The greater $\mu_A(x)$, the greater the truth of the statement that element x belongs to set **A**.

Fuzzy sets are often defined through membership functions to the effect that every element is allotted a corresponding grade of membership in the fuzzy set. Let us note fuzzy set **C**. The membership function that determines the grades of membership of individual elements x in fuzzy set **C** must satisfy the following inequality:

$$0 \leq \mu_C(x) \leq 1 \quad \forall x \in \mathbf{X} \tag{3.8}$$

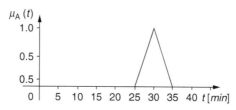

FIGURE 3.12   Membership function $\mu_A(t)$ of fuzzy set **A**.

Let us note fuzzy set **A**, which is defined as "travel time is approximately 30 hours." Membership function $\mu_A(t)$, which is subjectively determined is shown in Figure 3.12.

A travel time of 30 hours has a grade of membership of 1 and belongs to the set "travel time is approximately 30 hours." All travel times within the interval of 25–35 h are also members of this set because their grades of membership are greater than zero. Travel times outside this interval have grades of membership equal to zero.

Let us note fuzzy sets **A** and **B** defined over set X. Fuzzy sets **A** and **B** are equal (**A** = **B**) if and only if $\mu_A(x) = \mu_B(x)$ for all elements of set X.

Fuzzy set **A** is a subset of fuzzy set **B** if and only if $\mu_A(x) \leq \mu_B(x)$ for all elements x of set X. In other words, **A** $\subset$ **B** if, for every x, the grade of membership in fuzzy set **A** is less than or equal to the grade of membership in fuzzy set **B**.

The intersection of fuzzy sets **A** and **B** is denoted by **A** $\cap$ **B** and is defined as the largest fuzzy set contained in both fuzzy sets **A** and **B**. The intersection corresponds to the operation "and." Membership function $\mu_{A \cap B}(x)$ of the intersection **A** $\cap$ **B** is defined as follows:

$$\mu_{A \cap B}(x) = \min\left\{\mu_A(x), \mu_B(x)\right\} \tag{3.9}$$

The union of fuzzy sets **A** and **B** is denoted by **A** $\cup$ **B** and is defined as the smallest fuzzy set that contains both fuzzy set **A** and fuzzy set **B**. The membership function $\mu_{A \cup B}(x)$ of the union **A** $\cup$ **B** of fuzzy sets **A** and **B** is defined as follows:

$$\mu_{A \cup B}(x) = \max\left\{\mu_A(x), \mu_B(x)\right\} \tag{3.10}$$

Fuzzy logic systems arise from the desire to model human experience, intuition, and behavior in decision-making. Fuzzy logic (approximate reasoning, fuzzy reasoning) is based on the idea of the possibility of decison-making based on imprecise, qualitative data by combining descriptive linguistic rules. Fuzzy rules include descriptive expressions such as small, medium, or large used to categorize the linguistic (fuzzy) input and output variables. A set of fuzzy rules, describing the control strategy of the operator (decision-maker) forms a fuzzy control algorithm, that is, approximate reasoning algorithm, whereas the linguistic expressions are represented and quantified by fuzzy sets.

The basic elements of each fuzzy logic system are rules, fuzzifier, inference engine, and defuzzifier. The input data are most commonly crisp values. The task of a fuzzifier is to map crisp numbers into fuzzy sets. Fuzzy rules can conveniently represent the knowledge of experienced operators used in control. The rules can be also formulated by using the observed decisions (input/output numerical data) of the operator. A fuzzy rule (fuzzy implication) takes the following form:

If x is **A**, then y is **B**

where **A** and **B** represent linguistic values quantified by fuzzy sets defined over universes of discourse X and Y. The first part of the rule "x is **A**" is the premise or the condition preceding the second part of the rule "y is **B**" which constitutes the consequence or conclusion.

Let us consider a set of fuzzy rules containing three input variables x_1, x_2, and x_3 and one output variable y.

Rule 1: If x_1 is $\mathbf{P}_{11}$ and x_2 is $\mathbf{P}_{12}$ and x_3 is $\mathbf{P}_{13}$, then y is $\mathbf{Q}_1$,

or

Rule 2: If x_1 is $\mathbf{P}_{21}$ and x_2 is $\mathbf{P}_{22}$ and x_3 is $\mathbf{P}_{23}$, then y is $\mathbf{Q}_2$,

or

Rule k: If x_1 is $\mathbf{P}_{k1}$ and x_2 is $\mathbf{P}_{k2}$ and x_3 is $\mathbf{P}_{k3}$, then y is $\mathbf{Q}_k$.

The given rules are interrelated by the conjunction *or*. Such a set of rules is called a disjunctive system of rules and assumes the satisfaction of at least one rule. It is assumed that membership functions of fuzzy sets $\mathbf{P}_{k1}$ and $\mathbf{P}_{k3}$ ($k = 1, 2, ..., K$) are of a triangular shape, whereas membership functions of fuzzy sets $\mathbf{P}_{k2}$ and $\mathbf{Q}_k$ ($k = 1, 2, ..., K$) are of a trapezoidal shape. Let us note Figure 3.13 in which our disjunctive system of rules is presented.

Let the values i_1, i_2, and i_3, respectively, taken by input variables x_1, x_2, and x_3, be known. In the considered case, the values i_1, i_2, and i_3 are crisp. Figure 3.13 also represents the membership function of output **Q**. This membership function takes the following form:

$$\mu_Q(y) = \max_k \left\{ \min\left[\mu_{P_{k1}}(i_1), \mu_{P_{k2}}(i_2), \mu_{P_{k3}}(i_3)\right] \right\}, \quad k = 1, 2, \supset, \qquad (3.10a)$$

whereas fuzzy set **Q** representing the output is actually a fuzzy union of all the rule contributions $\mathbf{Y}_1$, $\mathbf{Y}_2$, ..., $\mathbf{Y}_k$, that is:

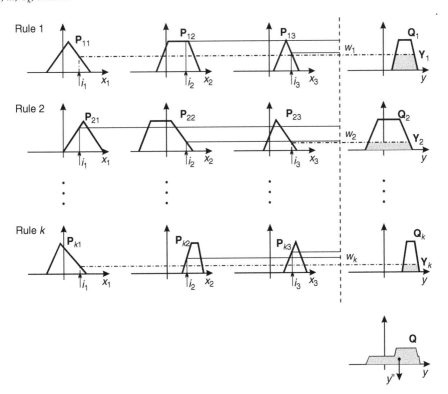

FIGURE 3.13 Graphical interpretation of a disjunctive system of rules.

$$Q = Y_1 \ U \ Y_2 \ U \ ... \ U \ Y_k \quad (3.11)$$

It is clear that

$$\mu_Q(y) = \max\left\{\mu_{Y_1}(y), \mu_{Y_2}(y), \supset, \mu_{Y_k}(y)\right\} \quad (3.12)$$

Consider rule 1, which reads as follows:

If x_1 is P_{11} and x_2 is P_{12} and x_3 is P_{13}, then y is Q_1.

The value $\mu_{P11}(i_1)$ indicates how much truth is contained in the claim that i_1 equals P_{11}. Similarly, values $\mu_{P12}(i_2)$ and $\mu_{P13}(i_3)$, respectively, indicate the truth value of the claim that i_2 equals P_{12} and i_3 equals P_{13}. Value w_1, which is equal to

$$w_1 = \min\left\{\mu_{P_{11}}(i_1), \mu_{P_{12}}(i_2), \mu_{P_{13}}(i_3)\right\} \quad (3.13)$$

indicates the truth value of the claims that, simultaneously, i_1 equals P_{11}, i_2 equals P_{12} and i_3 equals P_{13}.

As the conclusion contains as much truth as the premise, after calculating value w_1, the membership function of fuzzy set Q_1 should be transformed. In this way, fuzzy set Q_1 is transformed into fuzzy set Y_1 (Fig. 3.13). Values $w_2, w_3, ..., w_k$ are calculated in the same manner leading to the transformation of fuzzy sets $Q_2, Q_3, ..., Q_k$ into fuzzy sets $Y_2, Y_3, ..., Y_k$.

As this is a disjunctive system of rules, assuming the satisfaction of at least one rule, the membership function $\mu_Q(y)$ of the output represents the outer envelope of the membership functions of fuzzy sets Y_1, $Y_2, ..., Y_k$. The final value y^* of the output variable is arrived at upon defuzzification, that is, choosing one value for the output variable. In most applications an analyst or decision-maker looks at the grades of membership of individual output variable values, and chooses one of them according to the following criteria: "the smallest maximal value," "the largest maximal value," "center of gravity," "mean of the range of maximal values," and so on (Fig. 3.14).

3.7.2 Trucks Dispatching by Fuzzy Logic

Transportation companies receive a great number of requests every day from clients wanting to send goods to different destinations. Each transportation request is characterized by a large number of attributes, including the most important: type of freight, amount of freight (weight and volume), loading and unloading sites, preferred time of loading and/or unloading, and the distance the freight is to be transported. Transportation companies usually have fleets of vehicles consisting of several different types of vehicle. In addition to the characteristics of the transportation request, when assigning a specific type of vehicle to a specific transportation request, the dispatcher must also bear in mind the total number of

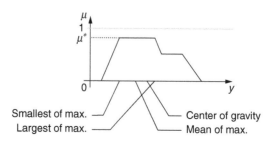

FIGURE 3.14 Defuzzification methods.

available vehicles, the available number of vehicles by vehicle type, the number of vehicles temporarily out of working order, and vehicles undergoing technical examinations or preventive maintenance work. When meeting transportation requests, one or more of the same type of vehicle might be used. In other cases, several different types of vehicles might be used. Depending on the characteristics of the transportation request and the manner in which the transportation company operates, vehicle assignments to transportation requests can be made several times a day, once a day, once a week, and so on. Without loss of generality, we considered the case when dispatching is carried out every day based on the principle "today for tomorrow." In other words, dispatchers have a set amount of time (one day) to match available vehicles to transportation tasks that are to begin the following day.

Assigning vehicles to planned transportation tasks is a daily problem in every transportation company. In most cases, dispatchers responsible for assigning the vehicles rely primarily on their experience and intuition in the course of decision-making. Experienced dispatchers usually have built-in criteria ("rules") which they use to assign a given amount of freight to be sent a given distance to a given vehicle with given structural and technical-operational characteristics (capacity, ability to carry freight certain distances, and so on).

A good dispatcher must have suitable abilities and skills, and his training usually requires a long period. The problem we consider is not one requiring "real-time" dispatching (which is needed to dispatch ambulances, fire department vehicles, police patrol units, taxis, dial-a-ride systems, and so on). However, the large number of different input data and limited time to solve the problem of assigning vehicles to requests can certainly create stressful situations for the dispatcher. These reasons support the need to develop a system that will help the dispatcher to make decisions.

3.7.2.1 Statement of the Problem

Let us consider the vehicle assignment problem within the scope of the following scenario. We assume that a transportation company has several different types of vehicle at its disposal. The number of different vehicle types is denoted by n. Individual vehicle types differ from each other in terms of structural and technical-operational characteristics. We also assume that the transportation company has a depot from which the vehicles depart and to which they return after completing their trip.

Let us consider a delivery system in which different types of freight are delivered to different nodes. We also assume that after serving a node, the vehicle returns to the depot. The reasons for such a delivery tactic are often because of the fact that different types of freight cannot be legally delivered in the same vehicle, and that different types of freight belong to different clients of the transportation company. As the vehicle returns to the depot after serving a node, we note that the routes the vehicle is to take are known. As shown in Figure 3.15, we are dealing with a set of routes in the form of a star, with each route containing a node to be served. Let us denote by m the total number of transportation requests to be undertaken the following day. Let us also denote by T_i the i-th transportation request, $(i = 1, 2, ..., m)$. Every transportation request T_i is characterized by four parameters (v_i, Q_i, D_i, n_i), where v_i is the node where freight is to be delivered when executing transportation request T_i, Q_i is the amount of freight to be transported by request T_i, D_i is the distance freight is to be transported in request T_i (the distance between depot D and node v_i), and n_i is the number of trips along route $\{D, v_i, D\}$ that can be made by one vehicle during the time period under consideration (one day).

In order to simplify the problem, we will assume that the number of possible trips n_i that can be made along route $\{D, v_i, D\}$ is independent of the vehicle type.

We shall denote by C_j the capacity of vehicle type j taking part in the service $(j = 1, 2, ..., n)$. The number of available type j vehicles is denoted by N_j. We also assume that

$$n_i \geq 1 \qquad i = 1, 2, ..., m \tag{3.14}$$

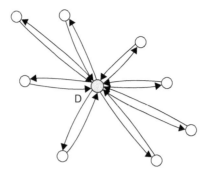

FIGURE 3.15 Depot D and nodes to be served.

Based on the discussed relation, we conclude that the vehicle can serve any node within the geographical region under consideration at least once a day and return to the depot.

Depending on the values of D_i and Q_i and the capacity C_j of the vehicle serving node v_i, one or more trips will be made along route $\{D, v_i, D\}$ during the day being considered. One type or a variety of vehicle types can take part in the delivery to node v_i. Let us first consider the case when only one type of vehicle takes part in serving any node. The more complicated case when several different types of vehicle serve a node is considered later. We would also note that in some cases there is the possibility of the transportation company not being able to serve all nodes with its available transportation capacities.

The standard VRP consists of designing a route to be taken by the vehicles when serving the nodes. In most articles devoted to the classical routing problem, it is assumed that the capacity of the serving vehicle is greater than or equal to demand in any node. In our case, the routes to be taken by the vehicles are known (Fig. 3.15). We shall denote by f_{ij} the number of trips (frequency) to be made by a type j vehicle when executing transport request T_i. It is clear that $f_{ij} \geq 0$ ($i = 1, 2, ..., m, j = 1, 2, ..., n$).

The problem we considered is to determine the value of f_{ij} ($i = 1, 2, ..., m, j = 1, 2, ..., n$) so that the available vehicles are assigned to planned transportation tasks in the best possible way.

3.7.2.2 Proposed Solution to the Problem

The total number of vehicles N available to the dispatcher at the moment he assigns vehicles is

$$N = \sum_{j=1}^{n} N_j \tag{3.15}$$

As already mentioned, the problem considered is the assignment of N available vehicles to m transportation requests. This belongs to the category of OR problems known as assignment problems.

Some transportation requests are "more important" than others. In other words, some clients have signed long-term transportation contracts, and others randomly request transportation that will engage transportation capacities for longer or shorter periods of time. In some cases there is no absolutely precise information about the number of individual types of vehicle that will be ready for operation the following day. Bearing in mind the number of operating vehicles and the number of vehicles expected to be operational the following day, the dispatcher subjectively estimates the total number of available vehicles by type. Some vehicle types are more "suitable" for certain types of transportation tasks than others. Naturally, vehicles with a 5 t capacity are more suitable to deliver goods within a city area than those with a 25 t capacity. On the other hand, 25 t vehicles are considerably more suitable than 5 t or 7 t vehicles for long-distance freighting.

As we can see, the vehicle assignment problem is often characterized by uncertainty regarding input data necessary to make certain decisions. It should be emphasized that the subjective estimation of individual parameters differs from dispatcher to dispatcher, or from decision-maker to decision-maker.

The number of available vehicles of a specific type might be "sufficient" for one dispatcher, while another dispatcher might think this number "insufficient" or "approximately sufficient." Also, one dispatcher might consider a certain type of vehicle "highly suitable" regarding a certain distance, while other dispatchers might consider this type of vehicle "suitable" or "relatively suitable." Clearly, a number of parameters that appear in the vehicle assignment problem are characterized by uncertainty, subjectivity, imprecision, and ambiguity. This raises the need in the mathematically modeling phase of the problem to use methods that can satisfactorily treat uncertainty, ambiguity, imprecision, and subjectivity. The approximate reasoning model presented in the following section is an attempt to formalize the dispatcher's knowledge, that is, to determine the rules used by dispatchers in assigning vehicles to transportation requests.

Approximate reasoning model for calculating the dispatcher's preference when only one type of vehicle is used to meet every transportation request

It can be stated that every dispatcher has a pronounced subjective feeling about which type of vehicle corresponds to which transportation request. This subjective feeling concerns both the suitability of the vehicle in terms of the distance to be traveled and vehicle capacity in terms of the amount of freight to be transported.

Dispatchers consider the suitability of different types of vehicles as being "low" (LS), "medium" (MS), and "high" (HS) in terms of the given distance the freight is to be transported. Also, capacity utilization (the relationship between the amount of freight and the vehicle's declared capacity, expressed as a percentage) is often estimated by the decision-maker as "low" (LCU), "medium" (MCU), or "high" (HCU).

The suitability of a certain type of vehicle to transport freight different distances, and its capacity utilization can be treated or represented as fuzzy sets (Figs. 3.16 and 3.17).

Vehicle capacity utilization is the ratio of the amount of freight transported by a vehicle to the vehicle's capacity. The membership functions of the fuzzy sets shown in Figures 3.16 and 3.17 must be defined individually for every type of vehicle.

The decision-maker assigns transportation requests to individual types of vehicle bearing in mind above all the distance to be traveled and the capacity utilization of the specific type of vehicle. When dispatching, the decision-maker–dispatcher operates with certain rules. Based on conversations with dispatchers who deal with the vehicle assignment problem every day, it is concluded that the decision-maker has certain preferences:

"Very strong" preference is given to a decision that will meet the request with a vehicle type having "high" suitability in terms of distance and "high" capacity utilization.

Or

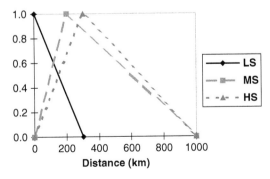

FIGURE 3.16　Membership functions of fuzzy sets: LS is low, MS is medium, and HS is high suitability in terms of distance.

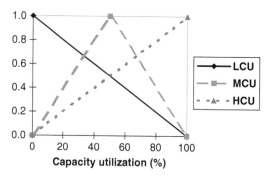

FIGURE 3.17 Membership functions of fuzzy sets: LCU is low, MCU is medium, HCU is high vehicle capacity utilization.

"Very weak" preference is given to a decision that will meet the request with a vehicle that has "low" suitability regarding distance and "low" capacity utilization.

The strength of the dispatcher's preference can be "very strong," "strong," "medium," "weak," and "very weak." Dispatchers most often use five terms to express the strength of their preference regarding the meeting of a specific transportation request with a specific type of vehicle. These five preference categories can be presented as corresponding fuzzy sets **P1**, **P2**, **P3**, **P4**, and **P5**. The membership functions of the fuzzy sets used to describe preference strength are shown in Figure 3.18. Preference strength will be indicated by a preference index, **PI**, which lies between 0 and 1, where a decrease in the preference index means a decrease in the "strength" of the dispatcher's decision to assign a certain transportation request to a certain type of vehicle.

For every type of vehicle, a corresponding approximate reasoning algorithm is developed to determine the dispatcher's preference strength in terms of meeting a specific transportation request with the type of vehicle in question. The approximate reasoning algorithms for each type of vehicle differ from each other in terms of the number of rules they contain and the shapes of the membership functions of individual fuzzy sets. For example, for a vehicle with a capacity of 14 t, the approximate reasoning algorithm reads as shown in Table 3.4.

Using the approximate reasoning by max–min composition, every preference index value is assigned a corresponding grade of membership. Let us denote this value by P_{ij}. This value expresses the "strength" of the dispatcher's preference that the i-th transportation request be met by vehicle type j. Similar approximate reasoning algorithms were developed for the other types of vehicle.

Calculating the dispatcher's preference when several types of vehicle are involved in meeting requests

Up until now, we have only considered the vehicle assignment problem when one type of vehicle is used to meet every transportation request. Some transportation companies often use several different types of vehicle to meet a specific transportation request. When meeting requests with several different types of vehicle, every request can be met in one or several different ways. For example, if the amount of freight in the i-th request equals $Q_i = 18\,t$ and if we have

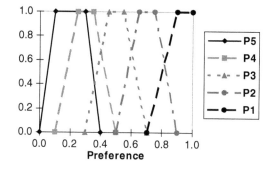

FIGURE 3.18 Membership functions of fuzzy sets: **P1** is very strong, **P2** is strong, **P3** is medium, **P4** is weak, **P5** is very weak preference.

TABLE 3.4 Approximate Reasoning Algorithm for a
Vehicle with a Capacity of 14 t

		Capacity Utilization		
		LCU	MCU	HCU
Suitability	LS	P5	P4	P3
	MS	P3	P2	P2
	HS	P2	P1	P1

two types of vehicle whose capacities are 5 t and 7 t, respectively, there are four possible alternatives to meeting the i-th request, shown in Table 3.5.

The first of the possible alternatives to meet any transportation request is the one in which only one type of vehicle is used, the vehicle with the greatest capacity. Every other alternative differs from the previous to the effect that there is a smaller share of vehicles with a higher capacity and a greater share of vehicles with a smaller capacity. The last possible alternative uses vehicles with the smallest capacity.

Let us denote the following:

Q_{ijk} is the amount of freight from the i-th request transported by vehicle type j when request T_i is met using alternative k.

N_{ijk} is the number of type j vehicles that participate in meeting request T_i when request T_i uses alternative k.

It is clear that the total freight Q_i from transportation request T_i that is met using transportation alternative k equals the sum of the amount of freight of request T_i transported by individual types of vehicles, that is,

$$\sum_{j=1}^{n} Q_{ijk} = Q_i \tag{3.16}$$

The capacity utilization (expressed as a percentage) λ_{ijk} of vehicle type j that takes part in meeting transportation request T_i using alternative k can be defined as,

$$\lambda_{ijk} = \frac{Q_{ijk}}{C_j \, N_{ijk} \, n_i} 100 \, [\%] \tag{3.17}$$

Let us denote by P_k the dispatcher's preference to use service alternative k to meet transportation request T_i. It is clear that,

TABLE 3.5 Comparison of the Total Number of Ton-Kilometers Realized for the Four Different Ways of Assigning Vehicles to Transportation Requests

Possible Ways of Assigning Vehicles to Transportation Requests	Amount of Time Needed to Assign Vehicles to Planned Transportation Requests	Total Number of Realized Ton-kilometers	Percentage of Realized Ton-kilometers
I	2 hr 30 min	163,821	92.26%
II	2 hr 15 min	154,866	87.15%
III	40	152,727	86.01%
IV	40	170,157	95.83%

$$P_k = \frac{\sum_{j=1}^{n} N_{ijk}\, C_j\, n_i\, P_{ij}}{\sum_{j=1}^{n} N_{ijk}\, C_j\, n_i} \tag{3.18}$$

The corresponding dispatcher's preference P_{ij} must be calculated for every type of vehicle j taking part in meeting transportation request T_i. Preference values P_{ij} are calculated based on approximate reasoning algorithms.

Based on relation 3.18, the dispatcher preference to meet transportation request T_i with any of the possible service alternatives k can be calculated.

Heuristic algorithm to assign vehicles to transportation requests

The basic characteristics of every transportation request are the amount of freight that is to be transported and the distance to be traveled. Therefore, requests differ in terms of the volume of transportation work (expressed in ton-kilometers) to be executed, and in terms of the revenues and profits that every transportation request brings to the transportation company. It was also emphasized in our previous remarks that a company might have long-term cooperation with some clients, while other clients request the transportation company's services from time to time. Therefore, some transportation requests can be treated as being "more important," or "especially important requests," having "absolute priority in being carried out," and so on. All of this indicates that before assigning vehicles to transportation requests, the requests must first be sorted. The requests can be sorted in descending order by number of ton-kilometers that would be realized if the request were carried out, in descending order of the amount of freight in each request, in descending order of the requests' "importance" or in some other way. The manner in which the requests are sorted depends on the company's overall transportation policy. It is assumed that sorting of the transportation requests is made before vehicles are assigned to transportation requests.

The heuristic algorithm of assigning vehicles to transportation requests consists of the following steps:

Step 1: Denote by i the index of transportation requests. Let $i = 1$.

Step 2: Generate all possible alternatives to meet transportation request T_i.

Step 3: Denote by $k(i)$ the index of possible alternatives to meet transportation request T_i. Let $k(i) = 1$.

Step 4: Analyze alternative $k(i)$. If available resources (number of available vehicles of a specific type) allow for alternative $k(i)$, go to Step 5. Otherwise go to Step 7.

Step 5: Determine the preference for every type of vehicle that takes part in implementing alternative $k(i)$ using an approximate reasoning by max–min composition.

Step 6: Calculate the dispatcher's preference to use alternative $k(i)$ to meet transportation request T_i. Use relation 3.18 to calculate this preference.

Step 7: Should there be any uninvestigated alternatives, increase the index alternative value by 1 ($k(i) = k(i) + 1$) and go to Step 4. Otherwise, go to Step 8.

Step 8: Should none of the potential alternatives be possible owing to a lack of resources, transportation request T_i cannot be met. The final value of the dispatcher's preference (when there is at least one alternative possible) equals the maximum value of the calculated preferences of the considered alternatives. In this case, transportation request T_i is met by the alternative that corresponds to the maximum preference value.

Step 9: Decrease the number of available vehicles for the types of vehicle that took part in meeting transportation request T_i by the number of vehicles engaged in meeting the request.

Step 10: If any transportation requests have not been considered, increase the index by i ($i = i + 1$) and return to Step 2.

3.7.2.3 Numerical Example

The developed algorithm was tested on a fleet of vehicles containing three different types of vehicle. Capacity per type of vehicle and their respective number in the fleet are: $Q_1 = 4.4\,t$ ($N_1 = 48$ vehicles), $Q_2 = 7.0\,t$ ($N_2 = 49$ vehicles), $Q_3 = 14\,t$ ($N_3 = 42$ vehicles).

Table 3.6 presents the characteristics of the set of 78 transport requests to be met. As can be seen from Table 3.6, each of the 78 transportation requests is characterized by amount of freight Q_i and distance D_i. The transportation work undertaken by the transportation company could be expressed in ton-kilometers (tkm). Based on the characteristics of the transportation requests, it is easy to calculate that the total number of the ton-kilometers to be carried out by the transportation company equals

$$\sum_{i=1}^{78} Q_i\, D_i = 177{,}570.3\ \text{tkm} \tag{3.19}$$

The quality of the solution obtained can be measured as the percentage of realized transportation requests and the percentage of realized ton-kilometers. As the transportation company's profit directly depends on the number of affected ton-kilometers, it was decided that the quality of the solution obtained should be judged on the basis of the total number of realized ton-kilometers. The solutions obtained from the developed model were compared with those obtained by an experienced dispatcher. Let us consider the following four ways of assigning vehicles to transportation requests:

1. An experienced dispatcher assigned vehicles to the transportation requests. The dispatcher was not given any instructions regarding the manner in which the assignments should be made.
2. An experienced dispatcher assigned vehicles to the transportation requests. The dispatcher was asked to assign only one type of vehicle to each transportation request.
3. Vehicles were assigned to transportation requests based on the developed algorithm, with only one type of vehicle being assigned to each transportation request.
4. Before assigning vehicles, the transportation requests were sorted by descending order of ton-kilometers. Vehicles were assigned to transportation requests using the developed algorithm, to the effect that one or several different types of vehicle took part in meeting each request.

The results obtained are shown in Table 3.7.

The developed model shows indisputable advantages compared to the dispatcher, particularly concerning the amount of time needed to assign vehicles to planned transportation requests. It might also be noted that the model sufficiently imitates the work of an experienced dispatcher. Using the model, it is possible to achieve results that are equal to or greater than the results achieved by an experienced dispatcher. Testing a large number of dispatchers and testing the model on a large number of different examples would confirm whether the model gives better results than the dispatcher in every situation.

Bibliography

Bodin, L. et al., Routing and Scheduling of Vehicles and Crews: The State of the Art, *Comput. Oper. Res.*, 10, 63–211, 1983.
Council of Logistics Management, http://www.clm1.org/mission.html, 12 February 1998.
Gillett, B. and Miller, L., A Heuristic Algorithm for the Vehicle Dispatch Problem, *Oper. Res.*, 22, 340–352, 1974.
Hillier, F.S. and Lieberman, G.J., *Introduction to Operations Research*, McGraw-Hill Science, Columbus, OH, 2002.

TABLE 3.6 Characteristics of 78 Transport Requests to Be Met

Request Number	Request Amount of Freight (Tons)	Distance (km)	Daily Number of Trips by One Vehicle	Request Number	Request Amount of Freight (Tons)	Distance (km)	Daily Number of Trips by One Vehicle
1	22.0	42.0	2	40	11.0	180.0	1
2	3.0	25.0	4	41	13.0	12.0	5
3	7.0	138.0	1	42	28.0	198.0	1
4	39.0	280.0	1	43	34.0	265.0	1
5	6.0	75.0	2	44	52.0	140.0	1
6	17.0	189.0	1	45	2.0	180.0	1
7	5.0	45.0	2	46	1.5	17.0	5
8	21.0	110.0	1	47	3.0	29.0	3
9	8.0	180.0	1	48	67.0	270.0	1
10	27.0	42.0	2	49	1.0	87.0	2
11	43.0	197.0	1	50	1.7	195.0	1
12	2.0	317.0	1	51	5.0	49.0	2
13	6.0	180.0	1	52	8.0	165.0	1
14	16.0	78.0	2	53	12.0	87.0	2
15	25.0	78.0	2	54	28.0	65.0	2
16	34.0	57.0	2	55	24.0	29.0	3
17	23.0	57.0	2	56	21.0	12.0	5
18	12.0	129.0	1	57	17.0	369.0	1
19	9.0	32.0	3	58	19.0	100.0	2
20	21.0	21.0	4	59	17.0	120.0	1
21	7.0	180.0	1	60	18.0	140.0	1
22	7.0	87.0	3	61	31.0	190.0	1
23	4.0	49.0	2	62	3.0	120.0	1
24	26.0	127.0	1	63	8.0	108.0	2
25	22.0	240.0	1	64	4.0	140.0	1
26	19.0	220.0	1	65	3.0	17.0	5
27	14.0	100.0	2	66	9.0	98.0	2
28	15.0	121.0	1	67	4.4	78.0	2
29	38.0	27.0	4	68	4.4	78.0	2
30	41.0	129.0	1	69	4.2	112.0	1
31	8.0	160.0	1	70	3.5	5.0	6
32	9.0	180.0	1	71	27.0	15.0	5
33	16.0	70.0	2	72	12.0	5.0	6
34	21.0	161.0	1	73	7.5	98.0	2
35	32.0	180.0	1	74	18.7	210.0	1
36	42.0	120.0	1	75	6.5	180.0	1
37	16.0	132.0	1	76	21.0	600.0	1
38	12.0	12.0	5	77	13.5	120.0	1
39	9.0	27.0	4	78	4.9	120.0	1

TABLE 3.7 Alternatives to Meeting the *i*-th Request

	Number of Vehicles in Service	
Alternative Number	7 t	5 t
1	3	0
2	2	1
3	1	3
4	0	4

Holland, J., *Adaptation in Natural and Artificial Systems*, University of Michigan Press, Ann Arbor, 1975.

Klir, G. and Folger, T., *Fuzzy Sets, Uncertainty and Information*, Prentice-Hall, Englewood Cliffs, NJ, 1988.

Larson, R. and Odoni, A., *Urban Operations Research*, Prentice-Hall, Englewood Cliffs, NJ, 1981.

Taha, H., *Operations Research: An Introduction* (8th Edition), Pearson Prentice Hall, Upper Saddle River, NJ, 2006.

Teodorović, D. and Vukadinović, K., *Traffic Control and Transport Planning: A Fuzzy Sets and Neural Networks Approach*, Kluwer Academic Publishers, Boston-Dordrecht-London, 1998.

Winston, W., *Operations Research*, Duxbury Press, Belmont, CA, 1994.

Zadeh, L., Fuzzy Sets, *Inform. Con.*, 8, 338–353, 1965.

Zimmermann, H.-J., *Fuzzy Set Theory and Its Applications*, Kluwer, Boston, 1991.

4

Logistics Metrics

4.1 Introduction 4-1
4.2 Logistics Data 4-3
Attribute Data in the Logistics Area • Variable Data
in the Logistics Area
4.3 Statistical Methods of Process Monitoring 4-4
Seven Tools of SPC • Control Charts in the Logistics Area
• Error Types • AT&T Run Rules • Types of Control
Charts • Construction of Control Charts
4.4 Logistics Performance Metrics 4-7
4.5 Case Study 4-9
Point-of-Use/Pull System • Performance Metrics
References 4-19

Thomas L. Landers
University of Oklahoma

Alejandro Mendoza
University of Arkansas

John R. English
Kansas State University

4.1 Introduction

With the growth of logistics and supply chain management (SCM), there is an urgent need for performance monitoring and evaluation frameworks that are balanced, integrated, and quantitative. Gerards et al. define logistics as "the organization, planning, implementation and control of the acquisition, transport and storage activities from the purchase of raw materials up to the delivery of finished products to the customers" [1]. SCM is defined by the Council of Supply Chain Management Professionals (CSCMP) [2] as follows: "(SCM) encompasses the planning and management of all activities involved in sourcing and procurement, conversion, and all logistics management activities. Importantly, it also includes coordination and collaboration with channel partners which can be suppliers, intermediaries, third-party service providers, and customers. SCM integrates supply and demand management within and across companies." Current frameworks of performance evaluation within most organizations are sets of known performance measures or metrics (PMs) that have evolved over time. CSCMP [2] defines performance measures as "indicators of the work performed and the results achieved in an activity, process or organizational unit. Performance measures should be both nonfinancial and financial."

Monitoring the performance of a given process requires a well-defined set of metrics to help us establish goals within organizations. Managers need guidance in identifying useful performance metrics, their associated units, unique data characteristics, monitoring techniques, and benchmarks against which such metrics can be compared.

A metric is a standard measure that assesses an organization's ability to meet customers' needs or business objectives. Many performance metrics are ratios relating inputs and outputs, thus permitting assessment of both effectiveness (the degree to which a goal is achieved) and efficiency (the ratio of the resources utilized against the results derived) in accomplishing a given task [3].

Metrics generally fall into two categories: (*i*) performance metrics and (*ii*) diagnostics metrics [4]. PMs are external in nature and closely tied to outputs, customer requirements, and business needs for the process. A diagnostic metric reveals the reasons why a process is not performing in accordance to expectations and is internal in nature. The CSCMP standards for delivery processes [5] stress key performance indicators (KPI) to be monitored by summary tools, such as scorecards or dashboards.

There is a growing body of knowledge and publications on topics of performance measurement and benchmarking for logistics operations. Frazelle and Hackman [6] developed a warehouse performance index using data envelopment analysis. Frazelle [7] has continued to report warehouse metrics and best practices. In 1999, the Council of Logistics Management (now the CSCMP) published a business reference book [8] on the topic. Several articles provide good reviews of performance measurement in logistics [9–11]. Other articles have proposed performance measurement frameworks, including identification and clustering of metrics [12–20].

Two major themes have emerged in the field of performance measurements for business processes in general and logistics processes in particular. The first is to maintain breadth of measurement across functions and objectives. Kaplan and Norton [21] proposed the Balanced Scorecard approach, with metrics in multiple categories (e.g., financial, operational efficiency, service quality, and capability enhancement). The Warehouse Education and Research Council (WERC) periodically reports on performance measurement in distribution centers [22]. Secondly, performance measurement should span the full supply chain. The Perfect Order Index (POI) [23] has emerged as a preferred best practice for measuring full-stream logistics and includes as a minimum the following attributes: on-time, complete, damage-free, and properly invoiced. POI requires discipline and integration of information systems across supply chain partners [23].

A typology measuring relative sophistication of logistics management approaches has been developed by A. T. Kearney [15]. This typology divides companies into four different stages. In Stage I, companies use very simple measures that are expressed in terms of dollars, where information usually comes from the financial organization using very few accounting ratios. In Stage II, companies begin to use simple measures of distribution in terms of productivity to evaluate performance. The use of measures is normally in response to a given problem. In Stage III, companies are proactive and have set meaningful goals for operations. The sophistication of performance measurements is very high. In Stage IV, companies integrate performance data with financial data and are thus able to integrate functional goals.

Comparability of measures, errors in the measurement systems, and human behavior are some of the issues in establishing and monitoring PMs. The marginal benefit of information gathered must exceed marginal costs. Trimble [4] points out that the PMs must be "SMART": Specifically targeted to the area you are measuring, the data must be Measurable (accurate and complete), Actionable (easy to understand), Relevant, and the information inferred from the data must be Timely.

Euske [24] provides a five-step process for developing a measurement system:

1. Establish the problem or goal and its context.
2. Identify the attributes, inputs, and outputs to be evaluated.
3. Analyze the way the measures are obtained.
4. Replace unsatisfactory measures with ones that fulfill the requirements.
5. Perform a cost-benefit analysis to assess the benefit of using a given measurement system.

Lockamy and Cox [18] establish three primary categories of performance measurements: customer, resource, and finance. Within each category, functions are identified. The customer category contains the marketing, sales, and field services functions; the resources category is made up of production, purchasing, design engineering, and transportation functions; and the finance category includes cost, revenue, and investment functions. The PMs for each of these three categories and their associated functions are typically assumed to be independent of one another. As discussed in Byers et al. [25] and implemented by Harp et al. [26], it is necessary to construct performance metrics that monitor performance vertically throughout an organization as well as integrating performance horizontally across the organization giving rise to balanced, full-stream logistics measurements [25].

Boyd and Cox [27] apply this integrating requirement to a case study. Specifically, they implement a negative branch approach (a cause and effect approach developed by Goldratt [28]) to analyze the value-added impact of existing PMs within a pressboard manufacturing process. Through the case study, they clearly demonstrate that performance metrics should not be blindly selected. Specifically, an effective performance metric framework facilitates continuous improvement for the organization.

In summary, performance metrics are data collected from a process of transformation from inputs into outputs to evaluate the existing status of a process. Performance metrics are systematically related to norms and other data. Transformations may include production processes, decision processes, development process, logistics processes, and so on.

4.2   Logistics Data

The monitoring of logistics systems is critical to measuring the quality of service. Data for logistics performance metrics are similar to traditional categories of data in other quality control applications. Quality control data are categorized into two types: attribute and variable. Variable data are measurements that are made on a continuous spectrum. For example, cycle times for receiving materials and issuance of stock are variable data as used in most organizations for a given service type. Alternatively, attribute data are classifications of type. For example, a package either meets or fails to meet packaging standards. Extending this concept further, if 100 packages are selected at random, the proportion of packages meeting inventory accuracy would also be considered as attribute data.

4.2.1   Attribute Data in the Logistics Area

Table 4.1 presents a set of logistics performance metrics of attribute type. Each metric has been either used or recommended for use within a given organization as described in the third column of Table 4.1. In the subsequent discussion, a framework suitable for mostly all logistics systems is presented that more completely enumerates logistics metrics.

4.2.2   Variable Data in the Logistics Area

Table 4.2 provides examples of the current and planned use of variable data within logistics environments. The logistics function is a complex process in which sub-operations are intertwined and may be

TABLE 4.1   Logistics Attribute Data

PM#	Performance Metric	Source
1	Data entry accuracy (total track frequency)	United Parcel Service [29]
2	Preservation and packaging	Defense Logistics Agency [26]
3	Inventory accuracy	Defense Logistics Agency [26]
4	Resolutions complete	Defense Logistics Agency [26]
5	Customer complaints	Defense Logistics Agency [30]
6	Damage freight claims	J.B. Hunt [30]
7	Carrier on-time pickup	Lucent Technologies [30]
8	% Location accuracy	Whirlpool [30]
9	% Empty miles	J.B. Hunt [30]
10	Picks from forward areas	Lucent Technologies [30]
11	Pick rate	Global Concepts [30]
12	% Perfect orders	Global Concepts [30]

TABLE 4.2   Logistics Variable Data

PM#	Performance Metric	Source
1	Cycle time for receipt of material	Defense Logistics Agency [26]
2	Cycle time for issuance of stock	Defense Logistics Agency [26]
3	Cost of nonconformance	Arkansas Best Freight [30]
4	Cost of maintenance	Lucent Technologies [30]
5	Transportation cost	J.B. Hunt [30]
6	Inventory on hand	Lucent Technologies [30]
7	Customer inquiry time	Defense Logistics Agency [30]

confounded. The performance must be considered in view of the process of natural variation. For example, consider cycle time for the receipt of material as used by the Defense Logistics Agency (DLA). The DLA records periodic cycle times and reports average cycle times to the appropriate management. The cycle time varies from one study period to the next for a given service (e.g., binable, high-priority items). The cycle time may occasionally exceed requirements, whereas at other times, it may fall significantly below the requirement. Personnel directly involved with the process know to expect variation from one period to the next, but management usually becomes concerned when requirements are either missed or exceeded. Performance requirements must be considered in the context of expected variation. Consequently, logistics functions should be monitored such that the process is controlled and evaluated in accordance to its natural variation.

In any process, whether it be manufacturing, logistics, or other service, natural variation is present, and must be properly addressed. The sub-processes should be controlled to within the range of their natural variation. Only when nonrandom patterns exist should operators adjust the process, because reaction to random behavior inevitably increases process variation. Patterns should be judged as nonrandom only based upon sound statistical inference. Statistical process control (SPC) provides the framework for statistical inference. SPC builds an environment in which it is the desire of all employees and supply chain partners associated with the process to strive for continuous improvements. Without top-level support, SPC will fail. The following section presents the tools suitable for logistics processes.

4.3 Statistical Methods of Process Monitoring

Statistical process control is a powerful collection of problem-solving tools useful in achieving process stability and improving process capability through the reduction of variability. The natural variability in a process is the effect of many small unavoidable causes. This natural variability is also called a "stable system of chance causes" [31]. A process is said to be in statistical control when it operates under only chance causes of variation. On the other hand, unnatural variation may be observed and assigned to a root cause. These unnatural sources of variability are referred to as assignable causes. Assignable causes can range from improperly adjusted machines to human error. A process or service operating under assignable causes is said to be out of SPC.

4.3.1 Seven Tools of SPC

Statistical process control can be applied to any process and relies on seven major tools, sometimes called the magnificent seven [31]:

1. Histogram
2. Check sheet
3. Pareto chart
4. Cause and effect diagram
5. Defect concentration diagram

6. Scatter diagram
7. Control chart

Histogram: A histogram represents a visual display of data in which three properties can be seen (shape, location or central tendency, and scatter or spread). The typical histogram is a type of bar chart with the vertical bars ordered horizontally by value of a variable. The vertical scale measures frequencies.

Check sheet: A check sheet is a very useful tool in the collection and interpretation of data. For example, a check sheet may capture data for a histogram. Events are tallied in categories. A check sheet should clearly specify the type of data to be collected as well as any other information useful in diagnosing the cause of poor performance.

Pareto chart: The Pareto chart is simply a frequency distribution (or histogram) of attribute data arranged by category. The Pareto chart is a very useful tool in identifying the problems or defects that occur most frequently. It does not identify the most important defects; it only identifies those that occur most frequently. Pareto charts are widely used for identifying quality-improvement opportunities.

Cause and effect diagram: The cause and effect diagram is a tool frequently used to analyze potential causes of undesirable problems or defects. Montgomery [31] suggests a list of seven steps to be followed when constructing a cause and effect diagram: (*i*) define the problem, (*ii*) form the team to perform the analysis, (*iii*) draw the effect box and the centerline, (*iv*) specify the major potential cause categories and join them as boxes connected to the centerline, (*v*) identify the possible causes and classify them, (*vi*) rank the causes to identify those that impact the problem the most, and (*vii*) take corrective action.

Defect concentration diagram: The defect concentration diagram is a picture of the process or product. The different types of defects or problems are drawn on the picture, and the diagram is analyzed to determine the location of the problems or defects.

Scatter diagram: The scatter diagram is used to identify the potential relationship between two variables. Data are plotted on an x-y coordinate system. The shape of the scatter diagram indicates the possible relationship existing between the two variables.

Control chart: The control chart is a graphical display of a quality characteristic that has been measured or computed from a sample versus the sample number or time.

4.3.2 Control Charts in the Logistics Area

To separate assignable causes from the natural process variation, we make use of control charts. Control charts are the simplest procedure of on-line SPC (Fig. 4.1). These charts make possible the diagnosis and correction of many problems, and help to improve the quality of the service provided. Control charts also help in preventing frequent process adjustments that can increase variability. Through process improvements, control charts often provide assurance of better quality at a lower cost. Therefore, a control chart is a device for describing in a precise manner exactly what is meant by statistical control [27].

A control chart contains a centerline that represents the in-control average of the quality characteristic. It also contains two other horizontal lines called the upper control limit (UCL) and lower control limit (LCL). If a process is in control, most sample points should fall within the control limits. These limits are typically called "3-sigma (3σ) control limits." Sigma represents the standard deviation (a measure of variability, or scatter) of the statistic plotted on the chart. The width of the control limits is inversely proportional to the sample size n.

Control charts permit the early detection of a process that is unstable or out of control. However, a control chart only describes how a process is behaving, not how it should behave. A particular control chart might suggest that a process is stable, yet the process may not actually be satisfying customer requirements.

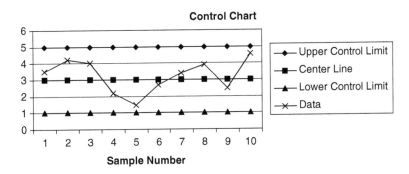

FIGURE 4.1   Control chart.

4.3.3   Error Types

In a control chart, the distance between the centerline and the limits controls decision-making based on error. There are two types of statistical error. A type I error, also known as a false alarm or producer's risk, results from wrongly concluding that the process is out of control when in fact it is in control. A type II error, also known as the consumer's risk, results from concluding that the process is in control when it is not. Widening the control limits in a control chart decreases the risk of a type I error, but at the same time increases the risk of a type II error. On the other hand, if the control limits are moved closer to the centerline, the risk of having a type I error increases while decreasing the risk of a type II error.

4.3.4   AT&T Run Rules

The identification of nonrandom patterns is done using a set of rules known as run rules. The classic Western Electric (AT & T) handbook [32] suggests a set of commonly used decision tools for detecting nonrandom patterns on control charts. A process is out of control if any one of the following applies:

1. One data point plots outside the 3-sigma limits (UCL, LCL).
2. Two out of three consecutive points plot outside the 2-sigma limits.
3. Four out of five consecutive points plot outside the 1-sigma limits.
4. Eight consecutive points plot on one side of the centerline.

The rules apply to one side of the centerline at a time. For example, in the case of rule 2, the process is judged out of control when two out of three consecutive points falling beyond the 2-sigma limits are on the same side of the centerline.

4.3.5   Types of Control Charts

Quality is said to be expressed by variables when a record is made of an actual measured quality characteristic. The $\bar{x}$, R, and S control charts are examples of variables control charts. When samples are of size one, individual (I) charts are suggested for monitoring the mean, and moving range (MR) charts are suggested for monitoring the variance. On the other hand, when a record shows only the number of articles conforming or nonconforming to certain specified requirements, it is said to be a record by attributes. The p, c, and u charts are examples of control charts for attribute data. One other important control chart is the moving centerline exponentially weighted moving average (EWMA), which is very effective in monitoring data that are not independent.

4.3.5.1 Control Charts for Variable Type Data

When dealing with variable data, it is usually necessary to control both the mean value of the quality characteristic and its variability. To monitor the mean value of the product, the $\bar{x}$ control chart is often used. Process variability can be monitored with a control chart for the standard deviation called the S chart, or a control chart for the range, called the R chart. The R chart is the more widely used. The $\bar{x}$ and R (or S) charts are among the most important and useful on-line SPC techniques. When the sample size, n, is large, $n > 12$, or the sample size is variable, the S chart is preferred to the R chart for monitoring variability.

4.3.5.2 Control Charts for Attribute Data

It is known that many quality characteristics cannot be represented numerically. Items inspected are usually classified as conforming or nonconforming to the specifications of that quality characteristic. This type of quality characteristic is called an attribute. Attribute charts are very useful in most industries. For example, from the logistics perspective, it is often necessary to monitor the percentage of units delivered on-time, on-budget, and in compliance with specifications.

The p-chart is used to monitor the fraction nonconforming from a manufacturing process or a service. It is based on the binomial distribution (number of successes in n trials) and assumes that each sample is independent. The fraction nonconforming is defined as the ratio of nonconforming items in a population to the total number of items in that population. Each item may have a number of quality characteristics that are examined simultaneously. If any one of the items being scrutinized does not satisfy the requirements, then the item is classified as nonconforming. The fraction nonconforming is usually expressed as a decimal, although it is occasionally expressed as the percent nonconforming.

There are many practical situations in which working directly with the total number of defects or nonconformities per unit or the average number of nonconformities per unit is preferred over the fraction nonconforming. The c-chart assumes that the occurrence of nonconformities in samples of constant size is rare. As a result, the occurrence of nonconformity is assumed to follow the Poisson probability distribution. The inspection unit must be the same for each sample.

4.3.5.3 Control Chart for Moving Centerline Exponentially Weighted Moving Average

The use of variable control charts implies the assumption of normal and independent observations. If the assumption of normality is violated to a moderate degree, the $\bar{x}$ control chart used to monitor the process average will work reasonably well due to the central limit theorem (law of large numbers). However, if the assumption of independence is violated, conventional control charts do not work well. Too many false alarms disrupt operations and produce misleading results. The moving centerline exponentially weighted moving average (EWMA) is effectively a one-step-ahead predictor to monitor processes when data are correlated. The moving centerline EWMA chart is also recommended for use in the logistics arena for a performance metric that is subject to seasonal variation.

4.3.6 Construction of Control Charts

Table 4.3 (variable type) and Table 4.4 (attribute, or fraction nonconforming type) summarize the parameters and equations for commonly used control charts applicable to logistics performance measurement.

4.4 Logistics Performance Metrics

The authors have developed a logistics performance measurement methodology through centers in the National Science Foundation Industry/University Cooperative Research Center program: Material Handling Research Center (MHRC), The Logistics Institute (TLI), and Center for Engineering Logistics

TABLE 4.3 Construction of Control Charts (Variable Type)

	Estimators of Mean	Estimator of Variation		Control Limits
$\bar{x}$ and R charts	Fixed sample size $$\bar{\bar{x}} = \frac{\bar{x}_1 + \bar{x}_2 + \cdots + \bar{x}_m}{m}$$	$$\bar{R} = \frac{R_1 + R_2 + \cdots + R_m}{m}$$	$\bar{x}$ chart	$UCL = \bar{\bar{x}} + A_2\bar{R}$ $Center\ Line = \bar{\bar{x}}$ $LCL = \bar{\bar{x}} - A_2\bar{R}$
		$R = x_{max} - x_{min}$	$\bar{R}$ chart	$UCL = D_4\bar{R}$ $Center\ Line = \bar{R}$ $LCL = D_3\bar{R}$
$\bar{x}$ and S charts	Fixed sample size $$\bar{\bar{x}} = \frac{\bar{x}_1 + \bar{x}_2 + \cdots + \bar{x}_m}{m}$$	$$\bar{S} = \frac{1}{m}\sum_{i=1}^{m} S_i$$	$\bar{x}$ chart	$UCL = \bar{\bar{x}} + A_3\bar{S}$ $Center\ Line = \bar{\bar{x}}$ $LCL = \bar{\bar{x}} - A_3\bar{S}$
	Variable sample size $$\bar{\bar{x}} = \frac{\displaystyle\sum_{i=1}^{m} n_i\bar{x}_i}{\displaystyle\sum_{i=1}^{m} n_i}$$	$$\bar{S} = \left[\frac{\displaystyle\sum_{i=1}^{m}(n_i - 1)S_i^2}{\displaystyle\sum_{i=1}^{m} n_i - m}\right]^{\frac{1}{2}}$$	S-chart	$UCL = B_4\bar{S}$ $Center\ Line = \bar{S}$ $LCL = B_3\bar{S}$
Individual measurements (sample size 1)	$$\bar{x} = \frac{x_1 + x_2 + \cdots + x_m}{m}$$	$$\overline{MR} = \frac{\displaystyle\sum_{i=1}^{m}\left|x_i - x_{i-1}\right|}{m - 1}$$	$\bar{x}$ chart	$UCL = \bar{x} + 3\dfrac{\overline{MR}}{d_2}$ $Center\ Line = \bar{x}$ $LCL = \bar{x} - 3\dfrac{\overline{MR}}{d_2}$
			$\overline{MR}$ chart	$UCL = D_4\overline{MR}$ $Center\ Line = \overline{MR}$ $LCL = D_3\overline{MR}$

and Distribution (CELDi). A workshop of invited industry leaders in logistics produced the initial framework [30].

Figure 4.2 presents the framework for the generic design of performance measures necessary for monitoring logistics support functions within most organizations. Clearly, there is overlap among each of the four groups, and as observed in Boyd and Cox [27], it is suggested that each PM be heavily scrutinized for its added value.

There are four primary groups of PMs presented in the framework: financial, quality, cycle time, and resource. In the design of a metrics framework, it is necessary to maintain balance and integration across each of these groups. These four primary groups represent a holistic view of the design of PMs necessary to evaluate and monitor the performance of most logistics support functions. The financial group represents the necessary dimension of evaluating short- and long-term profits to ensure the strong financial position of an organization. The quality group represents the dimension of evaluating an organization's quality of meeting customer expectations (external and internal). The cycle time group represents the necessity of evaluating process velocity and consistency. Finally, the resource dimension accounts for the necessary provision of process resources and the utilization and efficiency of processes.

The framework provides a high-level, balanced, and integrated approach. Table 4.5 categorizes and describes the subgroups of performance metrics in each of the four major groups. Tables 4.6 through

TABLE 4.4 Construction of Control Charts (Fraction Nonconforming Type)

	Estimators of Central Tendency	Estimator of Variation	Control Limits
1. p-charts	Fixed sample size $$\bar{p} = \frac{\sum_{i=1}^{m} D_i}{mn} = \frac{\sum_{i=1}^{m} \hat{p}_i}{m}$$	$$\sigma_p^2 = \frac{p(1-p)}{n}$$ $$\sigma_{\hat{p}} = \sqrt{\frac{p(1-p)}{n}}$$	Fixed sample size $$UCL = \bar{p} + 3\sqrt{\frac{\bar{p}(1-\bar{p})}{n}}$$ $$CL = \bar{p}$$ $$LCL = \bar{p} - 3\sqrt{\frac{\bar{p}(1-\bar{p})}{n}}$$ *p*-chart
	Variable sample size $$\bar{p} = \frac{\sum_{i=1}^{m} D_i}{\sum_{i=1}^{m} n_i}$$		Variable sample size $$UCL_i = \bar{p} + 3\sqrt{\frac{\bar{p}(1-\bar{p})}{n_i}}$$ $$CL_i = \bar{p}$$ $$LCL_i = \bar{p} - 3\sqrt{\frac{\bar{p}(1-\bar{p})}{n_i}}$$ *p*-chart
2. Number nonconforming type	$$\bar{c} = \frac{\sum D_i}{\sum n_i}$$	$$\sqrt{\bar{c}}$$ *c*-chart	$$UCL = \bar{c} + 3\sqrt{\bar{c}}$$ $$Center\ Line = \bar{c}$$ $$LCL = \bar{c} - 3\sqrt{\bar{c}}$$
	$$\bar{u} = \frac{\bar{c}}{m}$$	$$\sqrt{\bar{u}/n}$$ *u*-chart	$$UCL_i = \bar{u} + 3\sqrt{\frac{\bar{u}}{n}}$$ $$CL_i = \bar{u}$$ $$LCL_i = \bar{u} - 3\sqrt{\frac{\bar{u}}{n}}$$

Table 4.9 summarize recommended performance metrics for the framework in Figure 4.2 and Table 4.5: financial metrics (Table 4.6), quality metrics (Table 4.7), cycle time metrics (Table 4.8), and resource metrics (Table 4.9).

4.5 Case Study

The following case study demonstrates SPC applied to logistics performance metrics. The application is cycle time and quality metrics for material flow in a point-of-use pull system.

In the ideal application for just-in-time (JIT) manufacturing, there is a single product in high-volume continuous demand. Synchronized JIT production results in components being delivered directly from the supplier to the point of use, just at the time of need. Components are received and handled in standard, reusable containers. One stage in assembly is completed just as the resulting work-in-process (WIP) is needed in the next stage. If there is a time delay between two successive operations, a small temporary buffer storage area is provided between the operations. These buffers are called kanbans and serve three purposes in JIT doctrine: (*i*) limit WIP inventory, (*ii*) maintain shop discipline and housekeeping, and (*iii*) provide process visibility. Removal of a workpiece from the kanban empties the buffer and serves as a pull signal for another unit to be produced and placed in the kanban.

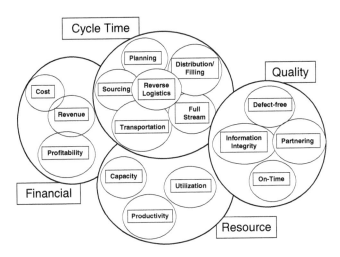

FIGURE 4.2   Performance metrics framework.

In the more typical case of high-mix, low-volume production, some WIP inventory, including component stocks, may be required. However, it is preferred to minimize the WIP by using the principles of JIT to the extent possible. An engineered storage area (ESA) supports point-of-use pull logic for material flow in the high-mix, low-volume shop. The two-bin system is perhaps the most simple and visible method of deploying component stocks into the workcenter ESA. The stock for a component is split into two storage bins. In the simplest form of two-bin system the quantities are equal. When the first bin is emptied, a replenishment order is initiated. If the stock level is reviewed continuously, then the second bin must contain a sufficient amount of material to meet production needs during the replenishment lead time. If the stock is reviewed periodically, then the second bin should contain sufficient stock to meet demand during lead time plus the periodic review interval. If there is variability in the supply or demand processes, then safety stock may be required, which increases the standard quantity in both bins.

TABLE 4.5   Description of Performance Metric Subgroups

Group	Subgroup	Description
Financial	Cost	Focus on cost elements
	Profitability	Consideration of both cost and revenue elements
	Revenue	Focus on revenue elements
Quality	Defect-free	Encompasses all elements of a perfect order outside of on-time and information integrity
	Information integrity	Measurement of information accuracy in the system
	On-time	Meeting partner/customer on-time commitments
	Partnering	Teaming of logistics players, including employees, in order to accomplish value-driven goals
Cycle Time	Distribution/filling	Focuses on distribution and filling time
	Full stream	Spans the entire supply chain
	Planning	Elapsed time related to planning and design
	Reverse logistics	Measurement of elapsed time related to returns
	Sourcing	Focuses on sourcing elapsed time
	Transportation	Measurement of transit time
Resource	Capacity	Related to the output capability of a system
	Productivity	Comparison of actual output to the actual input of resources
	Utilization	Comparison of actual time used to the available time

TABLE 4.6 Financial Metrics Subgroup

Metric	Data	Units	SPC
Annual cost of maintenance by operator	Total amount of money spent on maintenance in a fiscal year	$/operator-year	Pareto chart
Cost per operation	(Total cost)/(total number of operations)	$ per activity	x-bar and R charts or moving centerline EWMA and MR if the data are seasonal
Cost per piece	(Total cost)/(total number of pieces)	$ per piece	x-bar and R charts or moving centerline EWMA and MR if the data are seasonal
Cost per transaction	Cost of a transaction	$ per transactions	x-bar and R charts or moving centerline EWMA and MR if the data are seasonal
Cost per unit of throughput	Cost of a specific facility	$ per unit of throughput for a facility	x-bar and R charts or moving centerline EWMA and MR if the data are seasonal
Cost variance	Cost variance is the ratio between actual costs and standard costs. This ratio most likely varies with particular seasons	Percentage	Moving centerline EWMA and MR charts (assuming the data are seasonal)
Economic value added	(Net operating profit after taxes) – (capital charges)	$	x-bar and R charts
Gross profit margin	[(Sales) – (cost of good sold)]/(sales)	Percentage	Individual and MR charts
Increase in profile adjusted revenues per CWT*	[Positive trend in profile (segment of the market or a particular customer) adjusted revenues)]/[CWT]	$ per CWT	u chart with variable sample size
Inventory carrying Cost	Costs related to warehousing, taxes, obsolescence and insurance (total cost of warehousing, taxes, obsolescence, insurance)/(total cost)	Percentage	Individual and MR charts
Inventory on hand	Total cost of inventory on hand	Percentage	Individual and MR charts
Inventory shrinkage	Total money lost from scrap, deterioration, pilferage, etc.	$	Pareto chart
Logistics operating expenses	Inventory carrying cost + transportation + shipper expenses + distribution + administrative	$	Pareto chart
Material handling rate	(Material handling expense)/(material handling asset value)	Percentage	Individual and MR charts
Net profit margin	(Net profit after taxes)/(sales)	Percentage	Individual and MR charts or Trend charts
Operating expenses before interest and taxes	Cost of goods sold	$ per unit time	x-bar and R charts
Operating ratio	(Cost of goods sold + selling costs + general and administration cost)/(sales)	Percentage	SPC: Individual and MR charts or Trend charts
Payables outstanding past credit term	(Accounts payable past credit term)/(# accounts)	Percentage	P chart, variable sample size
Receivable days outstanding	(Accounts receivable)/(sales per day). Refers to the average collection period	Percentage	x-bar and R charts
Return on assets	(Net profits after taxes)/(total assets)	Percentage	Individual and MR charts or Trend charts
Return on investment	(Income)/(investment capital)	Percentage	Individual and MR charts
Revenue growth percentage	(Revenue at the end of a period) – (Revenue at the end of previous period). Change in revenue over time	Percentage	Individual and MR charts

continued

TABLE 4.6   Financial Metrics Subgroup (continued)

Metric	Data	Units	SPC
Transportation cost per unit (piece, CWT, mile)	(Total transportation costs)/(total number of units)	$ per unit	u chart, variable sample size
Trend	Growth or shrinking market share. Change in market share over time	Percentage	Individual and MR charts

*CWT = $ per 100 pounds of weight.

Bar code labeling and radio frequency communications promote the efficiency of an ESA. Each bin location is labeled with a barcode. When bin 1 is emptied, the user scans the barcode to trigger a replenishment cycle. As the replenishment is put away, the location barcode is again scanned to close out the cycle.

A printed wiring board (PWB) assembly operation in manufacturing of telecommunications switching equipment served as a case study for point-of-use pull material flow through an ESA. Figure 4.3 depicts the layout and flow.

The general framework in Figure 4.2 is utilized to design performance metrics specific to the point-of-use pull system in the PWB assembly shop. This custom system is compatible with available information systems at the company and can be used to move the organization to an environment that views logistics performance with respect to natural process variation. The facility is positioned to identify areas of excellence in current performance as well as opportunities for improvement.

TABLE 4.7   Quality Metrics Subgroup

Metric	Data	Units	SPC
Data entry accuracy	(# Errors)/(# transactions)	Percentage	p-Chart with variable sample size
Document accuracy	(# of orders with accurate documentation)/(total # orders)	Percentage nonconforming	p-Chart with variable sample size
Forecast accuracy	Mean absolute deviation or mean square error. This metric refers to the difference (error) between forecasted and actual	Percentage	x-bar and R charts
Inventory accuracy	(Parts in stock)/(parts supposed to be in stock). Refers to the total number of parts reported by the system of being in stock versus the actual number of parts present in stock	Percentage	p-Charts with variable sample size
Record accuracy	(Number of erroneous records)/(total number of records)	Percentage nonconforming	p-Chart with variable sample size
Tracking accuracy	(Entities in known status)/(total entities). This metric measures the accuracy of tracking job orders e.g., by lot control.	Percentage	p-Chart with variable sample size
On-time delivery	(On-time deliveries)/(total deliveries)	Percentage nonconforming	p-Chart with variable sample size
On-time entry into the system	(Orders with timely system entry)/(total orders)	Percentage nonconforming	p-Chart with variable sample size
On-time loading	(On-time loaded orders)/(total orders)	Percentage nonconforming	p-Chart with variable sample size
On-time marshalling	(Orders ready on time)/(total orders)	Percentage nonconforming	p-Chart with variable sample size
On-time pick up	(On-time pick-ups)/(total pick ups)	Percentage nonconforming	p-Chart with variable sample size
On-time put away	(Orders with timely put away)/(total orders)	Percentage nonconforming	p-Chart with variable sample size

TABLE 4.8 Cycle Time Metrics Subgroup

Metric	Data	Unit	SPC
Cycle sub-time, distribution/filling	Cycle sub-time. Cycle time at the distribution/filling segment	Time units	x-bar and R charts
Fill rate	(Number of lines filled)/(number of lines requested in order)	Percentage	p-Chart with variable sample size
Stock-to-non-stock ratio	(Material shipped)/(total material in stock). Percentage of material shipped by regular stock	Percentage	p-Chart with variable sample size
Cycle time (full stream)	Cycle time. Total or full stream cycle time. Elapsed time between order entry until cycle completion is visible in the computer system	Time units	x-bar and R charts
Days in inventory by item	(Units in inventory)/(average daily usage)	Days	x-bar and R charts
Cycle sub-time, planning/design	Cycle sub-time. Cycle time at the planning and design segment	Time units	x-bar and R charts
Cycle sub-time, reverse logistics	Cycle sub-time. Cycle time for the returns segment	Time units	x-bar and R charts
Cycle sub-time, sourcing	Cycle sub-time. Cycle time for the sourcing segment	Time units	x-bar and R charts
Point of use deliveries	(Number of deliveries)/(total deliveries)	Percentage nonconforming units	p-Chart with variable sample size
Supplier direct deliveries	(Total number of supplier deliveries)/(total number of deliveries)	Percentage	p-Chart with variable sample size
Throughput rate	(WIP)/(cycle time). WIP is the inventory between start and end points of a product routing	Units/time	x-bar, R charts
Cycle sub-time, transportation	Cycle sub-time. Cycle time for the transit segment	Time units	x-bar and R charts
Expedite ratio	(Number of shipments expedited)/(total number of shipments)	Percentage	p-Chart with variable sample size
Off-line shipments	(Number of off-line shipments)/(total number of shipments). Represents the percentage of off-line shipments	Percentage nonconforming units	p-Chart with variable sample size

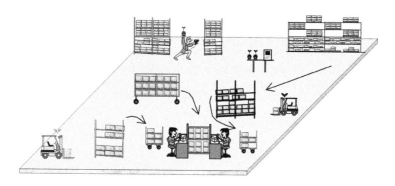

FIGURE 4.3 Case study in electronics manufacturing logistics.

TABLE 4.9 Resource Metrics Subgroup

Metric	Data	Units	SPC
Asset turnover	($ Sales)/($ assets)	Percentage	Individual and MR charts
Asset utilization	(Capacity used)/(capacity available)	Percentage	x-bar and R charts
Cube utilization (load factor)	(Cubic space used)/(cubic space available)	Percentage	x-bar and R charts
Downtime	(Total downtime)/(total available time)	Percentage	x-bar and R charts
Empty miles	(Total empty miles)/(total miles)	Percentage	u-Chart with variable sample size
Empty trailers/containers	(Total empty trailers/containers)/(total trailers/containers)	Percentage	p-Chart with variable sample size
Idleness	(Idle time)/(total available time)	Percentage	x-bar and R charts
Inventory turns	(Sales @ cost)/(average inventory @ cost)	Percentage	Individual and MR charts
Labor utilization	(Total labor used)/(total labor planned to use)	Percentage	Moving centerline EWMA
Material burden	(Good material)/(total material consumed)	Percentage	x-bar and R charts
Network efficiency	(Full enroute miles)/(total miles)	Percentage	u-Chart with variable sample size
Pack rate	(Orders packed)/(employee). Refers to the number of orders packed by a person in a given period of time (minutes, hours, days, etc.)	Packages per employee per unit of time	u-Chart with variable sample size
Pick rate	(Pieces)/(employee), (lines)/(employee), (orders)/(employee). Refers to the number of pieces or orders or lines picked by an employee in a given period of time	Pieces per employee per unit of time, lines per employee per unit of time, orders per employee per period of time	u-Chart with variable sample size
Productivity-on road	(Miles traveled by truck)/(number of days, weeks, etc.). Refers to the number of miles traveled by a truck in a period of time (day, week, etc.)	Miles per period of time	Individual and MR charts
Ratio of inbound to outbound	(Inbound transactions)/(outbound transactions)	Percentage	p-Chart with variable sample size
Receiving rates	(Number of pieces/orders/lines in a given time)/(# employees). Receiving of pieces or orders or lines per employee in a given period of time	Pieces or lines or orders per time unit per employee	u-Chart with variable sample size
Revenue or profit per square foot	(Revenue or profit)/(total space in square feet)	$ per square foot	u-Chart with variable sample size
Revenue per associate	(Total revenues)/(total number of associates)	$ per associate	u-Chart with variable sample size
Shipments per associate	(Number of shipments)/(number of associates)	Shipments per associate	u-Chart with variable sample size
Shipping rate	(Number of pieces/orders/lines in a given time)/(# employees). Shipping of pieces or order or lines per employee in a given period of time	Pieces or lines or orders per time unit per employee	u-Chart with variable sample size
Trailer turns	(Trips)/(period of time). Refers to the total number of trips by a trailer in a given period of time (day, week, month, etc.)	Turns per period of time	Individual and MR charts
Trailer/tractor ratio	(Number of trailers)/(number of tractors)	Trailers per tractor	u-Chart with variable sample size

4.5.1   Point-of-Use/Pull System

One of the processes at the facility is using point-of-use material presentation and pull-logic material flow. Management desires to monitor on-time delivery and accuracy of order filling for components supplied from the stockroom (or supplier) to the shop floor. The layout of the process and the available data resources are identified as follows.

Components are kept in a stock room that fills demand (in varying quantities) to seven different shops. Within each shop, there are associated delivery zones (dz). Each dz has an ESA, consisting of carton flow-rack stock points for components set up as a two bin system with working bins and reserve bins. Both bins have equal quantities of the same product as identified by their product number, or stock keeping unit (SKU). If the SKU from the working bin is depleted, an order is filled from the reserve bin. The box that has been used as a reserve is then moved forward to become the working bin.

When a reserve bin is moved forward to the working bin, a worker scans the barcode of the SKU and places a magnetic sticker on the bin beside the SKU barcode. The magnetic sticker indicates that the part needs to be restocked. The scan triggers a signal to the stockroom computer that notifies the stockroom personnel that the SKU should be restocked. The maximum desired timeframe for the SKU to be restocked is 4 h. When the SKU is delivered to the dz by the stockroom, a second scan is performed. This second scan triggers a signal to the stockroom indicating that the part has been delivered, and this signal marks the time of delivery. If the SKU is delivered within 4 h of the first scan, the action of delivery is considered on-time or good performance. On the other hand, if the SKU is not delivered within the timeframe, the delivery is considered past due. The elapsed time between scans is the total delivery time and is an important performance metric for this process.

If an SKU cannot be delivered from the stock room after the first scan, because it is out of stock, the system automatically sends an order signal to the outside supplier. When an order for the SKU is placed with a supplier, the part is expected to be delivered within five days. Deliveries within this timeframe are considered successful. If the delivery time exceeds five days, the performance is poor and the order is considered short and remains an open transaction. Every Monday the total number of parts short (i.e., open transactions) is collected. On Friday, the system is checked for the number of transactions that have been closed during the week. The difference between the two numbers (open transactions on Monday, closed transactions on Friday) is considered the total number of shortages. Number of shortages is a metric indicating the quality of the logistics system. The number of shortages divided by the number of SKUs within a dz is the performance metric of choice. Due to limitations of the data system, shortages occurring between Tuesday and Sunday are not reported until the following Monday. Therefore, there is a time lag in reporting shortages, and the reported shortages do not necessarily match orders being filled.

A planner is responsible for an assigned set of SKUs. There are 11 planners in the facility. If an SKU order should be placed with a supplier, the associated planner is responsible for the placement of the order and the final delivery of the part to the stock room. The performance of each planner is based on the number of open transactions. This performance metric should be monitored at the planner, shop, and facility levels.

Additionally, the size of the bins is related to the SKU volumetrics. If the packaging is modified by the vendor, the new package may not fit in its designated bin. Therefore, there is a need for monitoring the exceptions to standard packaging.

4.5.2   Performance Metrics

Three key metrics are identified, in view of available data resources, to monitor performance within the facility: delivery time, shortages, and standard packaging.

4.5.2.1   Delivery Time

Delivery time is the metric for monitoring time required to move SKUs from the stock room to the different dz. The target lead time for this operation is 4 h. The shop monitors the number of orders exceeding the 4 h requirement on a per-shift basis. Delivery time is transformed to attribute form at the facility. Attribute data implies that there are two possible events: success or failure. Failure in this case means that the lapsed time between the first scan (need for a SKU to be restocked) and the second scan (SKU restocked) exceeds 4 h. A p-chart with variable sample size is the preferred SPC method for tracking this attribute data. In the case of delivery time performance, the p-chart is used to monitor the percentage of deliveries made within 4 h. These data are collected automatically from the company's database. The formulas used to calculate a p-chart with variable sample size are given in Table 4.4. Figure 4.4 shows the performance of shop 1 for a particular month, and Figure 4.5 shows the performance of the first shift of shop 1 for the same month.

The centerline (CL) in Figure 4.4 is calculated as follows: the total number of nonconforming deliveries (173) is divided by the total number of samples (396), or 0.437. The data being plotted represents the fraction p of nonconforming deliveries. For the first sample the fraction nonconforming is equal to the total nonconforming deliveries for the sample (7) divided by the total number of deliveries for that sample (25). The 3-sigma control limits are calculated by placing control limits at three standard deviations beyond the average fraction nonconforming. For example, the control limits for sample one are:

$$LCL_1 = 0.437 - 3 * \sqrt{\frac{0.437 * (1 - 0.437)}{25}} = 0.437 - 0.297 = 0.139$$

If the LCL for any given sample is smaller than zero, then the value of the LCL is truncated to zero.

$$UCL_1 = 0.437 + 3 * \sqrt{\frac{0.437 * (1 - 0.437)}{25}} = 0.437 + 0.297 = 0.734$$

The resulting control limits and raw data are plotted in Figure 4.4.

The data are also used to monitor the performance of the shop on a per-shift basis. The calculation of the 3-sigma control limits is done in the same way as those for the shop performance. For example, for sample 15 on the first shift, the fraction nonconforming p is $3/15 = 0.2$. The fraction nonconforming, as well as the 3-sigma control limits, for each sample of the data on a per-shift basis are presented in Figure 4.5.

As can be seen in Figures 4.4 and 4.5, the points exceeding the UCL indicate a lack of stability in the process. The source of the nonrandom pattern should be determined and eliminated. The points plotting outside control limits require investigation, with the cause assigned and eliminated. Once the cause is eliminated, the associated points are no longer considered in the calculations, revised limits are

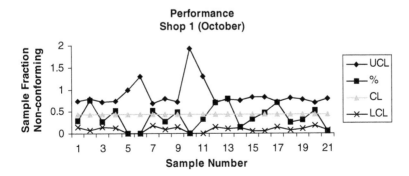

FIGURE 4.4   p-chart for attributes (shop 1).

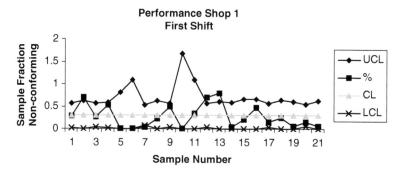

FIGURE 4.5   p-chart for attributes (shop 1, first shift).

calculated and the new plot is inspected for points plotting outside limits. Only extended in-control performance can be used to judge the capability of the process.

4.5.2.2   Shortages

The number of shortages is used to monitor the number of open transactions. The number of shortages is a performance metric that indicates the quality of the supply side of logistics systems and is readily available from data sources. The performance should be evaluated on planner, shop and facility levels. The later performance metric provides an aggregate view of all the combined shops, implying both the necessary horizontal as well as vertical dimensions of a balance PM system as suggested in Harp et al. [26].

The number of shortages, like the delivery time, is transformed to the attribute form for the facility. An open order must be closed within five days. Failure in this case is the failure to close an open order within the five-day time frame. A p-chart is recommended to track the percentage of open transactions. Since each SKU is assigned to different planners, a p-chart is allocated to each planner. Planner performance is based upon the percentage of open transactions to the total number of SKUs assigned to the planner. Figure 4.6 shows a p-chart used to monitor the number of shortages of planner 5. Figure 4.7 shows the p-chart used to monitor open transactions at the aggregate level.

Since each shop has a specific number of SKUs, a p-chart is also used to track the percentage of open transactions within a shop (ratio of open transactions to the total number of SKUs in a shop). In Figure 4.8, the p-chart is used to monitor the number of open transactions for shop 2. Furthermore, open transactions per shop should be monitored on an aggregate view as shown in Figure 4.9.

The points plotting outside the UCL indicate lack of stability in the process. Those points must be investigated and assigned to a cause that should be eliminated. After the points associated with this

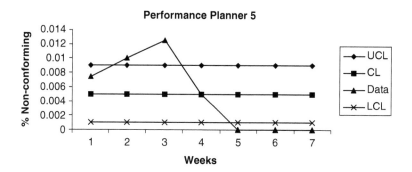

FIGURE 4.6   p-chart for attributes (planner 5).

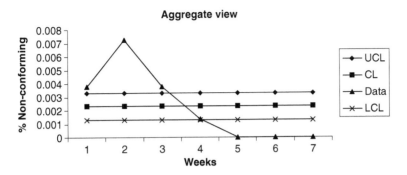

FIGURE 4.7   p-chart for attributes (aggregate).

cause are eliminated from the calculation, new revised limits are calculated and plotted. The new plot is inspected for stability.

4.5.2.3   Exceptions to Standard Packaging

Exceptions to standard packaging are also monitored for the process. This data presents the proportion of exceptions to standard packaging by shop.

Since management is interested only in the number of incorrect packaging incidents in relation to the total number of packages, a Pareto chart is recommended to monitor standard packaging. The Pareto chart for nonconforming packaging across all seven shops is shown in Figure 4.10. The data are categorized and ranked showing the cumulative percentage of incorrect packaging incidents by shop. The percentages are obtained by dividing the number of incorrect packaging incidents per shop by the total number of incidents. As a histogram showing the frequency of root causes, the Pareto chart is helpful in prioritizing corrective action efforts. The Pareto chart is used to identify major causes of phenomena like failures, defects, delays, etc. If a Pareto diagram is used to present a ranking of defects over time, the information is useful for assessing the trend of individual defects, frequency of occurrence, and the effect of corrective actions.

Intuitively, the shops with more SKUs will have a greater percentage of incorrect packaging incidents. However, as can be seen in Figure 4.10, shop 5 has the second greatest percentage of wrong packages even though it has the second smallest number of SKUs. The combination of Pareto charts and trend charts will provide the benefit of a better analysis tool, because the trend chart provides a tool for monitoring the process in view of its natural variation.

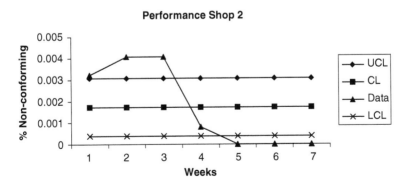

FIGURE 4.8   p-chart for attributes (shop 2).

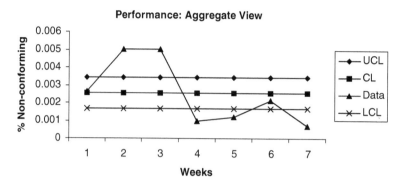

FIGURE 4.9   p-chart for attributes (aggregate).

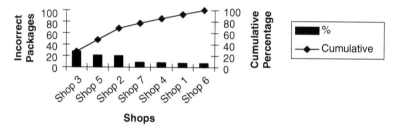

FIGURE 4.10   Pareto chart for nonconforming packaging (shops 1–7).

References

1. Gerards, G., ten Broeke, A.M., Kwaaitaal, A., vander Muelen, P.R.H., Spijkerman G., Vegter, K.J., Willemsen, J.Th.M. Performance Indicators in Logistics. *Approach and Coherence*, IFS Publications: UK, 1989.
2. CSCMP and Supply Chain Visions. *Supply Chain Management Process Standards: Enable Processes*, Council of Supply Chain Management Professionals: Oak Brook, IL, 2004.
3. Mentzer, J.T., Ponsford, B. An Efficiency/Effectiveness Approach to Logistics Performance Analysis. *Journal of Business Logistics*, vol. 12, no. 1, 1991, pp. 33–61.
4. Trimble, D. How to Measure Success: Uncovering the Secrets of Effective Metrics. Online Learning Center, Sponsored by ProSci, 1996, available at: http://www.prosci.com/metrics.htm. (accessed March 22, 2006).
5. CSCMP and Supply Chain Visions. *Supply Chain Management Process Standards: Deliver Processes*, Council of Supply Chain Management Professionals: Oak Brook, IL, 2004.
6. Frazelle, E.H., Hackman, S.T. The Warehouse Performance Index: A Single-Point Metric for Benchmarking Warehouse Performance, MHRC Final Report, #MHRC-TR-93-14, 1993.
7. Frazelle, E. *World Class Warehousing and Material Handling*, McGraw-Hill: New York, 2002.
8. Keebler, J.E., Durtsche, D.A. *Keeping Score: Measuring the Business Value of Logistics in the Supply Chain*, Council of Logistics Management: Oak Brook, IL, 1999.
9. Caplice, C., Sheffi, Y. Review and Evaluation of Logistics Metrics. *The International Journal of Logistics Management*, vol. 5, no. 2, 1994, pp. 11–28.
10. Caplice, C., Sheffi, Y. Review and Evaluation of Logistics Performance Measurement Systems. *The International Journal of Logistics Management*, vol. 6, no. 1, 1995, pp. 61–74.

11. Chow, G., Heaver, T.D., Henriksson, L.E. Logistics Performance: Definition and Measurement. *International Journal of Physical Distribution and Logistics Management*, vol. 24, no. 1, 1994, pp. 17–28.

12. Andersson, P., Aronsson, H., Storhagen, N.G. Measuring Logistics Performance. *Engineering Costs and Production Economics*, vol. 17, 1989, pp. 253–262.

13. Chan, F.T.S. Performance Measurement in a Supply Chain. *International Journal of Advanced Manufacturing Technology*, vol. 21, 2003, pp. 534–548.

14. Donsellar, K.V., Kokke, K, Allessie, M. Performance Measurement in the Transportation and Distribution Sector. *International Journal of Physical Distribution and Logistics Management*, vol. 28, no. 6, 1998, pp. 434–450.

15. A.T. Kearney, *Measuring and Improving Productivity in Physical Distribution*: The Successful Companies Physical Distribution Management: Chicago, IL, 1984.

16. Keebler, J.E., Manrodt, K.B. The State of Logistics Performance Measurement. *Proceedings of the Annual Conference*, Council of Logistics Management: Oak Brook, IL, 2000, pp. 273–281.

17. Legeza, E. Measurement of Logistics-Quality. *Periodica Polytechnica Series on Transportation Engineering*, vol. 31, no. 1–2, pp. 89–95.

18. Lockamy, A., Cox, J. *Reengineering Performance Measurement*, Irwin Professional Publishing: New York, 1994.

19. Mentzer, J.T., Konrad, B.P. An Efficiency/Effectiveness Approach to Logistics Performance Analysis. *Journal of Business Logistics*, vol. 12, no. 1, 1999, pp. 33–61.

20. Rafele, C. Logistics Service Measurement: A Reference Framework. *Journal of Manufacturing Technology Management*, vol. 15, no. 3, 2004, pp. 280–290.

21. Kaplan, R.S., Norton, D.P. *The Balanced Scorecard: Transforming Strategy into Action*, Harvard University Press: Cambridge, MA, 1996.

22. Hill, J. (Ed.) Using the Balanced Scorecard in the DC. WERC Sheet, Warehousing Education and Research Council, July 2005, pp. 3–4.

23. Novak, R.A., Thomas, D.J. The Challenges of Implementing the Perfect Order Concept. *Transportation Journal*, vol. 43, no. 1, 2004, pp. 5–16.

24. Euske, K.J. *Management Control: Planning, Control, Measurement, and Evaluation*, Addison-Wesley Publishing Co.: Menlo Park, CA, 1984.

25. Byers, J.E., Landers, T.L., Cole, M.H. A Framework for Logistics Systems Metrics. TLI Final Report, #TLI-MHRC-96-6, 1996.

26. Harp, C., Buchanan, J., English, J.R., Malstrom, E.M. Design of Group Metrics for the Evaluation of Logistics Performance. TLI Final Report, #TLI-MHRC-97-6, 1997.

27. Boyd, L.H., Cox, J.F. A Cause and Effect Approach to Analyzing Performance Measures. *Production and Inventory Management Journal*, Third Quarter, 1997, pp. 25–32.

28. Goldratt, E. *It's Not Luck*, North River Press: Great Barrington, MA, 1994.

29. Harp, C., Alsein, M., English, J.R., Malstrom, E.M. Quality Monitoring at UPS, TLI Final Report, #TLI-MHRC-97-6, 1997.

30. Mendoza, A., English, J.R., Cole, M.H. Monitoring and Evaluation of Performance Metrics. TLI October Workshop, Summary Report, Arkansas, 1997.

31. Montgomery, D.C. *Introduction to Statistical Quality Control*, 3rd Edition, John Wiley & Sons, Inc.: New York, 1996.

32. *Statistical Quality Control Handbook*, Western Electric Corporation: Indianapolis, IN., 1956.

5

Facilities Location and Layout Design

5.1 Introduction 5-1
5.2 Design Aggregation and Granularity Levels 5-3
5.3 Space Representation 5-7
5.4 Qualitative Proximity Relationships 5-9
5.5 Flow and Traffic 5-12
5.6 Illustrative Layout Design 5-17
5.7 Exploiting Processing and Spatial Flexibility 5-19
5.8 Dealing with Uncertainty 5-25
5.9 Dealing with an Existing Design 5-26
5.10 Dealing with Dynamic Evolution 5-28
5.11 Dealing with Network and Facility Organization ... 5-32
5.12 Design Methodologies 5-36
   Manual design • Heuristic Design • Mathematical
   Programming–Based Design • Interactive Design
   • Metaheuristic Design • Interactive Optimization–Based
   Design • Assisted Design • Holistic Metaheuristics
   • Global Optimization
5.13 Integrated Location and Layout Design
   Optimization Modeling 5-50
   Dynamic Probabilistic Discrete Location Model • Dynamic
   Probabilistic Discrete Location and Continuous Layout Model
5.14 Conclusion 5-55
Acknowledgments 5-56
References 5-57

Benoit Montreuil
University of Laval

5.1 Introduction

Organizations and enterprises around the world differ greatly in terms of mission, scale, and scope. However, all of them aim to deploy the best possible network of facilities worldwide for developing, producing, distributing, selling and servicing their products and offers to their targeted markets and clients. Underlying this continuous quest for optimal network deployment lies the facility location and layout design engineering that is the topic of this chapter. Each node of the network must be laid out as best as possible to achieve its mission, and similarly be located as best as possible to leverage network performance. There is a growing deliberate exploitation of the space-time continuum, which results in new facilities being implemented somewhere in the world every day while existing ones are improved upon or closed down. The intensity and pace of this flux are growing in response to fast and important market, industry and infrastructure transformations. Location and layout design is being transformed, from mostly being a cost-minimization sporadic project to being a business-enabling continuous process—a process embedded in a wider encompassing demand and supply chain design process, itself embedded in a business design process thriving for business differentiation, innovation, and prosperity. Location and layout design will always have significant impact on productivity, but it now is ever more recognized as having an impact on business drivers such as speed, leanness, agility, robustness, and personalization capabilities. The chapter grasps directly this growing complexity in its treatment of the location and layout domain, yet attempts to do so in a way that engineers will readily harness the exposed matter and make it theirs.

This chapter addresses a huge field of practice, education, and research. For example, the site www.uhd. edu/~halet, developed and maintained by Trevor Hale at the University of Houston currently provides over 3,400 location-related references. Location and layout design has been a rich research domain for over 40 years, as portrayed by literature reviews such as Welgama and Gibson (1995), Meller and Gau (1996), Owen and Daskin (1998), and Benjaafar et al. (2002). It is beyond the scope of this chapter to transmit all this knowledge. It cannot replace classical books such as those by Muther (1961), Reed (1961), and Francis et al. (1992) or contemporary books by Drezner and Hamacher (2002), and Tompkins et al. (2003).

The selected goal is rather to enable the readers to leapfrog decades of learning and evolution by the academic and professional community, so that they can really understand and act upon the huge location and layout design challenges present in today's economy. The strategy used is to emphasize selected key facets of the domain in a rather pedagogical way. The objectives are on one hand to equip the reader with hands-on conceptual and methodological tooling to address realistic cases in practice and on the other hand to develop in the reader's mind a growing holistic synthesis of the domain and its evolution.

To achieve its goals and objectives, the chapter is structured as follows. Sections 5.2 through 5.6 focus on introducing the reader to design fundamentals. Aggregation and granularity are discussed in Section 5.2. It is about managing the compromise between scale, scope, and depth that is inherent in any location or layout design study given limited resources and time constraints to perform the design project. Section 5.3 is about the essential element of any location and layout design study, that is, space itself, and how the designer represents it for design purposes. It exposes the key differences between discrete and continuous space representations, as well as the compromises at stake in selecting the appropriate representation in a given case. Sections 5.4 and 5.5 expose the impact of interdependencies on the design task. Section 5.4 focuses on the qualitative proximity relationships between entities to be located and laid out, as well as with existing fixed entities. Section 5.5 concentrates on the quantitative flow and traffic between these entities. Section 5.6 presents an illustrative basic layout design, exploiting the fundamentals introduced in the previous sections. The emphasis is not on how the design is generated. It is rather on the data feeding the design process, the intermediate and final forms of the generated design, and the evaluation of the design.

Sections 5.7 through 5.11 expand from the fundamentals by treating important yet more complex issues faced by engineers having to locate and lay out facilities so that the resulting design contributes as best as possible to the expected future performance of the organization or enterprise. Section 5.7 addresses how a designer can exploit the processing and spatial flexibility of the centers to be laid out and located, whenever such flexibility exists. Section 5.8 extends to describe how to deal with uncertainty when generating and evaluating designs. Section 5.9 deals with the fact that most design studies do not start from a green field, but rather from an existing design which may be costly to alter. Section 5.10 extends to dealing with the dynamic evolution of the design, which switches the output of the study from a layout or location set to a scenario-dependent time-phased set of layouts or locations. The design thus becomes more of a process than a project. Finally, Section 5.11 deals with the potential offered by network and facility organization, when the engineer has freedom to define the centers, their mission, their client–supplier relationships, their processors, and so on, as part of the design generation. Overall, Sections 5.2 through 5.11 portray a rich view of what location and layout design is really about. The aim is clear. A problem well understood is a problem half solved, while attempting to solve a problem wrongly assessed is wasteful and risky in terms of consequences.

Only in Section 5.12 does the chapter directly address design methodologies. This section does not attempt to sell the latest approaches and tools generated by research and industry. It rather openly exposes the variety of methodologies used and proposed by the academic and professional communities. The presentation is structured around a three-tier evolution of proposed design methodologies, starting from the most basic and ancient to the most elaborate and emerging. This section is conceived as an eye opener on the wealth of methodological avenues available, and the compromises involved in selecting one over the other depending on the case the engineer deals with. The following Section 5.13 provides a formal mathematical modeling of location and layout design optimization. It focuses on introducing two models which give a good flavor of the mathematical complexity involved and allow to formally integrate the location and layout facets of the overall design optimization. The chapter concludes with remarks about both the chapter and the domain.

5.2 Design Aggregation and Granularity Levels

Facilities location and layout are both inherently prone to hierarchical aggregation so as to best direct design attention and harness the complexity and scale of the design space. Figure 5.1 provides an illustration of hierarchical aggregation. The entire network of facilities of an enterprise is depicted on the top portion of Figure 5.1, as currently located around the world. The company produces a core module in Scandinavia. This core module is fed to three regional product assembly plants, respectively located in the United States, Eastern Europe, and Japan. Each of these assemblers feeds a set of market-dedicated distribution centers. The middle of Figure 5.1 depicts the site of the Eastern European Assembler, located on municipal lot 62-32. The plan distinguishes seven types of zones in the site. Facility zones are segregated into three types: administration, factory, and laboratory. Transportation zones are split into two types: road zone and parking and transit zone. There is a green zone for trees, grassy areas and gardens. Finally, there is an expansion zone for further expanding activities in the future. There are two factories on the site. The lower portion of Figure 5.1 depicts the assembly factory F2, itself comprised of a number of assembly, production, and distribution centers, as well as offices, meeting rooms, laboratories, and personal care rooms.

A modular approach to represent facility networks helps navigate through various levels of a hierarchical organization. In Figure 5.1, the framework introduced by Montreuil (2006) has been used. It represents the facilities and centers through their main role in the network: assembler, distributor, fulfiller, producer, processor, transporter, as well as a number of more specific roles. A producer fabricates products, modules and parts through operations on materials. A processor performs operations on clients' products and parts. An assembler makes products and modules by assembling them from parts and modules provided by suppliers. A fulfiller fulfils and customizes client orders from products and modules. A distributor stores, prepares and ships products, modules and parts to satisfy client orders. A transporter moves, transports, and handles objects between centers according to client orders. Montreuil (2006) describes thoroughly each type of role and its design issues. Using the same terms at various levels helps the engineer comprehend more readily the nature of the network and its constituents, and leverage this knowledge into developing better designs.

Depending on the scope of design decisions to be taken, the engineer selects the appropriate level of aggregation. However, he must always take advantage of in-depth knowledge of higher and lower levels of aggregation to leverage potential options, taking advantage of installed assets and fostering synergies.

The illustration has focused on hierarchical aggregation. In location and layout studies another type of aggregation is of foremost importance: physical aggregation. This is introduced here through a layout illustration, yet the logic is similar for multi-facility location. The layout of a facility can be represented with various degrees of physical aggregation for design purposes.

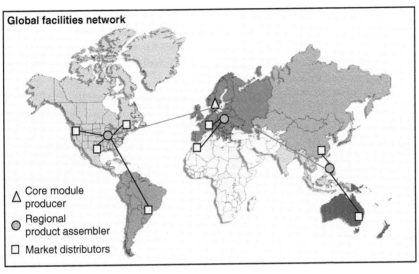

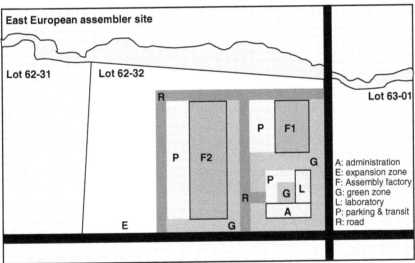

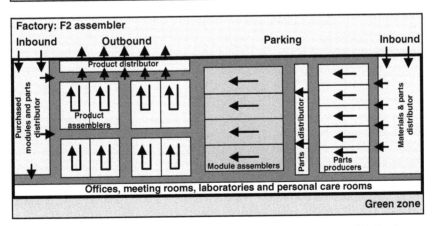

FIGURE 5.1   Hierarchical illustration of facilities network deployment, site layout, and facility layout.

The final deliverable is to be an implemented and operational physical facility laid out according to the design team specifications. The final form of these representations is an engineering drawing and/or a 3D rendering of the facility, with detailed location of all structural elements, infrastructures, walls, machines, etc., identifying the various centers sharing the overall space. For most of the design process, such levels of details are usually not necessary and are cumbersome to manipulate.

Figure 5.2 exhibits five levels of layout representation used for design purposes. The least aggregate first level, here termed processor layout, shows the location and shape of the building, each center, each aisle and each significant processor within each center (e.g., Warnecke and Dangelmaier 1982). The processor layout also locates the input and/or output stations of each center, the travel lane directions for each aisle and, when appropriate, the main material handling systems such as conveyor systems and cranes.

At the second level of aggregation lies the net layout which does not show the processors within each center (e.g., Montreuil 1991, Wu and Appleton 2002). The assumption when focusing the design process on the net layout is that prior to developing the entire layout for the facility, space estimates have been made for each center, leading to area and shape specifications, and that as long as these spatial specifications are satisfied, then the net layout embeds most of the critical design issues. The space estimation may involve designing a priori potential alternative processor layouts for each center. The transposition of the net layout to a processor layout for the overall facility is left as a detailed exercise where the layout of each center is developed given the shape and location decided through the net layout. Note that when the internal layout of the centers has influence on overall flow and physical feasibility, then basing the core of the design process on the net layout is not adequate.

At the third level of aggregation, the aisle set is not included anymore in the layout (e.g., Montreuil 1987, 1991). Instead, the space requirements for shaping each center are augmented by the amount of space expected to be used by aisles in the overall layout. For example, if by experience, roughly 15% of the overall space is occupied by aisles in layouts for the kind of facility to be designed, then the space requirements of each center are increased by 15%. This percentage is iteratively adjusted as needed. The layout depicting the location and shape of the centers is now termed a block layout.

At this third level, instead of including the aisle set explicitly, the design depicts the logical travel network (Chhajed et al. 1992). This network, or combination of networks, connects the I/O stations of the centers as well as the facility entry and exit locations. There may be a network representing aisle travel, or even more specifically people travel or vehicle travel. Other networks may represent travel along an overhead conveyor or a monorail. The network is superimposed on the block layout, allowing the easy alteration of one or the other without having to always maintain integrity between them during the design process, which eases the editing process. Links of the network can be drawn proportional to their expected traffic. When transposition of a block layout with a travel network into a net layout or a processor layout proves cumbersome due to the need for major adjustments, then such a level of aggregation may not be appropriate for design purposes.

At the fourth level of aggregation, the travel network is not depicted, leaving only the block layout and I/O stations (e.g., Montreuil and Ratliff 1988a). Editing such a block layout with only input/output stations depicted is easy with most current drawing packages. These stations clearly depict where flow is to enter and exit each center in the layout. Even though the I/O stations of each center can be located anywhere within the center, in practice most of the times they are located either at center periphery or at its centroid. The former is usually in concordance with prior space specifications. It is commonly used when it is known that the center is to be an assembly line, a U-shape cell, a major piece of equipment with clear input and output locations, a walled zone with access doors, etc. The latter centroid location, right in the middle of the center, is mostly used when the center is composed of a set of processors and flow can go directly to and from any of them from or to the outside of the center. It is basically equivalent to saying that one has no idea how flow is to occur in the center or that flow is to be uniformly distributed through the center.

The absence of travel network representation assumes that the design of the network and the aisle set can be straightforwardly realized afterward without distorting the essence of the network, and that flow travel can be easily approximated without explicit specification of the travel network. Normally, one of the two following assumptions justifies flow approximation. The first is that a free flow movement is representative, computed either through the rectilinear or Euclidean distance between the I/O stations between which a flow is expected to occur. Figure 5.3 illustrates these two types of free flow. Euclidean distance assumes that one can travel almost directly from one station to another while rectilinear distance assumes orthogonal staircase travel along the X and Y axes, like through a typical aisle set when one does not have

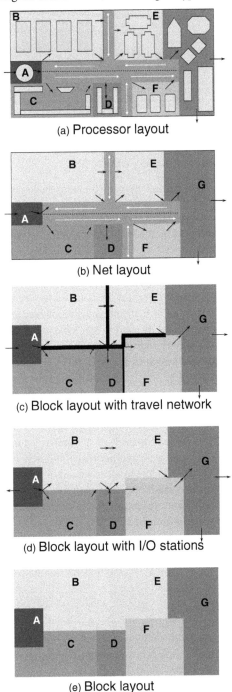

(a) Processor layout

(b) Net layout

(c) Block layout with travel network

(d) Block layout with I/O stations

(e) Block layout

FIGURE 5.2   Degrees of aggregation in layout representation for design purposes.

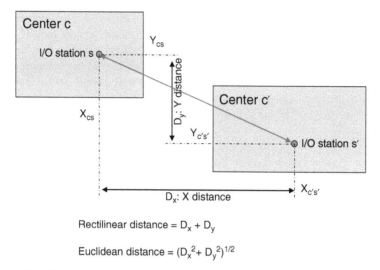

Rectilinear distance = $D_x + D_y$

Euclidean distance = $(D_x^2 + D_y^2)^{1/2}$

FIGURE 5.3 Free flow distance measured according to rectilinear or Euclidean distance.

to backtrack along any of the axes. The second alternative assumption is that flow travel is to occur along the center boundaries. Thus distances can be measured accordingly through the shortest path between the two I/O stations of each flow, along the contour network of the facility. This network is implicitly created by inserting a node at each corner of one center and/or the facility, and inserting a link along each center or facility boundary segment between the nodes. In Figure 5.2d, a flow from the northern output station of center B to the input station of center G would be assumed to travel from the output station of B southward along the west boundary of center B, then turning eastbound and traveling along the southern boundaries of center E, and keeping straight forward to reach the input station of center G.

At the fifth level of aggregation, only the block layout is drawn in Figure 5.2e. This is the simplest representation. On the one hand it is the easiest to draw and edit. On the other hand it is the most approximate in terms of location, shape, and flow. For the last 50 years, it has been by far the most commonly taught representation in academic books and classes, often the only one (e.g., Tompkins et al. 2003), and it has been the most researched. It is equivalent to the fourth level with all the I/O stations located at the centroid of their center. The underlying assumption justifying this level of aggregation is that the relative positioning and shaping of the centers embeds most of the design value and that this positioning and shaping can be done disregarding I/O stations, travel networks, aisle sets and processors, which are minor issues and will be dealt with at later stages. While in some settings this is appropriate, in many others such an aggregation can be dangerous. It may lead to the incorrect perception that the implemented layout is optimal because its underlying block layout was evaluated optimal at the highest level of aggregation, thus limiting and biasing the creative space of designers.

It is always a worthwhile exercise, when analyzing an existing facility, to draw and study it at various levels of aggregation. Each level may reveal insights unreachable at other levels, either because they do not show the appropriate information or because it is hidden in too much detail.

5.3 Space Representation

Location and layout is about locating and shaping centers in facilities or around the world. The design effort attempts to generate expected value for the organization through spatial configuration of the centers within a facility, or of facilities in wide geographical areas. Space is thus at the nexus of location and layout design. It is therefore not surprising that representation of space has long been recognized to be an important design issue. The essential struggle is between a discrete and a continuous representation of space.

Figure 5.4 allows contrasting both types of space representation for layout design purposes (Montreuil, Brotherton and Marcotte 2002). Leftmost is the simplest and freest continuous representation of space. In the top left, the facility is depicted as a rectangle within which the centers have to be laid out. In the bottom left, an example layout is drawn. To reconstruct this layout, a designer simply has to remember the shape of each center and the coordinates of its extreme points, as well as for the facility itself. Here centers have a rectangular shape, so one needs only to remember, for example, the coordinates of their respective southwest and northeast corners. As long as the shape specifications and spatial constraints are satisfied, the designer can locate and shape centers and the facility at will.

Third from the left in Figure 5.4 is a basic example of discrete space representation. Here, the top drawing represents space as an eight-by-eight matrix of unit discrete square locations. The size requirements of each center have to be approximated so that they can be stated in terms of number of unit locations. Shape requirements express the allowed assemblies or collages of these unit blocks. For example, the blocks are usually imposed to be contiguous. The length-to-width ratio and overall shape of the block assembly are also usually constrained. The design task is to best assign center blocks to discrete locations given the specified constraints. It is common for the discrete representation to force complex shapes for the centers in order to fit in the discrete facility matrix.

At first glance it seems hard to understand why one would want to use anything but a continuous space representation as it is more representative and natural. However, discrete space representation has a strong computational advantage, especially when a computer attempts to generate a layout using a heuristic. Manipulating continuous space and maintaining feasibility are much harder in continuous space for a computer. This is why the early layout heuristics such as CRAFT, CORELAP, and ALDEP (Armour and Buffa 1963, Lee and Moore 1967, Seehof and Evans 1967) in the 1960s used discrete space, and why many layout software applications still use it and researchers still advocate it. This trend is slowly getting reversed with more powerful heuristics, optimization models, and software. However, due to a long legacy, it is important for layout designers to master both types of representation.

Rightmost in Figure 5.4 is a more generalized nonmatrix discrete space representation. It corresponds to a facility that has a fixed overall structure characterized by a central loop aisle and centralized access on both western and eastern sides of the facility. The available space for centers becomes a set of discrete locations, each with specific dimensions. Such a kind of discrete representation is an interesting compromise, especially when space is well structured. For example, in a hospital the main aisle structure is often fixed and there are discrete rooms that cannot be easily dismantled or modified. With a discrete space representation, each room becomes a discrete space location. Even though a continuous representation can handle such cases a discrete representation can be adequate for design purposes.

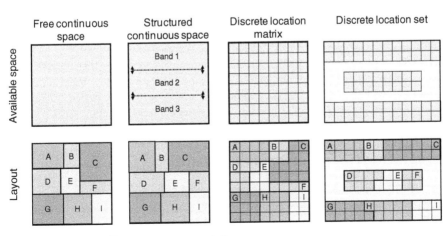

FIGURE 5.4   Continuous vs. discrete space representations.

Second from the left in Figure 5.4 is depicted a growing trend in layout software. It uses a continuous representation, yet it limits the layout possibilities through space structuring (Donaghey and Pire 1990, Tate and Smith 1995). Here the space structuring is expressed through the imposition of using three bands: a northern band, a central band, and a southern band. Within each band, centers have to be laid out side by side along the west–east axis. So the design process involves assigning centers to a band, specifying the order of centers within a band, sizing the width of each band (here its north–south length) and then shaping each center within its band. The advantage of such a representation is that with a simple layout code, the entire layout can be regenerated easily, provided simple assumptions exist, such as the sum of center areas equals the facility area. Here the code (1:A,B,C; 2:D,E,F; 3:G,H,I) enables reconstruction of the layout, provided that the facility shape is fixed. It states that northern zone one includes centers A, B, and C ranked in this order from west to east; and similarly for the other zones. Bands are one type of space structuring. Zones and space filling curves are other well-known methods (Meller and Bozer 1996, Montreuil et al. 2002a). Besides computational advantages, an interesting feature of space structuring is that it has the potential to foster simpler layout structures.

Space representation issues also involve the decision to explicitly deal with the 3D nature of facilities or to use a limited 2D representation. Simply, one should recognize that a facility is not a rectangle, for example, but rather a cube. Then one must decide whether he treats the cube as such or reduces it to a rectangle for layout design purposes. The height of objects, centers, and facilities becomes important when height-related physical constraints may render some layouts infeasible and when flow of materials and people involve changes in elevation. The most obvious situation is when one is laying out an existing multi-story facility with stairways and elevators. In most green field situations, the single-story vs. multi-story implementation is a fundamental decision. In some cases, it can be taken prior to the layout design study; in other settings it is through the layout design study that the decision is taken.

In forthcoming eras when space factories and nano factories are to be implemented, the 3D space representation will become mandatory. In space factories, the lack of gravity permits exploitation of the entire volume for productive purposes, objects moving as well up and down as from left to right. In nano factories, the forces influencing movement of nano objects are such that their travel behavior becomes complex. For example, nano objects may be attracted upward by other nearby objects.

As a final edge on the discrete vs. continuous space representation choice comes the notion of space modularity. To illustrate the notion, consider a facility where space is organized as the concatenation of $10 \times 10 \times 10$ ft^3 cubes. Centers and aisles are assigned to groups of such cubes, charged an occupancy rate per cube. Such a modular space organization may prove advantageous in certain settings in a stochastic dynamic environment (refer to subsequent Section 5.10). In such cases, then either a zone-based continuous representation or a discrete representation can be equivalently used.

The choice between discrete and continuous space representation is also a core decision in facilities location decision-making. Using a discrete representation requires selection in the early phases of the decision process the set of potential locations to be considered. The task is then to optimize the assignment of facilities to locations. When using a continuous representation, the decision-maker limits the boundaries of the space to be considered for potential location for each facility. Then the task is to optimize the coordinates of each facility. These coordinates correspond to the longitude and latitude of the selected location, or approximate surrogates. The compromises are similar as in layout design. Making explicit the characteristics of each potential location is easier with a discrete representation, yet this representation limits drastically the set of considered locations.

5.4 Qualitative Proximity Relationships

When spatially deploying centers in a facility or locating facilities around the world, there exist relationships between them that result in wanting them near each other or conversely far from each other. Such relationships can be between pairs of facilities or between a center and a fixed location. Each relationship exists for a set of reasons which may involve factors such as shared infrastructures, resources and personnel, organizational interactions and processes, incompatibility and interference, security and safety, as well as material and resource flow. Each case may generate specific relationships and reasons for each one.

These relations can be expressed as proximity relationships, which can be used for assessing the quality of a proposed design and for guiding the development of alternative designs. A proximity relationship is generically composed of two parts: a desired proximity and an importance level. Figure 5.5 shows a variety of proximity relationships between the 12 centers of a facility. For example, it states that it is important for centers MP and A to be near each other for flow reasons. It also states that it is very important that centers D and G be very far from each other for safety reasons. Such relationships can also be expressed with fixed entities. For example, in Figure 5.5, there are relationships expressed between a center and the outside of the facility. This is the case for center G: it is critical that it be adjacent to the periphery of the facility.

The desired proximity and the importance level can both be expressed as linguistic variables according to fuzzy set theory (Evans et al. 1987). In Figure 5.5, the importance levels used are vital, critical, very important, important, and desirable. The desired proximity alternatives are adjacent, very near, near, not far, far, and very far. Other sets of linguistic variables may be used depending on the case.

On the upper left side of Figure 5.5, the proximity relationships are graphically displayed, overlaid on the proposed net layout of the facility. Each relationship is drawn as a line between the involved entities. Importance levels are expressed through the thickness of the line. A critical relationship here is drawn as a 12-thick link while an important relationship is 3-thick. A vital relationship is 18-thick and is further highlighted by a large X embedded in the line. Gray or color tones can be used to differentiate the desired proximity, as well as dotted line patterns. Here a dotted line is used to identify a *not* distance variable such as *not far* or *not near*. In the color version of Figure 5.5, desired proximity is expressed through distinctive colors. For example, *adjacent* is black while *very far* is red. Such a graphical representation helps engineers to rapidly assess visually how the proposed design satisfies the proximity relationships. For example, in Figure 5.5, it is clearly revealed that centers E and PF do not respect the *very near* desired proximity even though it is deemed to be *critical*. Using graphical software it is easy to show first only the more important relationships, then gradually depict those of lesser importance.

Even though just stating that two centers are desired to be near each other may be sufficient in some cases, in general it is not precise enough. In fact, it does not state the points between which the distance is measured, and using which metrics. In Figure 5.5 are depicted the most familiar options within a facility. For example, inter-center distances can be measured between their nearest boundaries, their centroid, or their pertinent I/O stations. Distances can be measured using the rectilinear or Euclidean metrics, or by computing the shortest path along a travel network such as the aisle network. The choice has to be made by the engineer based on the logic sustaining the relationship. In the wide area location context, distances are similarly most often either measured as the direct flight distance between the entities or through the shortest path along the transport network. This network can offer multiple air-, sea-, and land-based modes of transportation.

When evaluating a design it is possible to come up with a proximity relationship-based design score. Figure 5.5 illustrates how this can be achieved. When starting to define the relationships, each importance level can be given a go/no-go status or a weight factor. In Figure 5.5, a *vital* importance results in an infeasible layout if the relationship is not fully satisfied. A critical importance level is given a weight of 64 while a desirable importance level has a weight of one. For each desired proximity variable a graph can be drawn to show the relationship satisfaction given the distance between the entities in the design. For example, in the upper right side of Figure 5.5, it is shown that the engineers have stated that a *not far* relationship is entirely satisfied within a 9-m distance and entirely unsatisfied when the distance exceeds 16 m. At a distance of 12 m it is satisfied at 50%. It is important to build consensus about the importance factors and proximity-vs.-distance satisfaction levels prior to specifying the relationships between the entities. Given a design, the distance associated with every specific relationship is computed. It results in a relationship satisfaction level. For example, it is important that centers A and C be near each other, as measured through the distance between their I/O stations assuming aisle travel. The computed distance is 12 m, which results

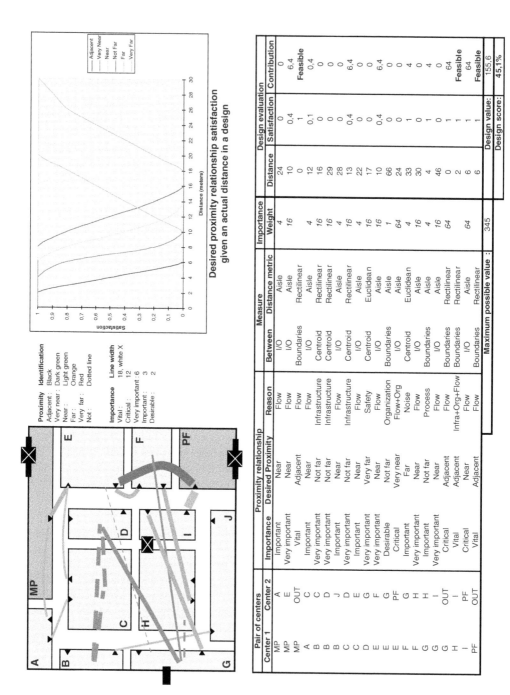

FIGURE 5.5 Qualitative proximity relationships based on evaluation of a layout design.

in a satisfaction level of 10%. Since the weight associated with such an important relationship is four, the contribution of this relationship to the design score is 0.4, whereas the upper bound on its contribution is equal to its weight of four. When totaling all relationships, the score contributions add up to a total of 155.6. The ideal total is equal to the sum of all weights, which in this case is 345. Therefore, the design has a proximity relationship score of 45.1%. This leaves room for potential improvement.

Simplified versions of this qualitative proximity relationships representation and evaluation scheme exist. For decades, the most popular has been Muther's AEIOUX representation (Muther 1961), where the only relationships allowed are: A for absolutely important nearness, E for especially important nearness, I for important nearness, O for ordinarily important nearness, U for unimportant proximity, and X for absolutely important farness. In most computerized implementations using this representation, a weight is associated to each type of relationship and proximity is directly proportional to the distance between the centroids of the related centers. Simpler to explain and compute, such a scheme loses in terms of flexibility and precision of representation.

In general, the reliance on qualitative relationships requires rigor in assessing and documenting the specific relationships. In an often highly subjective context, the relationship set must gain credibility from all stakeholders, otherwise it will be challenged and the evaluation based on the relationship set will be discounted. This implies that the perspectives of distinct stakeholders must be reconciled. For example, one person may believe a specific relationship to be very important while another may deem it merely desirable. Some may be prone to exaggerate the importance while others may do the inverse. It is also important to realize that some relationships may be satisfied with other means than proximity. For example, two centers may be desired to be far from each other since one generates noise while the other requires a quiet environment. If noise proofing isolation is installed around the former center, then the pertinence of the proximity relationship between the two centers may disappear.

5.5   Flow and Traffic

In most operational settings, the flow of materials and resources is a key for evaluating and optimizing a layout or location decision. It is sometimes sufficient to treat it through qualitative relationships as shown in Section 5.4. However, in most cases it is far more valuable to treat flow explicitly. Flow generally defines the amount of equivalent trips to be traveled from a source to a destination per planning period. There are two basic flow issues at stake here associated with implementing a design. First is the expected flow travel or flow intensity. Second is the flow traffic. The former is generically computed by summing over all pairs of entities having flow exchanges, the product of the flow value between them and their travel distance, time or cost, depending on the setting. Flow travel has long been used as the main flow-related criterion for evaluating alternative layout and location designs (Francis et al. 1992). The goal is for the relative deployment of entities to be such that travel generated to sustain the flow is as minimal as possible. The second flow issue, flow traffic, measures the load on the travel network, through intensity of flow through each of its nodes, links, and associated aisle segments and routes. Congestion along links and at nodal intersections is aimed to be minimized by the design (Benjaafar 2002, Marcoux et al. 2005).

Table 5.1 provides the flow matrix for the case of Figure 5.5. For example, it depicts that it is expected that there will be 125 trips per period from the output station of center A to the input station of center C. The matrix also illustrates a key issue when dealing with flow: the differences and complementarities between loaded travel and empty travel. Loaded travel corresponds to trips made to transport materials, and in general resources, from their source location to their target destination location. A forklift transferring a pallet of goods from location A to location B is an example of loaded travel. Empty travel occurs when the forklift reaches location B, deposits the transferred pallet, and becomes available for transporting something else while there is currently nothing to be transported away from location B. The forklift may wait there until something is ready for transport if the expected delay is short, but in many cases it will move to another location with a load to be transported away from it, causing empty travel. A ship

TABLE 5.1 Illustrative Flow Estimation Matrix for the Case of Figure 5.5

From/To	MP	A	B	C	D	E	F	G	H	I	J	PF	Loaded	Empty	Total From
MP		150	50			300							500	*0*	500
A		*175*	25	125	25								175	*175*	350
B			*130*	15	10					100	5		130	*130*	260
C				*215*	40	175							215	*215*	430
D		25	10	40	*100*						25		100	*100*	200
E			45			*475*	300		15			430	790	*475*	1265
F							*300*		300				300	*300*	600
G									350				350	*0*	350
H								15	*300*	300			300	*315*	615
I						*170*				*500*		500	500	*670*	1170
J				35	25					20	*125*	45	125	*125*	250
PF	*500*					*145*		*335*					0	*980*	980
Loaded	0	175	130	215	100	475	300	0	315	670	125	980	**3485**		
Empty	*500*	*175*	*130*	*215*	*100*	*790*	*300*	*350*	*300*	*500*	*125*	*0*		3485	
Total to	500	350	260	430	200	**1265**	600	350	615	**1170**	250	**980**			**6970**

Entries: Trips/period.
Shaded and italics: Empty.
Bold: High relative value.
Loaded trip entries: From the output station of source center to the input station of destination center.
Empty trip entries: From the input station of source center to the output station of destination center.

transporting containers from China to Canada is another example of loaded travel, while the same ship traveling empty to pick up containers in Mexico is an example of empty travel.

In the flow matrix of Table 5.1, empty travel is written in italics. For example, it shows that 500 empty trips are to be expected from center PF to center MP. To be precise, the empty trips are from the input station of center PF to the output station of center MP, bringing transporters to enable departures from center MP. Similarly, Table 5.1 depicts that 175 trips per period are expected from the input station of center A to the output station of center A. Table 5.1 indicates a total flow of 7,320 trips per period, split equally between 3,485 loaded trips and 3,485 empty trips per period. By a simple usage of bold characters, Table 5.1 highlights the most important flows for layout analysis and design.

Table 5.2 provides the distance to be traveled per trip assuming vehicle-based aisle travel in the layout of Figure 5.5. The provided distances are between the I/O stations of the centers having positive flow.

The travel matrix of Table 5.3 is derived by multiplying the flow and distance for each corresponding matrix entry of Tables 5.1 and 5.2. For example, travel from the output station of center G to the input station of center I is estimated to be 16,100 m per period. The expected total travel is 151,410 m per period. So by itself the G to I flow represents roughly 11% of the total travel. Table 5.3 presents an interesting evaluation metric, which is the average travel. It simply divides the total travel by the total flow. Here it allows one to state that the average travel is 22 m per trip, with 24 m per loaded trip and 20 m per empty trip. An engineer can rapidly grasp the relative intensity of travel with such a metric. Here 22 m per trip is a high value in almost every type of facility, readily indicating a strong potential for improvement. The lower portion of Table 5.3 depicts total flow, total travel, and average travel for each center. This highlights that centers E, I, and PF each have a total travel higher than 40,000 m per period, that centers MP, D, and J each have an average travel around 30 m per trip and that center G has an average travel of 40 m, making these centers the most potent sources of re-layout improvement.

Given the flows of Table 5.1 and the layout of Figure 5.5, traffic can be estimated along each aisle segment and intersection. Assuming shortest path travel, Figure 5.6 depicts traffic estimations. The aisles forming the main loop contain most of the traffic. Only one small flow travels along a minor aisle, east of centers C and H. In fact, it reveals that most of the minor aisles between centers could be deleted without

TABLE 5.2 Distance Matrix for the Layout of Figure 5.5

From/To	MP	A	B	C	D	E	F	G	H	I	J	PF
MP		24	30			10						
A		8	6	12	28							
B			16	26	42						28	56
C				4	16	22						
D		46	50	40	24						50	
E			48			16	10		42			24
F							10		30			
G								20		46		
H								16	8	38		
I						12				14		6
J				64	48					22	26	10
PF		42				24		34				

Entries: Trips/period.
Shaded and italics: Empty.
Loaded trip entries: Distance from the output station of source center to the input station of destination center.
Empty trip entries: Distance from the input station of source center to the output station of destination center.

forcing longer travel. The main south and east aisles get most of the traffic. The most active corners are the I-PF-J and D-E-MP intersections. However, the smooth distribution of traffic does not emphasize hot spots for congestions. Further analyses based on queuing theory (Kerbache and Smith 2000, Benjaafar 2002) or relying on discrete event simulations (Azadivar and Wang 2000, Huq et al. 2001, Aleisa and Lin 2005) would be required to estimate congestion effects in more depth.

Table 5.1 provides expected flows for the illustrative case. In practice, the engineer has to estimate these flows. There are basically two ways used to do so. The first is to track actual flows occurring in the actual facility during a sampling period and to extrapolate the expected flows from the sampling results, taking into account overall expected trends in demand. In technologically rich settings, precise tracking of actual flow can be achieved through the use of connective technologies such as GPS, RFID, or bar coding, using tags attached to the vehicles and/or objects being moved. In other settings, it requires people to perform trip samplings.

The second way to estimate flows is to rely on product routing and demand knowledge for estimating loaded trips and to rely on approximate analytical or simulation-based methods for estimating empty trips. Illustratively, Table 5.4 provides the planned inter-center routing and expected periodic demand for each of a set of 18 products. From these can be estimated the loaded flows of Table 5.1. For example, in Table 5.1 there is a flow of 25 loaded trips per period from center A to center D. From Table 5.4, the A to D flow is estimated through adding trips from A to D in the routings of products 5, 7, 8, and 9.

Whereas the loaded flow estimation is here rather straightforward, the estimation of empty flow requires assumptions on the behavior of vehicles when they reach an empty status and on the dispatching policy of required trips to individual vehicles. In Table 5.1, the empty flow is estimated using the two following simple assumptions. First, vehicles reaching the input station of a center are transferred in priority to the output station of that center to fulfill the needs for empty vehicles. Second, centers with exceeding incoming vehicles aim to transfer the exceeding vehicles to the nearest center having a lack of incoming loaded vehicles to fulfill its need for departing vehicles. A transportation model is used to allocate empty vehicle transfers according to center unbalances, as originally advocated by Maxwell and Muckstadt (1982). There exists a variety of alternative methods for empty travel estimation (see e.g., Ioannou 2007). It is important for the method to reflect as precisely as possible the behavior expected in the future layout implementation.

The illustrated approach for estimating flows from product routing and demand permits highlighting of three fundamental issues. First, the computations divide the expected demand by the transfer lot in order to estimate the trips generated by a product routing segment. However, in practice the transfer lot is

TABLE 5.3 Travel Evaluation for the Layout of Figure 5.5 Based on the Flow Matrix of Table 5.1

From/To	MP	A	B	C	D	E	F	G	H	I	J	PF	Loaded	Empty	Total	Average
MP		3600	1500			<u>3000</u>							8100	0	8100	16
A		*1400*	150	1500	700								2350	1400	3750	11
B			*2080*	390	420					2800		280	3890	2080	5970	23
C				*860*	640	3850							4490	860	5350	12
D		1150	500	1600	*2400*						1250		4500	2400	6900	35
E			2160			***7600***	3000		630			**10320**	16110	7600	23710	19
F							*3000*		**9000**				9000	3000	12000	20
G										**16100**			16100	0	16100	46
H								*240*	*2400*	8600	2800		11400	2640	14040	23
I						*2040*				***7000***		3000	3000	9040	12040	10
J				2240	1200					440	*3250*	<u>450</u>	4330	3250	7580	30
PF	*21000*					*3480*		***11390***					0	35870	35870	37
Loaded	0	4750	4310	5730	2960	6850	3000	0	9630	27940	4050	14050	83270			24
Empty	21000	1400	2080	860	2400	13120	3000	11630	2400	7000	3250	0		68140		20
Total	21000	6150	6390	6590	5360	19970	6000	11630	12030	34940	7300	14050			151410	22
Average	42	18	25	15	27	16	10	33	20	30	29	14	24	20	22	
Total flow⊠	1000	700	520	860	400	2530	1200	700	1230	2340	500	1960			6970	22
Total travel⊠	29100	9900	12360	11940	12260	43680	18000	27730	26070	46980	14880	49920			151410	
Average travel⊠	29	14	24	14	31	17	15	40	21	20	30	25			22	

Entries: Trips/period, ⊠meters/period.
Shaded and Italics: Empty.
Bold: High relative value.
Underline: Low value given high flow.

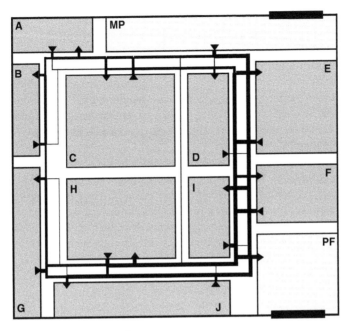

FIGURE 5.6 Expected traffic in the current layout.

often dependent on the distance to be traveled and the type of handling system used. This illustrates the typical chicken-and-egg phenomenon associated with layout design and material handling system design, requiring iterative design loops to converge toward realistic estimates.

Second, in the lean manufacturing paradigm, large transfer lots are perceived as an inefficiency hideout (Womack and Jones 1998), leading to the proposition that the layout be designed assuming a will to use transfer lots of one. The optimal layout assuming the stated transfer lots may well be different from the

TABLE 5.4 Product Routings and Expected Product Demand

Product	Demand	Transfer Lot	Trips/ Period	Inter-Center Routing										
1	2500	20	125	IN	MP	A	C	E						
2	560	16	35	IN	MP	B	J	C	D	C	E			
3	5250	15	350	IN	G	I	PF	OUT						
4	225	15	15	IN	MP	B	J	D	J	I				
5	120	12	10	IN	MP	A	D	A	B	C	E			
6	3000	10	300	IN	MP	E								
7	50	10	5	IN	MP	A	D	A	B	C	D	C	E	
8	30	6	5	IN	MP	A	D	A	B	D	B	PF	OUT	
9	30	6	5	IN	MP	A	D	A	B	D	B	J	D	J PF OUT
10	875	5	175	E	F	H	I							
11	650	5	130	E	PF	OUT								
12	25	5	5	E	B	J	D	J	I					
13	25	5	5	E	B	J	PF	OUT						
14	75	5	15	E	H									
15	625	5	125	E	F	H	I	PF	OUT					
16	175	5	35	E	B	J	PF	OUT						
17	1500	5	300	E	PF	OUT								
18	125	5	25	I	PF	OUT								

IN and OUT, respectively, refer to inbound from suppliers and outbound to clients.

The transfer lot expresses the number of units planned to be transported concurrently in each trip.

optimal layout assuming unitary transport. This illustrates the interaction between layout design and operating system planning, requiring a fit between their mutual assumptions.

Third, in the illustration, all trips are hypothesized equivalent. In practice, a forklift trip carrying a standard pallet is not equivalent to a forklift transporting a 10-m long full metal cylinder with a diameter of 30 cm, the latter being much more cumbersome and dangerous. Compare a forklift trip with a walking individual transporting a hammer. This is why the flow definition used the notion of equivalent trips, requiring the engineer to weigh the different types of trips so that the layout compromises adequately, taking into consideration their relative nature. Muther (1961) proposes a set of preset weights to standardize trip equivalence computations.

It is important whenever possible to transpose the travel and traffic estimations in terms of operating cost and investment estimations. This is often not a straightforward undertaking. Flow travel and traffic influence differently the operating cost and investment in a facility depending on whether handling involves trips by humans and/or vehicles or it involves items moving along a fixed system such as a conveyor. When using a conveyor to travel between two points, there is a fixed cost to implement the conveyor, and there is often negligible cost involved in actually moving specific items on the conveyor. Flow traffic along a conveyor influences investments in a staircase fashion. As traffic gets closer to the upper bound manageable by a given technology, faster technology is required that costs more to acquire and install.

When trip-based travel is used, then flow travel increases translate more directly into cost and investment increases. First, each vehicle spends costly energy as it travels. Second, as travel requirements augment, the number of required vehicles and drivers generally augments in a discrete fashion. Third, higher traffic along aisle segments and intersections may require implementing multiple lanes, extending the space required for aisles and affecting the overall space requirements. Fourth, when using trip-based travel, the time for each trip is the sum of four parts: the pickup time, the moving time, the waiting time, and the deposit time. The pickup and deposit times are mostly fixed, given the handling technology selected for each trip and the items to be maneuvered. They range mostly from a few seconds to a minute. The only ways to reduce them are by improving the technology and its associated processes, and by avoiding making the trip. The latter can be achieved when the I/O stations are laid out adjacent to each other, or when the flow is reassigned to travel along a fixed infrastructure. The moving time depends both on the path between the entities and the speed and maneuverability of the handling technology used. The waiting time occurs when traffic becomes significant along aisles and at intersections. In small facilities, it is often the case that pickup and deposit times dominate moving and waiting times because of short distances.

In location decisions, cost estimation relative to flow travel and traffic involves making assumptions or decisions relative to the transportation mode to be used (truck, plane, boat, etc.), fleet to be owned or leased, routes to be used and contracts to be signed with transporters and logistic partners. Congestion is not along an aisle or at an intersection in a facility. It is, rather, along a road segment, a road intersection, at a port, at customs, etc. The geographical scope is generally wider, yet the logical issues are the same.

5.6 Illustrative Layout Design

In order to provide an example of layout design, an engineer has been mandated to spend a day trying to develop an alternative design for the case used in the previous sections. He was simply provided with the case data and given access to spreadsheet and drawing software. The case data includes the relationships of Figure 5.5, the flows of Table 5.1, and the space requirements of Table 5.5. For safety reasons, it has also been required that at least four distinct aisles provide access to the exterior of the facility.

The engineer has first developed the design skeleton of Figure 5.7. This design skeleton is simply a flow graph. The engineer has drawn nodes for each center. The node diameter is proportional to the area requirements for the center. The loaded flows have been drawn as links whose thickness is proportional to flow intensity. The engineer has placed the nodes relative to each other and the exterior so as to approximately minimize travel and to respect roughly the qualitative proximity relationships. He has also decided to split facility input between two main entrances IN1 and IN2.

TABLE 5.5   Space Requirements for the Illustrative Case

Center	Minimal Area Requirements	Length-to-Width Maximum Ratio	Fixed Shape	Fixed Relative Location of I/O Stations	Can Be Mirrored
A	18	2.0	N	N	Y
B	16	4.0	Y	Y	N
C	64	2.0	N	N	Y
D	24	3.0	N	N	Y
E	48	2.0	N	N	Y
F	30	2.0	N	N	Y
G	26	6.5	N	N	Y
H	54	2.0	N	N	Y
I	21	3.0	N	N	Y
J	39	5.0	Y	N	Y
MP	51	6.0	N	N	Y
PF	42	2.0	N	N	Y
Building		3.0	N	N	Y

Then the engineer has transformed the design skeleton into an actual layout with three self-imposed objectives: (*i*) stick as much as possible to the design skeleton relative placement, (*ii*) minimize space by avoiding unnecessary aisles, and (*iii*) keep the shape of centers and building as simple as possible. The engineer has personally decided to put priority on minimizing flow travel, qualitative proximity relationships being a second priority. While developing the design, the engineer has iteratively estimated empty travel, using two simple self-imposed rules: (*i*) give priority of destination choice to empty trips from input stations of centers with higher inbound loaded flow and (*ii*) avoid assigning more than roughly half the empty trips out of a station to any specific destination. This is a looser estimation than that made for the current design, while being defendable as a viable operating strategy to deal with empty travel. Figure 5.8 depicts the resulting alternative design preferred by the engineer.

Table 5.6 provides the flow matrix resulting from the engineer's empty travel estimation. Examination of the empty trip allocations show, for example, that the engineer has assigned empty travel out of the

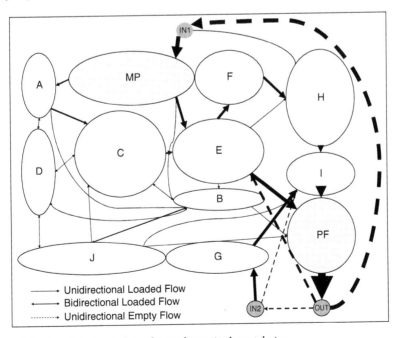

FIGURE 5.7   A flow based design skeleton for an alternative layout design.

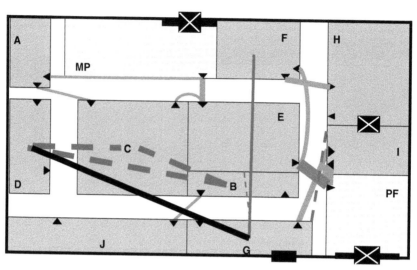

FIGURE 5.8   Alternative layout 1 with superimposed qualitative relationships.

critical PF center to nearby centers E, G, and I, while the empty flows out of low-inbound-traffic center D have been assigned to more dispersed centers G and J, as well as to the output station of center D itself. Overall it has the same amount of empty trips. They are simply reshuffled differently given the proposed layout.

From an expected performance perspective, Table 5.7 shows that the expected total travel for the alternative design is now estimated at 84,765 m per period and the average travel is now at 12 m per trip, a 44% reduction over the current design. Table 5.8 provides its proximity relationship score of 48.9%, an 8.4% improvement over the current design. From a space perspective, the alternative layout slightly reduces the space requirements for the building. Its area shrinks from 441 to 435 square feet. This is mostly because of the reduction of unnecessary aisles in the alternative design.

5.7   Exploiting Processing and Spatial Flexibility

A key issue in location and layout design has become to exploit the flexibility offered by new technologies and means of operations. Processing flexibility allows processors to be allocated a variety of products to be treated, each with a given performance rating. Spatial flexibility occurs when management accepts multiple centers or processors, either identical or having intersecting capabilities, to be distributed through the facility. The combination of processing and spatial flexibility has the potential to improve significantly the design performance. Simple examples can be seen in everyday life. Switching from a single centralized toilet area or break area in a facility to multiple smaller areas spread through the facility has significant impact on people movement. A chain adding another convenience store in a city both helps it reach new customers through better convenience and reshuffles its clientele among the new and existing stores.

Exploiting flexibility makes the design process more difficult as it involves treating the flows as variables, rather than mere inputs, and dealing with capacity. The flows indeed become dependent on the relative locations and performance of entities. Thus, to evaluate a design, one has to estimate how in future operations the flows will be assigned given the design and the operating policies. Given the estimated flow, one can then apply the travel and traffic scoring methods shown in Section 5.5.

For illustrative purposes, assume that in the case used in the previous sections, each center is devoted to a single process and is composed of a specific number of identical processors, as shown in processor layout 1 in Figure 5.9. For simplicity purposes, assume also that the processing times for each process are product independent as stated in Table 5.9.

TABLE 5.6 Flow Estimation for Layout Alternative

From/To	MP	A	B	C	D	E	F	G	H	I	J	PF	Loaded	Empty	Total From
MP		150	50			300							**500**	*0*	500
A		*70*	*25*	125	25+15						*90*		175	*175*	350
B			*65*	15	10						100	5	130	*130*	260
C	50	*105*		*215*	40+60	175							215	*215*	430
D		25	*45*	40	*25*						25+35		100	*100*	200
E	260					300	300		15	350		430	**790**	*475*	**1265**
F	150					150		40	300				300	*300*	600
G	40						65						350	*0*	350
H						150	15		110	180		500	300	*315*	615
I							60	170	190	20		45	**500**	*670*	**1170**
J			65	35	25					125			125	*125*	250
PF						490	300	*170*	300	*320*	125		**0**	*980*	**980**
Loaded	0	175	130	215	100	790	300	0	300	500	125	980	3485		
Empty	**500**	*175*	*130*	*215*	*100*	*475*	*300*	*350*	*315*	*670*	*125*	*0*		3485	
Total to	500	350	260	430	200	**1265**	600	350	615	**1170**	250	980			6970

Entries: Trips/period.
Shaded and italics: Empty.
Bold: High relative value.
Squared: Both loaded and empty travel.
Loaded trip entries: From the output station of source center to the input station of destination center.
Empty trip entries: From the input station of source center to the output station of destination center.

TABLE 5.7 Travel Evaluation for Layout Alternative

From/To	MP	A	B	C	D	E	F	G	H	I	J	PF	Loaded	Empty	Total	Average
MP		**3600**	2000			1200							**6800**	_0_	6800	14
A			_1500_	1250	_320_						_2250_		2800	_3080_	5880	17
B		_560_		630	440			_390_			800	110	1980	_1300_	3280	13
C	_1000_	_1470_	910			1050				_1440_			1930	_3910_	5840	14
D		550		800				_1840_			_2540_		2440	_4180_	6620	**33**
E	_1040_		540	_2150_			**6600**		255			**3440**	**10835**	_3190_	**14025**	11
F	_3000_							_1200_	2100				2100	_4200_	6300	11
G										5600			5600		5600	16
H	_760_				725	_1800_	_1050_			**3600**			3600	_3020_	6620	11
I					450		_1200_		_1210_			5000	5000	_4760_	9760	8
J				945					_1520_	1020		2025	4715	_1785_	6500	**26**
PF						_3920_		_1700_		_1920_			0	_7540_	7540	8
Loaded	0	4150	4480	3625	2095	2250	6600	0	2355	10220	1450	10575	47800			14
Empty	_5800_	_2030_	_1495_	_2150_	_1710_	_5720_	_2250_	_5130_	_2730_	_3360_	_4140_	_0_		36965		11
Total	5800	6180	5975	5775	4255	7970	8850	5130	5085	13580	5590	10575			84765	12
Average	12	18	23	13	21	6	15	15	8	12	22	11	14	11	12	
Total flow*	1000	700	520	860	400	2530	1200	700	1230	2340	500	**1960**			6970	
Total travel*	12600	12060	9255	11615	10875	21995	15150	10730	11705	23340	12090	18115			**84765**	
Average travel*	13	17	18	14	27	9	13	15	10	10	24	9			12	

Entries: Trips/period, * meters/period.
Shaded and italics: Empty.
Bold: High relative value.
Underline: Low value given high flow.

TABLE 5.8 Evaluation of Layout Alternative Based on Qualitative Relationships

Pair of Centers		Proximity Relationship			Measure		Importance		Design Evaluation	
Center 1	Center 2	Importance	Desired Proximity	Reason	Between	Distance Metric	Weight	Distance	Satisfaction	Contribution
MP	A	Important	Near	Flow	I/O	Aisle	4	24	0	0
MP	E	Very important	Near	Flow	I/O	Aisle	16	4	1	16
MP	In1	Vital	Adjacent	Flow	Boundaries	Rectilinear		0	1	Feasible
A	C	Important	Near	Flow	I/O	Aisle	4	10	0.45	1.8
B	C	Very important	Not far	Infrastructure	Centroid	Rectilinear	16	22	0	0
B	D	Very important	Not far	Infrastructure	Centroid	Rectilinear	16	37	0	0
B	J	Important	Near	Flow	I/O	Aisle	4	8	0.85	3.4
C	D	Very important	Not far	Infrastructure	Centroid	Rectilinear	16	15	0.1	1.6
C	E	Important	Near	Flow	I/O	Aisle	4	6	1	4
D	G	Very important	Very Far	Safety	Centroid	Euclidean	16	35	1	16
E	F	Very important	Near	Flow	I/O	Aisle	16	22	0	0
E	G	Desirable	Not far	Organization	Boundaries	Aisle	1	8	1	1
E	PF	Critical	Very near	Flow + Org	I/O	Aisle	64	8	0.2	12.8
F	G	Important	Far	Noise	Centroid	Euclidean	4	32	1	4
F	H	Very important	Near	Flow	I/O	Aisle	16	7	0.95	15.2
G	H	Important	Not far	Process	Boundaries	Aisle	4	18	0	0
G	I	Very important	Near	Flow	I/O	Aisle	16	16	0	0
G	OUT	Critical	Adjacent	Flow	Boundaries	Rectilinear	64	0	1	64
H	I	Vital	Adjacent	Infra + Org + Flow	Boundaries	Rectilinear		0	1	Feasible
I	PF	Critical	Near	Flow	I/O	Aisle	64	10	0.45	28.8
PF	OUT	Vital	Adjacent	Flow	Boundaries	Rectilinear		0	1	Feasible
						Maximum possible value:	345		Design value:	168.6
									Design score:	48.9%

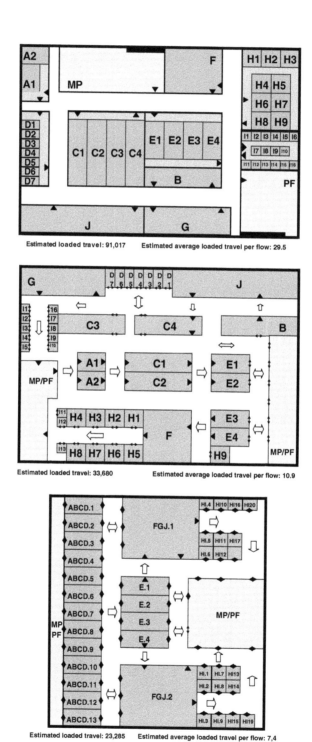

FIGURE 5.9 Processor layouts of alternatives 1, 2 with spatial flexibility and 3 with added processing flexibility.

TABLE 5.9   Elemental Process and Specialized Processor Specifications

Operation Type	A	B	C	D	E	F	G	H	I	J
Unit time (minutes)	0.6	0.5	0.9	5	0.3	0.2	0.1	5	1.5	0.7
Number of processors	2	1	4	7	4	1	1	9	13	1
Processor size	3*2	2*9	2*7	1*2	2*5	6*5	2*13	2*2	1*1	3*13
Expected net utilization (%)	93	68	90	88	80	31	55	91	86	96

A period is set to a 20-hour workday.
Processor efficiency is estimated at 80%.

As in Section 5.6, an engineer was again asked to generate an alternative layout exploiting this knowledge and the potential for spatially dispersing identical processors instead of grouping them in a single functional center. He was allowed to use flexible centers responsible for both inbound materials and outbound products. He generated alternative layout 2 shown in Figure 5.9. First, given the relatively small size of the case, he has decided not to create centers and has, rather, developed the design directly at the processor level. Second, he has indeed exploited flexibility allowed to disperse processors. He has strictly separated groups of processors of types H and I. He has contiguously laid out processors of types C and E, yet has oriented them so as to better enable efficient travel for distinct products. Third, the spatial dispersion exploited is not extreme. In fact, it is limited to a fraction of the overall design.

Assume now that there exist flexible processors capable of performing multiple processes. In fact, here assume there are three flexible processor types respectively termed ABCD, FGJ, and HI capable of performing the processes embedded in their identifier. In order for the example to focus strictly on exhibiting the impact of flexibility, first the processing times are identical as in Table 5.9 and, second, the flexible processors have space requirements such that the overall space they jointly need is the same as the original specialized processors. The engineer was again required to generate an alternative layout exploiting this flexibility as well as spatial dispersion. He has designed the significantly different alternative layout 3 in the lower part of Figure 5.9.

The design scores provided under the layouts of Figure 5.9 illustrate vividly the potential of exploiting spatial dispersion and flexibility. Alternative 1 is used as a comparative basis. It has an estimated loaded travel score of 91,017. Alternative 2, exploiting spatial dispersion, has an estimated loaded travel score of 33,680, slicing 63% off alternative 1's travel. Alternative 3 reduces further the estimated loaded travel score to 23,285, which slices 31% off alternative 2's travel.

The scores have been estimated by assuming that the factory operating team will favor the products with a high number of equivalent trips when assigning products to processors. Heuristically, the engineer has first assigned the best paths to products P3, P6, and so on, taking into consideration processor availability and processing times. For example, in alternative 2, product P3 getting out of center G is given priority for routing to processors I1 to I9 and then to the nearby MP/FP center.

When locating facilities around the world or processors within a facility, exploiting flexibility leads to what are known as location-allocation problems (Francis et al. 1992). The most well-known illustration is the case where distribution centers have to be located to serve a wide area market subdivided as a set of market zones or clients. There are a limited number of potential discrete locations considered for the distribution centers. Each distribution center is flexible, yet has a limited throughput and storage capacity which can be either a constraint or a decision variable.

The assignment of market zones to specific distribution centers is not fixed a priori. The unit cost of designating a zone through a distribution center located at a given discrete location is precomputed for each potential combination, given the service requirements of each market zone (e.g., 24-h service). The goal is to determine the number of distribution centers to be implemented, the location and capacity of each implemented center, and the assignments of zones to centers. This can be done for single product cases and for multiple product cases. The same logic applies for flexible factories aimed to be spread around a wide market area so as to serve its production to order needs.

5.8   Dealing with Uncertainty

Explicitly recognizing that the future is uncertain is becoming ever more important in location and layout design. Such designs aim to be enablers of future performance. A design conceived assuming point estimates of demand may prove great if the future demand is in line with the forecast. However, it can prove disastrous if the forecast is off target (Montreuil 2001). Stating intervals of confidence around demand estimates may be highly beneficial to the engineer having to generate a design. For example, consider the demand estimates provided in Table 5.4. The demand for product P1 is forecast to be 2,500 per period. It makes quite a difference if the forecaster indicates that within 99% the demand is to be between 2,400 and 2,600, between 2,000 and 2,700, or between 0 and 7,500. Applied to all products, it significantly influences flow, capacity usage, and required resources to sustain desired service levels.

The case when the product mix is known, the demand for each product is known with certainty, as well as their realization processes, is fast becoming an exceptional extreme. Therefore, the engineer must gauge the level of uncertainty concerning each of these facets, and ensure that he develops a design that will be robust when faced with the uncertain future.

As proposed by Marcotte et al. (2002), Figure 5.10 depicts a graph where each dot corresponds to a design. Each design has been evaluated under a series of scenarios. The graph plots each design at the coordinates corresponding to the mean and standard deviation of its score over all scenarios. The ideal design has both lowest mean and standard deviation. However, as shown in Figure 5.10, often there is not such a single dominating design. In fact, an efficient robust frontier can normally be composed by a series of designs that are not dominated by any other design through its combination of mean and standard deviation. In the case of Figure 5.5, there are five such designs. The leftmost design along the frontier has the lowest mean and the highest standard deviation, whereas the rightmost design has the highest mean and the lowest standard deviation. The mean and standard deviation for each of the five dominating designs are respectively (2,160; 315), (2,240; 235), (2,260; 215), (2,760; 180), and (3,600; 155) from left to right. The choice between the five designs becomes a risk management compromise. A more adventurous management is to opt for designs on the left, while more conservative management is to opt for designs on the right. For example, if a two-sigma robustness is desired, this means that the comparison should be around the sum of the mean and two standard deviations. Here this results in looking for the minimum between (2,790; 2,710; 2,690; 3,120; 3,910). This means that the third from left design on the robustness frontier is the most two-sigma robust design. In fact, the leftmost design is the most one-half-sigma robust. The three leftmost designs are equivalent at one-sigma, and then the third from left is the most robust at two-sigma, three-sigma, and four-sigma, making a sound choice for a wide variety of risk attitudes.

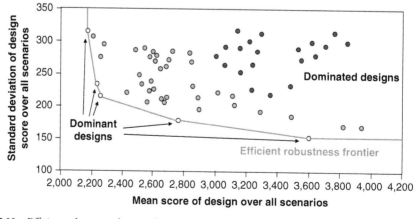

FIGURE 5.10   Efficient robustness frontier for a set of designs subject to stochastic scenarios.

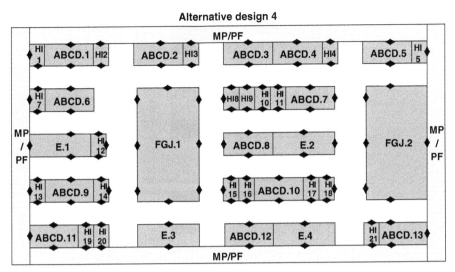

FIGURE 5.11   Alternative layout designed for high uncertainty, given the same set of processors as alternative 3 in Figure 5.9.

The *a priori* acknowledgment of uncertainty should lead engineers to generate designs appropriate for the uncertainty level. As an illustration, an engineer has been requested to generate a layout using the same processors as the third alternative in Figure 5.9, but for a situation where there is complete uncertainty relative to (*i*) the product mix, (*ii*) the demand and (*iii*) the realization process for the products. He has generated a design exploiting the holographic layout concept (Montreuil and Venkatadri 1991, Marcotte et al. 1995, Montreuil and Lefrançois 1996, Lamar and Benjaafar 2005), which differs significantly from all previous alternatives. In high uncertainty contexts, the holographic layout concept suggests strategic spreading of copies of the identical processors through the facility space so that from any type of processor there are nearby copies of every other type (Fig. 5.11). This distribution ensures a multiplicity of short paths for a variety of product realization processes. This can be verified from Figure 5.10 by randomly picking series of processor types and attempting to find a number of alternative distinct short paths visiting a processor of each type through the facility in the randomly generated order.

5.9   Dealing with an Existing Design

The vast majority of layout and location decisions have to take into consideration the fact that there exists an implemented current design that will have to be transformed to become the selected next design. In some cases, it is an insignificant matter to re-layout or to redeploy facilities. In such cases the next design can be developed without explicit consideration of the actual design.

However in most cases, reshuffling an actual implementation is not that easy. At the extreme, some processors cannot be moved. They have become monuments in the facility. An example is a papermaking machine in a paper factory: once installed, you do not move it. Between the two extremes of move at no cost and move at infinite cost lies an infinite spectrum of situations.

Figure 5.12 indicates graphically an interesting way to approach re-layout studies when there are non-negligible moving costs. Iteratively, the engineer should generate alternative designs which take as fixed all entities having at least a specified level of moving cost. On the top portion of Figure 5.12 is displayed the current design, here assumed to be layout 2 from Figure 5.9, displaying through gray tones the expected moving cost associated with each processor. At the first level, the engineer erases only the entities that have negligible moving costs. At the second level, he erases all entities with nonimportant moving costs.

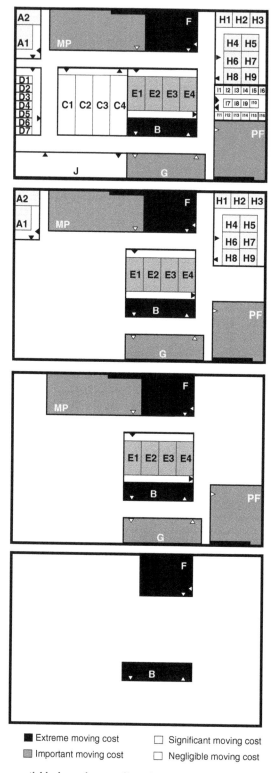

FIGURE 5.12 Design space available dependent on allowed move costs.

At the third level he erases all entities that do not have extreme moving costs. At the fourth and final level, he erases all entities.

At each level, the engineer generates a variety of alternative designs. This results in a pool of alternative designs with distinct estimated moving costs. This allows the engineer to really size the impact of moving costs. Also, when he presents them to the management, it has the potential to generate pertinent managerial discussions, beyond the current layout decision to be taken, which may open avenues in the future. Examples include aiming to implement easy-to-relocate processors, and avoiding putting monuments in the center of action of the facility.

Designs should be compared based on the expected operating cost, but also adding to it are the design transformation expenditures. Moving costs generally involve a fixed cost whenever the entity is even slightly moved. There is often a low level cost whenever the entity is moved within a nearby limited space from its current location. The cost then increases significantly when the move is outside this nearby region. The cost can be fixed as soon as there is a displacement or it can be proportional to the distance being moved. Move costs are sometimes not computable separately, entity per entity: they depend on the set of moves to be concurrently undertaken. It is interesting to assess that the cost of transforming layout 2 into layout 3 in Figure 5.9 would be astronomical given the moving cost specifications of Figure 5.12.

The second aspect relative to redesign is the timing of moves and its impact on current operations and overall implementation cost associated with transforming the current design into the prescribed design. In many settings, the space is so tight that in order to make some moves feasible, some other space has to be created a priori. This creates a cascading effect of interdependent moves which can have impact on the transformation feasibility and cost (Lilly and Driscoll 1985). Also in many settings the operations cannot be stopped for significant durations while the transformation occurs, sometimes except if stocks can be accumulated ahead of time. Some transformations may make it easy to continue operations during the moves. Others may make it very cumbersome and costly. Therefore, it is important to generate a time-phased moving plan that is proven feasible and whose cost is rigorously estimated.

5.10   Dealing with Dynamic Evolution

With the acute shrinking of product life cycles as well as the increasing pace of technological and organizational innovation, in most situations facilities should not anymore be located and laid out assuming a steady state perspective as was generally done in the past. Layout and location dynamics, explicitly considering the time-phased evolution of facilities and networks, is thus also becoming a key issue for the engineer (Rosenblatt 1986, Montreuil 2001). He has to recognize that as the current design is about to be transformed into the proposed design, this proposed design will have a finite existence. It will also have to be transformed into a subsequent design at a later time. The same will occur to this subsequent design and all subsequent others, in a repeating cycle over the entire life of the facility in layout cases or the network of facilities in location cases.

Only when relaying out or redeploying facilities involves insignificant efforts can the engineer optimize the next design strictly for the near future expectations, as (*i*) there will be negligible costs in transforming the current design into the next design and (*ii*) it will later be easy to reshuffle this next design into subsequent designs as needed. In most cases, however, there are significant costs involved in dynamically altering designs. Thus, the engineer has to explicitly deal with the dynamic evolution of his designs. This implies for him developing a dynamic plan as illustrated in Figure 5.13, which shows a four-year layout plan for a facility. Figure 5.13 uses gray shadings to distinguish processors in terms of expected moving cost.

In this age of high market turbulence, the complexity of dealing with the dynamic nature of the design task is confounded by the fact that all demand, process, flow, and space requirements are estimates based on forecasts and that these forecasts intrinsically are known to be ever more prone to error as one looks further into the future. For example, what will be the demand for a product family tomorrow, next month, next quarter, next year, in three years?

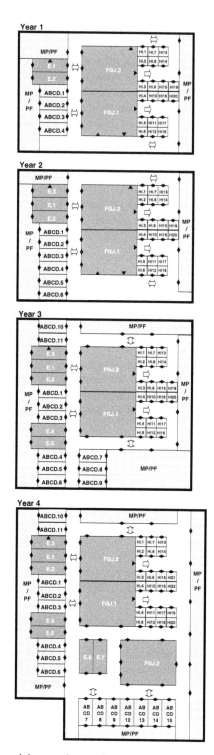

FIGURE 5.13   Myopically generated dynamic layout plan.

TABLE 5.10 Multi-Year Demand Forecasts

Product Number	Expected Daily Demand per Year			
	Y1	Y2	Y3	Y4
1	1000	2000	3000	2500
2	0	0	270	560
3	6000	5750	5500	5250
4	300	275	250	225
5	0	0	0	120
6	0	1000	2000	3000
7	0	0	0	50
8	0	0	0	30
9	0	0	0	30
10	1000	1000	950	875
11	800	750	700	650
12	25	25	25	25
13	0	0	0	25
14	0	0	0	75
15	200	300	450	625
16	0	50	100	175
17	200	500	1000	1500
18	125	110	100	125

Tables 5.10 and 5.11 illustrate this phenomenon for the products of Table 5.4. In these tables, the demand stated in Table 5.4 becomes the expected average daily demand in the fourth future year. There are also forecasts for the first three years preceding this fourth year. Table 5.10 shows that the demand for some products is forecasted to be expanding while the demand for others is forecasted to be shrinking. For each forecast of Table 5.10, Table 5.11 provides the estimated standard deviation over the forecasted mean. For example, in year 1 the average daily demand for product 3 is forecasted to be 6,000 units with a stan-

TABLE 5.11 Multi-Year Uncertainty of Average Daily Demand Forecasts

Product Number	Standard Deviation of Expected Average Daily Demand			
	Y1	Y2	Y3	Y4
1	150	360	675	750
2	0	0	30	83
3	300	345	413	525
4	20	22	25	30
5	0	0	0	25
6	0	200	500	1000
7	0	0	0	12
8	5	0	0	8
9	2	0	0	3
10	50	60	71	88
11	200	225	263	325
12	5	6	8	10
13	0	0	0	5
14	0	0	0	20
15	40	72	135	250
16	0	15	38	88
17	140	420	1050	2100
18	6	6	7	12

TABLE 5.12 Multi-Year Expected Average Processor Requirements

| Processor Type | Expected Average Processor Requirements | | | |
	Y1	Y2	Y3	Y4
ABCD	4	5	9	13
E	2	2	3	4
FGJ	2	2	2	2
HI	19	19	19	20
Total	27	28	33	39

dard deviation of 300 units, while in year 3 the average daily forecast is down to 5,500 units, yet with a high standard deviation of 413. Such information may come from analyzing historical forecast performance in forecasting demand for such a family, respectively, one year and three years ahead (Montgomery et al. 1990). This means that within two standard deviations (two-sigma) or 98% probability using normal distribution estimation, in current year zero the average daily demand for P3 is expected to be between 5,400 and 6,600 units in year 1, and between 4,674 and 6,326 units in year 3.

Using the process requirements of Table 5.4 and assuming the flexible processors introduced in the lower part of Figure 5.9, these forecasts permit computation of estimates for the average expected number of processors of each type, provided in Table 5.12. Also, they allow robust estimations for processor requirements, such as the two-sigma robust estimates provided in Table 5.13. In year 3, for processor type ABCD, the average estimate is 9 units, while the two-sigma robust estimate is 11 units. Overall the robust estimate adds up to a total of 28 processors in year 1 to a total of 47 in year 4.

Figure 5.13 provides a four-year layout plan generated in year 0 by an engineer. To help understand the compromises involved, the engineer was asked to first generate a design for year 1 based on the estimates for year 1. He had to then transform this year-1 design into a year-2 design, taking into consideration the expected flows for year 2 and the cost of transforming the year-1 design into the year-2 design. He had to repeat this process for years 3 and 4. Clearly, this is a rather myopic approach because in no time was he considering the overall forecasted flows and processor requirements for the entire four-year planning horizon. Analyzing the plan, it is clear that the engineer's decision in year 1 to lay out the two FGJ processors adjacent to each other has defined a developmental pattern that has had repercussions on the designs he has produced for year 2 to year 4. Even though possible, he has not planned to move any of the processors E and FGJ once laid out in their original location, which has created a complex flow pattern in year 4, as contrasted with the elegant simplicity of the lower layout of Figure 5.9. Formally evaluating the dynamic plan requires evaluation of each design statically as described in the previous sections, and computation of the expected transformational costs from year to year. The evaluation requires the generation of demand scenarios probabilistically in line with the forecast estimates of Tables 5.10 to 5.13. Due to space constraints, the results of such an evaluation for the plan of Figure 5.13, and the generation and evaluation of alternative plans that take a more global perspective, are left as an exercise to the reader.

TABLE 5.13 Multi-Year Two-Sigma Robust Processor Requirements

| Processor Type | 2-Sigma Robust Processor Requirements | | | |
	Y1	Y2	Y3	Y4
ABCD	4	6	11	15
E	2	3	5	7
FGJ	2	2	2	3
HI	20	20	20	22
Total	28	31	38	47

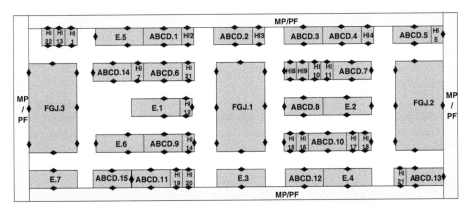

FIGURE 5.14 Illustrating the steady robust, immune-to-change, design strategy by expanding the template of Figure 5.11 to transpose it into a design for year 4 given the forecasts of Tables 5.10 to 5.13.

In practice, there are two main strategies to deal with dynamics. The first is to select processors and facilities that enable easy design transformation, and to try to dynamically alter the design so as to always be as near to optimal as possible for the forthcoming operations. Figure 5.13 can be seen as an example of this strategy. The second is to develop a design that is as robust as possible, as immune to change as possible, a design that requires minimal changes to accommodate in a satisfactory manner a wide spectrum of scenarios (Montreuil and Venkatadri 1991, Montreuil 2001, Benjaafar et al. 2002). Figure 5.14 provides an example of this strategy by simply expanding the robust design of Figure 5.11 to be able to deal with the estimated requirements for year 4. It is left as an exercise to assess how to subtract processors from Figure 5.10 to deal with the lower expected requirements for years 1 to 3.

In the above examples, a yearly periodicity has been used for illustrative purposes. In practice, the rhythm of dynamic design reassessment and transformation should be in line with the clock speed of the enterprise, in synchronization with the advent of additional knowledge about the future and the lead time required for processor and facility acquisitions and moves. Even decades ago, some companies were already reconfiguring their shop floor layouts on a monthly basis, for example, in light assembly factories dedicated to introducing new products on the market, assembling them until demand justified mass production.

In the illustrative example of Figure 5.13, the planning horizon has been set to four years. Again, this depends on the specific enterprise situation. It can range from a few days in highly flexible, easy-to-alter designs to decades in rigid designs in industries with low clock speeds.

5.11 Dealing with Network and Facility Organization

Layout and location design studies often take the organization of the facility network as a given, yet organizational design has a huge impact on spatial deployment optimization. The organization of the network states for each center and/or facility its specific set of responsibilities. This bounds the type of products, processes, and clients the center is to deal with. According to Montreuil and Lefrançois (1996) the responsibility of an entity is defined by a set of combinations of markets, clients, outbound products, processes, processors, inbound products and suppliers, specified quantitatively and through time. For example, a center can be responsible for manufacturing all plastic products offered by the enterprise to the Australian market. Another center can be responsible for assembling up to 10,000 units a year of a specific product. Through the responsibility assignment process, the organizational design also defines the customer–supplier relationships among centers and facilities.

In some cases the organizational design is not complete when the layout or location design process is launched, depending on the output of this process to finalize the design. This is the case, for example, with location-allocation problems. The organizational design states, for example, that the logistic network is to comprise only distribution centers that are to be the sole source for their assigned market segments. The set of market segments is defined geographically and in terms of demand. Depending on the actual location and sizing of distribution centers, the assignment of segments to centers can be performed, completing the organizational design.

Adapted from Montreuil et al. (1998), Table 5.14 provides a responsibility-based typology of centers and facilities. First, types of centers are segregated by their defining orientation. The options are product, process, project, market, and resource orientations. A product-oriented organization defines the responsibility of the center in terms of a set of products. In contrast, a process organization does not state responsibilities in terms of products; it is, rather, in terms of processes. The same logic holds for the three other orientations.

For each orientation, Table 5.14 provides a set of types of centers, stating for each its type of responsibility. Product-oriented organizations are segregated into three types: product, group, and fractal. A product

TABLE 5.14 Responsibility-Based Center Typology

Center Orientation	Center Type	Responsibility Set	Responsibilty in Terms of Demand Satisfaction
Product		Set of products	All or a fraction
	Product	Single product	All or a fraction
	Group	Specific group or family of products	All or a fraction
	Product fractal	Most products; generally multiple centers are replicated to meet demand	A fraction
Process		Set of processes	All or a fraction
	Function	A single function, elementary process or operations	Generally all, yet can be a fraction
	Process	A composite process composed of linked elementary processes	All or a fraction
	Holographic	A set of elementary processes; generally multiple centers are distributed to meet demand	A fraction
	Process fractal	Most processes; generally multiple centers are replicated to meet demand	A fraction
Project		Set of projects	All or a fraction
	Order or contract	A specific order, contract or, in general, project	Generally all, except for very large cases
	Repetitive project	Projects of the same that repeatedly occur through time	All or a fraction
	Program	A long-term program involving a large number of planned deliveries	Generally all
Market		Set of markets and/or clients	Generally all
	Client	A specific client	Generally all
	Client type	A set of clients sharing common characteristics and requirements	Generally all
	Market	A market or market segment, defined by geography or any other means	Generally all
Resource		Set of resources to be best dealt with	Generally all
	Inbound product	Set of inbound products needing to be processed	Generally all
	Supplier	Set of suppliers whose input has to be processed	Generally all
	Team	Set of people whose capabilites have to be best exploited and needs have to be best met	Generally as much as possible given their capacity, capabilites and preferences
	Processor	Set of processors (equipment, workstation, etc.) to be exploited as best as possible	Generally as much as possible given their capacity and capabilites

Source: Adapted from Montreuil et al. in *Material Handling Institute*, Braun-Brumfield Inc., Ann Arbor, Michigan, 1998, 353–379.

center is devoted to a single product. A group center is devoted to a group or family of products. Note that *product* is here a generic term which encompasses materials, components, parts, and assemblies as well as final products. Table 5.14 also indicates that a product or group center may be made responsible for only a fraction of the entire demand for that product or group. For example, it can be decided that there are to be two product centers mandated to manufacturing a star product. The former is to be responsible for the steady bulk of the demand while the latter is to deal with the more fluctuating portion of demand above the steady quantity assigned to the former. Similarly, instead of assigning the fluctuating portion to another product center, it can assign it to a group center embracing similar situations. The possibilities are endless. A product-oriented fractal organization offers a different perspective. It aims to have a number N of highly agile centers, each capable of dealing with most products, assigning to each fractal center the responsibility of 1/N of the demand of each product. This allows operations management to dynamically assign products to centers in function of the dynamic repartition of demand among the products (Venkatadri et al. 1997, Montreuil et al. 1999). Implementing a product organization has tremendous impact on flow through the network and the constitution of each center in the network. Product centers rarely have flow of products between them, except when one provides products that are input to the other. There is more complex flow within the center as one switches from a product center to a group center and then to a fractal center. Also, when only product or group centers are used, most of the specific customer–supplier relationships are predefined. Whenever fractal centers are used, then workflow assignments become dynamic operational decisions.

Process orientations are segregated into four types: function, process, fractal, and holographic. Function, process, and fractal types are the process-oriented equivalent of the product-oriented product, group, and fractal types. For example, a process-oriented fractal center is responsible for being able to perform most elementary processes, with 1/N of the overall demand for these processes (Askin 1999). Again, adopting a process orientation has significant impact on workflow patterns. For example, function centers have minimal flow between the processors constituting them and have significant flows with other centers. Illustratively, an injection center has minimal flow between the injection molding machines, except for the sharing of molds, tools, and operators. In fact, when a network is composed only of function centers, a product with P processing steps will have to travel between P distinct function centers. In such cases the relative layout of centers becomes capital in order to contain the impact on inter-center material handling/transport. Holographic organization generates a number of small centers responsible for a limited set of related processes. Most centers are replicated and strategically distributed throughout the network or facility. In fact, the robust flexible layouts of Figures 5.11 and 5.14 are exploiting a holographic organization where each processor is conceived as a small center, instead of a function organization as in the layout of Figure 5.5.

Project-oriented organizations lead to center types that are defined in terms of orders, contracts, projects, or programs. A manufacturer bids for and then wins the bid for a major contract involving a set of products and processes to be performed in given quantities according to a negotiated delivery schedule. When its managers decide to implement a facility strictly devoted to delivering this contract, the resulting facility is of the contract type. Similarly, when a factory within an automotive network is awarded a multi-year program to manufacture all engine heads of a certain type for the European market and when it devotes a center to this production, its organization now has a program center. Repetitive project centers are centers well conceived and implemented to realize specific types of projects that come up repetitively. This is common in the aeronautical industry where, for example, large centers are well equipped to perform a variety of overhauls, maintenance or assembly of airplanes depending on the flow of projects signed by the enterprise.

Resource-oriented organizations can be segregated into four types of centers. Inbound product centers and supplier centers are respectively specialized to perform operations on certain types or groups of inbound products or on all products incoming from a set of suppliers. Processor and team centers are similar conceptually, designed to exploit the capabilities and capacity of a set of processors and humans, respectively. A center grouping all the CNC machines in a factory is an example of processor center.

Table 5.14 opens a wealth of organizational design options. First, each network can be composed of any combination of centers from the various types. Second, the types provided have to be perceived as building blocks which allow the design of composite or hybrid types of centers, such as a center devoted to performing a set of processes on a group of products. Third, it can be used recursively. Higher-level facilities or centers have to be organized according to a pure or hybrid type. These can be composed of a network of internal lower level centers. Each of these has to be organized, not restricted to the same type as its parent.

To illustrate the impact of network organization, for the illustrative case leading to the layouts starting in Figure 5.5, there has been the implicit assumption that the organizational design states that all the products and processes have to be performed within the same centralized facility. When this constraint is removed and further market information is provided, a network organization such as depicted in Figure 5.15 is quite possible. In the network of Figure 5.15, a global factory is proposed to manufacture products P1, P2, as well as P4 to P9. Another global factory is specialized to manufacture product P3. Three market-specific product group facilities are to be implemented. These will all make products P10 to P18. Each will be dedicated to serving a specific market: America, Europe, or Asia. Each market is to be assigned a number of regional distribution centers fed by the global P3 factory and the three P10-to-P18 factories. Now, instead of having to locate and lay out a single global facility, the design task involves locating interacting factories and distribution centers, and to organize, size and lay out each of these.

Here the organizational emphasis has been put on the centers, stressing the importance of their specific responsibilities and their customer–supplier relationships. Figure 5.16 depicts clearly another important network organization facet: the type of organization structure of the network. Figure 5.16 provides a sample of seven types of structures resulting from organizational design of the network.

The first is termed a fixed product structure. Here the idea is that the product is brought to one location and does not move until departing the system. The processors and humans are the ones moving to, into, and away from the stationary product. The second structure type is a parallel network, where all flow is leading inbound products into one of the centers and then out of the system. The third is a flow line where each center is fed by a supplier and itself feeds a client center, this being repeated until the product gets out of the system. Centers store and/or perform operations on the product.

The fourth structure is a serial-parallel network, typical of a flow shop. This structure combines the flow line and the parallel network. It is conceived as a series of stages. At each stage, there are parallel centers

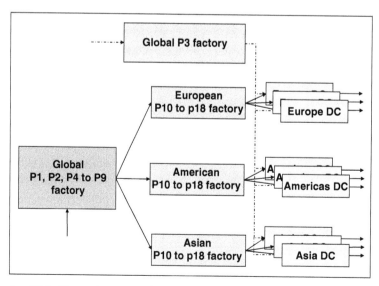

FIGURE 5.15 A multi-facility product oriented organization of the illustrative case.

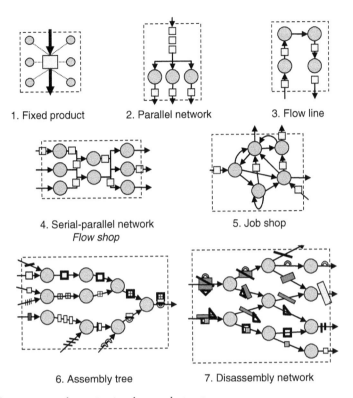

1. Fixed product   2. Parallel network   3. Flow line

4. Serial-parallel network   5. Job shop
Flow shop

6. Assembly tree   7. Disassembly network

FIGURE 5.16 Illustrative set of organizational network structures.

jointly responsible for delivering the stage responsibility. The fifth structure is a job shop network characterized by a profusion of inter-center flows that have no dominant serial or parallel pattern.

Whereas structures one to five can be mainly mono-echelon, the sixth and seventh example structures are multi-echelon in nature. Indeed, they explicitly deal with the fact that products are needed constituents of other higher-level products and organize the network around these bill-of-materials relationships among products. The sixth structure is an assembly tree. Here each center feeds a single center which later performs operations on the delivered products and/or assembles the delivered products into higher-level products. The seventh structure is a disassembly network. Instead of assembling products, it disassembles them. Instead of being restricted to a directed tree, it is conceived as a more flexible directed network. Here the main difference is that a center may have more than one client center, while maintaining the no backtracking constraint of the tree structure. One can easily think of a disassembly tree structure or an assembly network structure.

Network structures have direct influence on flow patterns and therefore on layout and location decisions. In fact, it can be said that the organizational combination of responsibility assignment and network structure sets the stage for layout and location studies. However, more important in a highly competitive economy is the fact that integrating the organizational, location, and layout design processes offers the potential for designing networks with higher overall performance potential.

5.12 Design Methodologies

The previous sections have focused on the essence of the location and layout design representation, stressing key facets and issues. This section focuses on methodologies used for generating the designs.

It is important to start by humbly stating that currently there is no generic automated method capable of dealing with all issues covered in the previous sections and of providing optimal, near-optimal, heuristically optimized or even provably feasible designs. It is also important to state that most issues brought forward in the previous sections are inherent parts of most location and layout design studies. Indeed, facilities end up being located and laid out every day around the world, resulting in feasible yet imperfect networks which have to be adjusted to improve their feasibility and performance as their implementation and operation reveal their strengths and weaknesses and their growing inadequacy to face evolving demands.

In this section, the emphasis is not on trying to document reported methodologies pertinent for each type of situation as defined through combinations of facets introduced in the previous sections. For example, there will be no specific treatment of stochastic dynamic layout of flexible processors in continuous space, or of deterministic static location of unlimited capacity facilities in discrete space. The combinations are too numerous. References have already been provided through the previous sections, which propose either surveys of methods or introduce appropriate methods.

The section, rather, takes a macroscopic perspective applicable to most situations. It does so by mapping the evolution of the types of methodologies available to designers. Indeed, as depicted in Figure 5.17, location and layout design has evolved methodologically through the years into nine states that concurrently exist today, each with its application niches. The outer circle includes the oldest methods: manual, heuristic, and mathematical programming. The middle circle includes the more recent methods which have evolved from those in the outer circle: interactive, metaheuristic, and interactive optimization. Finally, the inner circle includes the most recent methods: assisted, holistic metaheuristic, and global optimization. The nine methodological alternatives are hereafter described.

5.12.1 Manual Design

The earliest and most enduring method is the manual method. Sheets of paper and cardboard, colored pencils, and scissors are the basic tools used. The engineer, based on his understanding of the qualitative proximity relationships, quantitative flow estimations, cost structures and constraints, gradually draws a series of designs, from which he picks a limited set of preferred alternatives to present to the management for decision. For example, layouts are sketched on paper. They are assembled on boards using pieces of

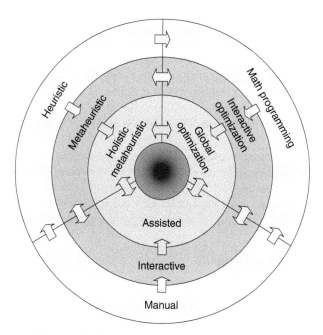

FIGURE 5.17 Evolution of design methodologies.

carton for each center, or using Lego-style building blocks. They are approximated using real-size flat panels on working floors through which the engineer can walk the design, or they are designed by really shuffling the actual layout until a satisfying design is implemented. In practice, the evaluation of each design is often very limited and coarse, even regularly limited to a multi-criteria ranking of alternatives, where each criterion is evaluated quite subjectively or approximately.

In the manual method, computers are used quite minimally. They are exploited for reporting the preferred designs and sometimes to evaluate the final set of designs.

The manual method has the advantages of being simple and expediting. It may work well when the design complexity is low and the degrees of freedom limited. It can rapidly prove tedious and limitative as the case size and complexity increase. However, for good or for bad, a large number of designs are still achieved this way in practice.

Starting in the 1960s, researchers have worked toward automating layout and location design. Two basic directions have been taken: simple heuristics and mathematical programming.

5.12.2   Heuristic Design

Researchers who generated heuristics for layout and location design have aimed to capture the power of the computer to generate satisfying designs by systematically searching the solution space using approximate yet rigorous methods. Kuenh and Hamburger (1963) and Nugent et al. (1968) are typical in heuristic location design while Armour and Buffa (1963) and Lee and Moore (1967) are typical in heuristic layout design. Two types of heuristics have been developed, with myriads of instances of each type and a multitude of hybrids combining both types. The two types are construction and improvement heuristics. As exemplified by ALDEP (Seehof and Evans 1967) and CORELAP (Lee and Moore 1967), a construction heuristic gradually iterates between selection and placement activities until a design is completed or infeasibility is reached. The selection activity decides on the next center to place in the design or, more generally, on the order according to which centers are to be inserted in the design under construction. The placement activity locates and shapes the selected center into the partial design.

The variety of construction heuristics comes from the multiple options for selecting the next center and for placing it into the partial design. Selection can use qualitative relationships or flow, can take into consideration or not those already placed, can be deterministic or randomized, and so on. The simplest way ranks the centers in decreasing order of flow or proximity relationships intensity and then selects them for placement in that order, placing them in the best available location given its space requirements and its flow or relationships with already placed centers. When deterministic selection is used, the heuristic generates a single design. When randomized selection is used, then the heuristic generates numerous designs, scores each of them, memorizes the best N designs and then reports them at the end of the randomized sampling.

Placement is the most difficult part of construction heuristics. So as to ease the generation of feasible designs, the earlier ones relied on a discrete space representation and did not support such restrictions as having to use existing constraining buildings. In general, as the heuristic advances in its iterations, the center of the design gets occupied, leaving mostly space available at the outskirts of the design, subject to ever more feasibility constraints to fit the centers in the design. In such heuristics, placement is eased when a combination of design code and filling pattern is imposed. For example, when layouts are coded as strings (e.g., A-MP-F-…-PF), centers can be iteratively inserted in the string code. Then the design can be generated by systematically placing the centers according to the string code. A layout heuristic can, for example, start to place the first center in the northwest corner of an existing facility, then move left with the second and third, until it reaches the eastern boundary. Then it can move one layer southward and head back westward, zigzagging until the design is completed.

Improvement heuristics such as CRAFT (Armour and Buffa 1963) start with a given initial design. Then they iteratively generate potential local improvements to the design, estimate the improvement potential, implement the preferred improvement, evaluate the potentially improved design, and set it as the incumbent design if it improves on the current best design, repeating this process until no further improvement is reachable or sufficient time has elapsed. Local improvements are typically two-way, three-way, or multiple-way exchanges. A two-way exchange of centers A and B consists of locating center A at B's current location and vice versa. In a three-way exchange, centers A, B, and C, respectively, take on the current location of centers B, C, and A, or of centers C, A, and B. As an example, a typical heuristic based on two-way exchanges scans iteratively all pairs of centers in a predetermined order. For each pair, it estimates the design score given the exchange of center locations. If the estimation reveals a potential score improvement over the current one, two options are possible. Either the heuristic implements immediately the exchange and scores the resulting design or it keeps on testing all two-way exchanges and implements only the best potential exchange.

For computational speed reasons, an improvement heuristic evaluates the potential of an exchange rather than immediately evaluating the altered design. For example, in layout design with centers of various sizes, the impact of exchanging centers A and B may not be obvious. The simplest case is when centers A and B are of equal size and all flows and proximity relationships are assumed to be between the centers' centroids. In such a case, simply interchanging the centroid locations in the travel or proximity relationship score computations is sufficient to estimate the real impact of the interchange. In most other cases, it is not so easy. For example, exchanging centers F and J in the layout of Figure 5.4 requires altering the location and shape not only of the involved pair, but also of several other nearby centers. Center J is larger than center F and it has a fixed shape with a large length-to-width ratio. While center F fits into the current location of center J, the converse is not true since center J does not fit in the current location of center F. Fitting J in the northeast region would require significant reshuffling. This is why many such heuristics, following the lead of their ancestor CRAFT (Armour and Buffa 1963), forbid interchanges involving non-adjacent different-size centers. In the illustrative case, even a simpler exchange such as centers D and I in Figure 5.6 requires dealing with the aisle segment between them, to reposition as best as possible their I/O stations, and to adjust the empty travel estimates. This is why, before realizing these tasks, a heuristic applied to this case would assume direct interchange of the centroid, I/O stations and boundaries, would not re-estimate the empty travel, and then would compute the selected design score (e.g., minimizing flow travel). Once an exchange is selected as the candidate for the current iteration, then the modifications are really made in the layout and the score is more precisely computed.

Most heuristics have been implemented with a number of simplifying restrictions and assumptions. For example, in layout design, most generate a block layout instead of a more elaborate design such as a net layout. They do not support I/O stations and aisle travel. They deal only with loaded travel minimization or qualitative proximity relationship maximization, assuming Muther's AEIOUX coding.

Typically a heuristic is coded in a software application that allows case data entry and editing, heuristic parameter setting, and graphical solution reporting. Most such software is developed by researchers solely to support the developed heuristic. They rarely allow choosing among a variety of heuristics. Capabilities for interactive editing of produced designs are usually quite limited, the emphasis being placed on automating the design task.

5.12.3 Mathematical Programming–Based Design

Researchers have long recognized that some simple instances of location and layout design can be modeled mathematically and solved optimally in short polynomial time. A well-known example is the location of a single new facility interacting in continuous space with a set of fixed facilities so as to minimize total travel given deterministic flows (Francis et al. 1992). Another well-known location example, solvable using the classical linear assignment model (Francis et al. 1992), involves the assignment of a set of facilities to a set of discrete locations so as to minimize total travel and implementation costs, provided that at most a single facility can be assigned to any specific location and that the assignment costs can be computed a priori for each potential facility-to-location assignment, being independent of the relative assignment of facilities.

As yet another example, take dedicated warehouse layouts in which each product is assigned to a fixed set of storage locations in which no other product can be located. It is well known that, assuming deterministic demand, products can be optimally assigned to storage locations according to the cube-per-order index when all products have the same inbound and outbound behavior in terms of dock usage (Heskett 1963, Francis et al. 1992). For example, they all come in a given dock and all go out using the same other dock. The cube-per-order method (*i*) computes the expected distance travelled by a product assigned to each storage location and then ranks the locations in nondecreasing order of expected distance, (*ii*) computes the cube-per-order index as the ratio of product storage space requirements over product throughput, and then ranks the products in nondecreasing order of this index, and finally (*iii*) iteratively assigns the first remaining location to the first non-fully assigned product, until all are assigned or no more space is available.

As a final example, given a continuous-space block layout with rectangular shaped centers, the optimal location of all I/O stations can be found in polynomial time if one aims to minimize rectilinear travel and if each station can be located anywhere within a predetermined rectangular zone (Montreuil and Ratliff 1988a).

When a design case fits exactly with a problem solvable in polynomial time, then its solution algorithm should be applied so as to get the optimal solution. Most cases do not readily fit exactly such easily solvable problems, yet if the gap is not too enormous, the case can be manually adapted to fit the problem and the optimal solution can be used as an approximate solution to the real situation, heuristically adjusted to reach satisfying feasibility. This can also be used for more complex (NP-Complete) mathematical programming problems that have been researched and for which there exist (*i*) good optimal solution algorithms exploiting techniques such as branch-and-bound, decomposition and branch-and-cut, or (*ii*) good generic heuristics capable of providing satisfying solutions.

Such an approach has led to the dominance of the quadratic assignment model in representing layout and location design problems for decades prior to the early 1990s. The model of the quadratic assignment problem (QAP) is defined as follows:

Minimize

$$\left[\sum_{\forall cl} a_{cl} A_{cl} + \sum_{\forall clc'l'} c_{clc'l'} \left(A_{cl} A_{c'l'} \right) \right] \tag{5.1}$$

subject to

$$\sum_{l \in L^c} A_{cl} = 1 \qquad \forall c \tag{5.2}$$

$$\sum_{c \in C^l} A_{cl} \le 1 \qquad \forall l \tag{5.3}$$

Variables:

A_{cl} Binary variable equal to 1 when center *c* is assigned to location *l*, or 0 otherwise

Parameters and sets:

a_{cl} Cost of assigning center *c* to location *l*
$C_{clc'l'}$ Cost of concurrently assigning center *c* to location *l* and center *c'* to location *l'*
C^l Set of centers allowed to be located in location *l*
L^c Set of locations in which center *c* is allowed to be located

The QAP forces the engineers to define a discrete location set, such that each center has to be assigned to a single location (constraint 5.2) and that a single center can be assigned to any location (constraint

5.3). In layout design, most of the cases researched in the scientific literature involve an M-row, N-column matrix of square unit-size locations and a set of at most M*N centers with fixed square unit-size shape. The QAP problem is among the most difficult combinatorial problems. For decades, cases with at most 10 locations could be solved optimally. Even today, the largest cases optimally solved involve up to 30 locations (Anstreicher et al. 2002). However, being such a well-known problem, the QAP has been a battling ground for researchers, leading to the availability of numerous generic heuristics and metaheuristics applicable for location and layout cases if the engineer is capable of modeling them as a QAP (e.g., Nourelfath et al. 2007).

The early advances in the manual, heuristic, and mathematical programming–based methodologies have led the way for the middle circle methodologies of Figure 5.17, described below.

5.12.4 Interactive Design

Interactive design follows directly the trail of manual design, with the difference lying in being adopted by engineers who are fluent with commercial spreadsheets such as Excel as well as with computer-aided drawing and design software (CAD) such as AutoCad, CATIA, SolidWorks and Visio, even presentation software such as PowerPoint, and with geographical information systems (GIS) such as MAPINFO or Google Earth. A spreadsheet is used for computing design scores and performing local analyses. For layout cases, the CAD software is used to draw and edit the designs, as well as to show the flows and relationships. For wide area location cases, the GIS software serves the same purpose.

Computer-aided drawing and design software has two main advantages. First, it is used for referential technical drawing of facilities in many organizations, used for keeping up to date the precise equipment, service, and utilities layout. The software and the drawings thus become freely available to the engineer for layout design purposes. Second, CAD software is often exploiting the notions of drawing object libraries and drawing layers, which speed up and ease the layout drawing effort. The main disadvantages of using CAD software are that (1) they are most often geared for precision drawing and may become cumbersome to use for design purpose, and (2) they do not understand layout design. An object is mostly a drawing object. A flow is simply a link from an object to another. The software does not embed knowledge and methods exploiting the fact that the object is a center and that the flow involves trips of products or resources between centers. The engineer must assume the sole responsibility for the representativeness of its drawn designs. The same types of advantages and disadvantages apply for GIS systems used for location purposes, adapted to a set of geographical sites rather than a set of facilities.

In the future, there will be more seamless integration of CAD and GIS software, allowing to show or edit a large-scale logistic network and to then swiftly dig into the facilities part of the network.

As generic technological capabilities increase, interactive design is enabled to achieve better representations in ever easier ways. For example, 3D drawings, renderings, and walks-throughs add significant value to an engineer involved in facilities layout. They allow dealing directly with multi-floor facilities, and in more generic terms, to exploit the cube rather than its rectangular surface. They allow a visual grasping of the facility layout, which is by far superior to 2D representations. This has been well known for decades. However, such capabilities are still very rarely used in practice because of the combination of software price, 3D drawing complexity and lack of computational power to deal with large-scale layouts. These three constraints are rapidly diminishing with new-generation software. As engineers will learn to exploit them generically, they will gradually use them more for facilities layout purposes.

Interactive design is widely used in practice, second only to manual design. Both suffer from the same threat: they depend heavily on the engineer. The tools are generic and do not understand layout or location and do not have any layout and location optimization capabilities. This is why the value of both manual and interactive design depends on the engineer's mastering of the layout and location issues and on his creativity in generating great designs.

5.12.5 Metaheuristic Design

Metaheuristics have evolved from heuristics for two main reasons. The first is an attempt to get out of the local optima trap in which heuristics get stuck. This has led to developing metaheuristics exploiting techniques such as simulated annealing (Meller and Bozer 1996, Murray and Church 1996), tabu search (Chittratanawat 1999, Abdinnour-Helm and Hadley 2000), genetic and evolutionary algorithms (Banerjee et al. 1997, Norman et al. 1998), ant colony algorithms (Montreuil et al. 2004) and swarm intelligence (Hardin and Usher 2005). The second reason is the researchers' attempt to go beyond solving the basic layout and location problems, to get away from enforcing myriads of simplifying assumptions and constraints. In location, this has led to metaheuristics for addressing complex problems (e.g., Kincaid 1992, Crainic et al. 1996, Cortinhal and Captivo 2004). In layout, researchers have attempted, for example, to integrate the automatic generation of block layouts with their travel network (e.g., Fig. 5.2c) (Norman et al. 1998). The combination of both reasons has had high stimulating impact on researchers.

Metaheuristics operate at least on two levels. The first level uses heuristics to develop a design subordinated to master decisions taken at the second level. This second level drives the overall heuristic search process, iteratively exploiting the heuristics of level 1 to scan the solution space. Complex implementations may have multiple levels, with the higher levels exploiting the lower levels in the same way as exemplified in the two-level illustration.

When trying to avoid the local optima trap, researchers have relied upon the exploitation of generic metaheuristic techniques. Genetic algorithms provide a fine example to understand how such metaheuristics are used in layout and location settings. Very shortly, genetic algorithms attempt to mimic genetic evolution leading to survival of the fittest. In layout design, members of the population are individual layouts. Used at the second level of the metaheuristic, the genetic algorithm iterates through rounds which each enact a number of immigrations, mutations, and crossovers from which is generated the next generation. At all iterations only the N best layouts are kept to form the population of the next generation.

The key to understanding how genetic algorithms work in layout is that they exploit the notions of layout code and space structuring, both introduced in Section 5.3. Remember that the code for the three-band layout of Figure 5.4 is (1:A,B,C; 2:D,E,F; 3:G,H,I). Given this code and the knowledge that the layout is restricted to be structured into three horizontal bands, the band layout of Figure 5.4 can be reconstructed. Hence, the second level of the metaheuristics is used to search the solution space in terms of layout codes while the first level uses a heuristic or an optimization model to generate a layout from the code generated in the second level.

At the second level, the activities are simple once focused to be performed using layout codes. For example, immigration is simply achieved through the randomized generation of a new layout code. At all iterations, the genetic algorithm randomly generates a number of immigrant codes.

A mutation of the (1:A,B,C; 2:D,E,F; 3:G,H,I) code can be achieved in many ways. For example, a center can be transferred from a band to another [e.g., D in (1:A,D,B,C; 2:E,F; 3:G,H,I)], a center can be moved from its current position in the string to another position while keeping the number of centers in each band intact [e.g., D in (1:A,B,C; 2:E,F,G; 3:H,I,D)], a pair of centers can exchange positions in the code [e.g., D and B in (1:A,D,C; 2:B,E,F; 3:G,H,I)], and the entire content of two bands can be exchanged [e.g., bands 1 and 2 in (1:D,E,F; 2:A,B,C; 3:G,H,I)]. At all iterations, the genetic algorithm randomly selects the layout codes to be mutated and the way each one is to be mutated.

A crossover involves two members of the population. As an example, consider the layout codes (1:A,B,C; 2:D,E,F; 3:G,H,I) and (1:D,H,I; 2:B,A,G; 3:E,C,F). An illustrative crossover could be formed by taking into priority the first band as in the first code, the second band as in the second code, the third band as in the first code, and then assigning any unassigned center to its current ordered position in the first or second code, picking from both codes in rotating order. This starts the crossover-generated code with (1:A,B,C). Second, it extends it as (1:A,B,C; 2:G). Third, it again extends it as (1:A,B,C; 2:G; 3:H,I). Fourth, it finalizes it by inserting the missing centers: (1:A,B,C; 2:D,G,F; 3:H,E,I). At each iteration, the genetic algorithm randomly selects the pairs of layout codes used for crossover purposes, and how the crossover is to be performed for each pair.

The layout code resulting from each mutation, crossover, and immigration is transferred to the level-one heuristic optimizer which generates a layout design respecting the layout code and the space structuring. This layout is scored according to the selected metric. The layout score serves for deciding which layouts are to form the next generation. The genetic algorithm keeps on searching until a time or iteration limit has been reached, or until no better layout has been generated since a specified number of iterations. The regular usage of randomization for generating layout codes, the multiplicity of ways layout codes can be generated, and the systematic screening of the score of the layout generated from each layout code augment the probability that the metaheuristic will not get stuck in local optima and thus potentially get nearer to optimality within a given solution time.

Without getting into as much detail, other metaheuristic techniques used are the following. The first and simplest to be tested has been simulated annealing, mostly used in conjunction with improvement heuristics. The second level of the metaheuristic simply dynamically adjusts the probability that the improvement heuristic at the first level will accept implementation of an exchange with negative impact on the performance of the current best design. The logic is as follows: When the heuristic finds better layouts at a good pace, the probability is kept low. When the heuristic begins to have trouble finding better layouts through local improvement, then the probability is increased, letting the improvement heuristic deteriorate temporarily its current best design so as to get away from the current local optimum region. Tabu search is another fruitful metaheuristic technique. It puts emphasis on forbidding to consider in the improvement algorithm moves that have been recently examined, speeding up the solution process by avoiding unnecessary repetitive loops examining the same potential layouts over and over.

Ant colony algorithms share with genetic algorithms the exploitation of layout code and space structuring. They differ in their second-level implementation. The underlying metaphor is to think of a resulting layout as the output of an ant looking for food. If the layout is good, then the ant leaves traces of pheromone at milestones along the path during its return trip. Milestones depend on the metaheuristic implementation: they can correspond to locating specific centers in some portion of the layout or to locating specific centers adjacent to each other. Other ants looking for food will trace a path which is influenced to some degree by the intensity of pheromone left at milestones by preceding ants, augmenting the probability that the ant will end up in hot spots for layout quality. At each iteration, the metaheuristic launches a number of ants whose job is to find a path toward a complete layout code. Then this layout code is evaluated by generating a layout based on this code, as is done with genetic algorithms. Dependent on the design score, various amounts of pheromone are deposited at key constructs within the design. As the metaheuristic proceeds, the aim is for the collectivity of ants to learn to avoid layout constructs which lead to bad layouts and to seek layout constructs that are often found in great designs. In order to avoid being trapped in local optima, the amount of pheromone at each construct decays with time and the selection by an ant of its next construct insertion given a partial code is made according to weighted randomization among the possible constructs available for insertion at the current code state.

The first and second reasons driving the development and use of metaheuristics are melted in various implementations. As an example, AntZone (Montreuil et al. 2004) is a metaheuristic that is based on ant colony techniques. AntZone generates block layouts with located I/O stations with the objective of minimizing rectilinear travel. Its exploits space structuring by having users select among different types of band layouts: 2H-bands; 3H-bands; 3V-bands; 1V-band + 3H-bands + 1V-band; etc. For example, the second-from-left layout of Figure 5.4 is constructed using 3H-bands. AntZone also lets the engineer specify a priori how many centers are allowed at maximum along each band and then it defines a flexible-size rectangular zone for each position along each band. A potential space structuring for the second layout from left in Figure 5.4 can be $[H_1:(Z_1,Z_2,Z_3,Z_4)/H_2:(Z_5,Z_6,Z_7,Z_8)/H_3:(Z_9,Z_{10},Z_{11},Z_{12})]$. A layout code then becomes an assignment of centers to zones. The layout code for the considered layout in Figure 5.4 is then simply (A,B,C,-,D,E,F,-,G,H,I,-). At the second level, the ant colony algorithm explores the solution space of layout codes. At the first level, a linear programming model generates the optimal block layout with located I/O stations, given a specified layout code.

Currently, most of the best-known solutions for large cases of the QAP, the block layout problem and their variants have been obtained using metaheuristics. Their advantage is their automatic capability of generating in reasonable time better designs than simpler heuristics. Their main disadvantage is their software implementation complexity, especially since most current implementations have been developed by research teams and are not widely available to practitioners.

5.12.6   Interactive Optimization–Based Design

In the early 1980s it became clear that trying to use mathematical programming for solving large realistic cases was out of reach in location and layout design involving interaction between facilities. Researchers started to look for sub-problems which could be solved optimally or near-optimally using heuristics. A design methodology emerged from this trend, termed interactive optimization-based design (Montreuil 1982). The concept is to let the engineer in the driver seat like in interactive design, while giving him access to a variety of focused optimizers supporting the various design tasks.

The earliest such methodologies used optimization to generate more advanced design skeletons than simple flow graphs and relationship graphs, from which the engineer had to interactively generate a design. The three best-known layout design skeleton-based methodologies, respectively, rely on the maximum-weighted planar adjacency graph (Foulds et al. 1985, Leung 1992), the maximum-weighted matching adjacency graph (Montreuil et al. 1987), and the cut tree (Montreuil and Ratliff 1988b).

The adjacency graph methods exploit three properties of any 2D layout. The adjacency graph property is that for any layout, one can draw an adjacency graph where each node is an entity in the layout (center, aisle segment, the outside, etc.) and each link corresponds to a pair of entities being adjacent to each other. The planar adjacency graph property states that the adjacency graph of a 2D layout is planar, meaning that it can be drawn without link crossings. Figure 5.18 illustrates these first two properties for the block layout of Figure 5.2e.

The matching adjacency graph property states that when assigning a value to each link equal to the boundary length shared by both entities defining the link, then the sum of all link values associated with a given entity is equal to the perimeter of that entity, defining the degree of the node representing the entity. For example, as shown in Figure 5.19, center A is adjacent to centers B and C and to the outside. The adjacent boundaries between A and these three entities are respectively 11.9, 10.3, and 8.8 m long. The sum of these adjacencies is 31, which is the perimeter of center A.

Every layout has a planar adjacency graph. If one could find the adjacency graph of the optimal layout, then the engineer could generate the optimal layout itself. For example, given the building and center space requirements, one can use the adjacency graph of Figure 5.18 as a design skeleton from which can

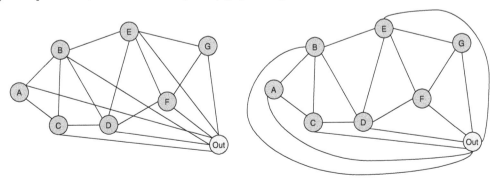

FIGURE 5.18   Illustrating the adjacency graph property and planar adjacency graph property using the block layout of Figure 5.2e.

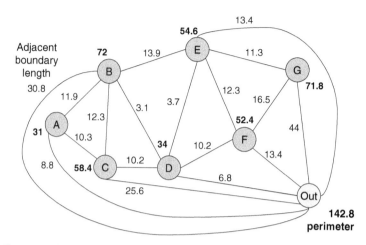

FIGURE 5.19   Illustrating the matching adjacency graph property using the block layout of Figure 5.2e.

be drawn the layout of Figure 5.2e with much ease. Given a weight for each potential link, the weighted maximum planar graph problem (Osman et al. 2003) aims to find the planar graph whose sum of link weights is maximal. In layout design, the weight for each link corresponds either to the flow between the centers or their qualitative proximity relationship importance expressed through the weight of their desired proximity type (e.g., adjacent: 100, very near: 50, not far: 2, very far: −50). A heuristic can be used to generate rapidly a near-optimal maximum-weighted planar graph. The engineer interactively draws the planar graph. Then he generates layouts respecting as much as possible the relative positioning of centers in the drawn graph and the adjacencies suggested by its links. This may be easy or rather difficult since not all planar graphs can be transformed in feasible layouts respecting the spatial requirements of each center and the building.

A similar approach is used when exploiting the matching adjacency graph property. The maximum-weighted b-matching problem (Edmonds 1965) can be solved optimally in polynomial time. This problem finds the graph, respecting the degree of each node while embedding links into the graph and stating a usage for each link respecting its imposed lower and upper usage bounds, which maximizes the sum over all links of the product of their usage and their value. In layout design, each node corresponds to a center; the value of each link is set as done earlier for the planar graph approach, yet here is divided by the upper usage bound; the degree of each node is bounded by desired lower and upper limits imposed on the center perimeter; a positive lower bound on a link forces the centers to be adjacent to a given extent; and, finally, the upper bound on a link indicates the maximum allowed adjacent boundary length between two centers. For example, the maximum adjacency between a 12 × 20 rectangular center and a 15 × 30 rectangular center is at most 20 m. The b-matching algorithm finds its optimal graph which is used by the engineer as a design skeleton representing the targeted adjacency graph. The engineer interactively generates a satisfying layout by iteratively drawing and adjusting a layout respecting the matching graph as much as possible, or resolving the b-matching model with adjusted link bounds to forbid or enforce specific adjacencies.

Cut trees are another type of design skeleton used in layout design. Cut trees can be computed from a flow graph in polynomial time (Gomory and Hu 1961). Figure 5.20 depicts the cut tree for the inter-center undirected loaded flow graph extracted from Table 5.6. Montreuil and Ratliff (1988b) prove that (*i*) the cut tree is the optimal inter-center travel network when the network links are all set to a unitary-length link and the travel network is restricted to have a noncyclic tree structure and (*ii*) if the centers have to be placed in two distinct facilities with a specific pair of centers forced to be separated from each other, then the cut tree will always indicate optimally which centers should be in each of the two facilities, assuming no restraining space constraints. For example, in Figure 5.20, if centers C and I have to be in distinct facili-

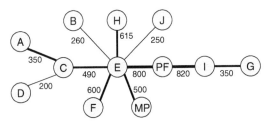

FIGURE 5.20   Inter-center cut tree based on the loaded flows of Table 5.6.

ties, then one has simply to find the single path between C and I in the cut tree, here C-E-PF-I, and then find the link with lowest value and cut it to find the optimal separation of centers. Here the lowest value link is C-E with a value of 490. Therefore, centers A, C, and D are best located in a facility and the remaining centers in the other facility. The 490 value indicates how much flow is to circulate between the facilities. In layout design, one seeks to decide what to put near each other and what to put far from each other.

As a design skeleton, the cut tree can guide an engineer into generating a layout. The cut tree can be molded at will to fit specific building constraints. The main rules are to systematically aim to locate centers so that higher value links and paths in the cut tree are as small as possible, and to avoid unnecessary link crossings. The cut tree can also be used for layout analysis, as shown in Figure 5.21, where the cut tree is overlaid on the current and alternative layouts of Figures 5.6 and 5.8, respectively. It is easy to see that the current design respects poorly the guidance of the cut tree, while the alternative layout, which has a significantly better travel score, does better even though it does not do it as best as could be.

Using design skeletons was the first stage of interactive optimization–based design. Montreuil et al. (1993b) later introduced a linear programming model for swiftly finding the optimal block layout, with located I/O stations, minimizing rectilinear flow travel given a set of flows and the relative positions of centers as inferred by the drawing of a design skeleton. This allows the engineer to manipulate the design skeleton, then to request a layout optimization based on the drawing of the design skeleton, and to examine a few seconds later the resulting layouts, iterating until he is satisfied with the design. Also developed were models and approaches for designing the travel network given a block layout with located I/O stations (e.g., Chhajed et al. 1992). Complementarily, a linear programming based model was introduced for optimizing the design of a net layout given a block layout and travel network (Montreuil and Venkatadri 1988, Montreuil et al. 1993a). The model optimally shrinks cell sizes from gross to net shapes, locates aisle segments appropriately and locates I/O stations so as to minimize aisle-based travel.

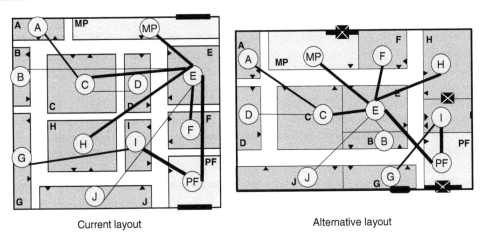

Current layout   Alternative layout

FIGURE 5.21   Cut tree over imposed on the current and alternative layouts of Figures 5.6 and 5.8.

The combination of these optimization models, used interactively by the engineer, allows him to generate designs that, even though they may not be globally optimal, benefit from optimized components and from the human capability to integrate them creatively. The main advantage of interactive optimization is that it enables the engineer while leaving him in the driver seat. The focused and integrated usage of optimization lets him address large cases efficiently. The main disadvantage lies in the current lack of wide and open accessibility of design software capable of sustaining such rich interactive optimization.

5.12.7 Assisted Design

Introduced as one of the three methodologies in the inner circle of Figure 5.17, assisted design has evolved mainly from interactive design and has been influenced by interactive optimization and metaheuristic design. The underlying hypothesis justifying the emergence of assisted design is that since layout and location design has high complexity, wide scope and large scale, it does not lend itself to fully automated design. Therefore, the underlying principles of assisted design are: (*i*) the engineer is to be at the core of the design process and (*ii*) it should have access to a design environment which enables him as best as possible to master the complexity, scope, and scale so as to efficiently generate high-quality creative designs. In assisted design, the focus is on (*i*) making sure the engineer is well trained into understanding the concepts, issues, and methods pertinent for his design task, (*ii*) providing him with an empowering assisted design environment, and (*iii*) training the engineer into being fluent in using the environment. As contrasted with interactive design which relies on mostly generic tools, adapted design relies on specialized, knowledge-intensive software tools which have been conceived for location and layout purposes. The tools may embed generic tools, but these are seamlessly integrated and they are parameterized for layout and location purposes. In order to make clearer what assisted design is really about, below are described examples of commercial and academic assisted layout design platforms.

On the commercial side, the Plant Design and Optimization Suite from Tecknomatix, a business unit of UGS (www.ugs.com), is currently the best-known application in line with assisted design. The suite loosely couples their Plant Simulation, FactoryCAD, FactoryFlow, Factory Mockup, eM-Sequencer, and Logistics software. Of most direct interest among these is the FactoryFlow software, whose core was developed in the late 1980s up to the mid-1990s. It is introduced by the company as a graphical material handling system that enables engineers to optimize layouts based on material flow distances, frequency and costs, and that allows factory layouts to be analyzed by using part routing information, material storage needs, material handling equipment specifications, and part packaging (containerization) information. Embedded in the Autocad software, www.Autodesk.com, FactoryFlow aims to assist the engineer through a series of interactive factory design features coupled with specific feasibility, material handling flow, and equipment capacity analysis tools.

On the academic side, the concept of assisted design has been investigated for a long time. In the early 1980s, Warnecke and Dangelmaier (1982) developed an early prototype. Later on, Montreuil and Banerjee (1988) and Montreuil (1990) investigated object-oriented technologies and knowledge representation toward intelligent layout design environments, while Goetschalckx et al. (1990) investigated integrated engineering workstations as a platform for rapid prototyping of manufacturing facilities. The WebLayout design platform (Montreuil et al. 2002b) is perhaps the most comprehensive effort to date in making available an open assisted design environment to the community. WebLayout is conceived as a Web-based platform enabling researchers, professors, and students from around the world to concurrently experiment, test, and learn the basic and latest concepts and methodologies in factory design. Figure 5.22 illustrates a design generated by a team of students using WebLayout: it includes the site, the factory building and structure, and the production area layout as well as the office and service area layout. WebLayout allows multiple levels of granularity, from multi-site block layouts to processor layouts. It supports factory organization using responsibility networks, enabling engineers to contrast for the same case factory designs based on function, process, holographic, product, group or fractal organizations, or any combination of the above. It accepts probabilistic demand distributions for the products and supports the analysis

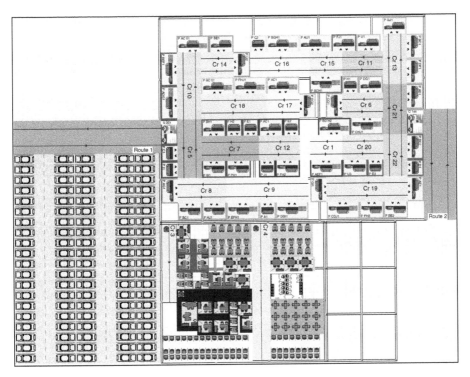

FIGURE 5.22 Example of design generated by a student team using WebLayout.

of their impact on production and handling resource requirements and overall expected economic performance. WebLayout is conceived from the ground up to allow various optimization, heuristic, analysis, evaluation, and validation tools to be readily integrated, so as to help engineers in various evolving ways to generate high-performance designs.

Even though they represent the current state of the art, both platforms are still primitive in terms of assisted design capabilities. There is still huge room for improvement and creativity, especially in a global, dynamic, and turbulent world where engineers have to transform their perception of location and layout design to become a process rather than a project. For example, none of the platforms currently have any significant dynamic layout capabilities, beyond letting the engineer enter an existing layout.

5.12.8 Holistic Metaheuristics

Holistic metaheuristics are significant extensions of first-generation metaheuristics. Driven by their fundamental intent to encompass a much more global design scope, they integrate a complex set of lower-level-focused heuristics, metaheuristic, and optimization model solvers.

A vivid example of such a holistic metaheuristic is HoloPro, conceived and developed to support holographic factory design in a wide spectrum of environments (Marcotte 2005, Marcotte and Montreuil 2004, 2005). Environments range from products, processes, and demand being known deterministically to being uncertain and all the way to being basically unknown. The design task addressed by HoloPro involves the automatic generation of (*i*) a set of processors to be implemented, (*ii*) a set of holographic centers, each with its embedded processors, (*iii*) a center and processor layout of the factory, (*iv*) expected work patterns for every center and processor, and (*v*) expected global and product-specific flow patterns. The metaheuristic is structured around a set of interacting agents, each responsible for a specific set of design tasks. Each agent relies on solving focused heuristics or optimization models.

Within a few minutes, it provides sets of holographic factory designs, complete with evaluation of their relative robustness when faced with expected uncertainties.

Detailed discussion of HoloPro is beyond the scope of this chapter. What is important to grasp is the fact that researchers have begun to address more holistic location and layout design tasks, attempting to provide automated approaches for generating optimized designs, and that the current preferred means for doing so is by developing and exploiting more complex, seamlessly integrated metaheuristics. However, this trend is barely in its infancy. It will require strong academic and industry commitment for it to grow. Indeed, getting involved in researching, developing, and maintaining such holistic metaheuristics is significantly more demanding in terms of academic, technological, and financial resources than previous generation approaches addressing much more localized, aggregated, and/or simplified problems.

5.12.9 Global Optimization

Complementary to assisted design and holistics metaheuristics lies global optimization. The main drive here is to develop optimization models of more comprehensive location and layout design tasks. For a long time, this drive has been mostly associated with a desire by researchers to formally define the problems they were addressing. They knew very well that the resulting models could not be solved beyond very simple cases and that they would have to rely on approximate techniques to really solve the problem for realistic-size cases. The very significant advances in performance of optimization solvers such as CPLEX (www.ilog.com) have pushed the frontier far enough that many problems considered unsolvable for realistic sizes have become amenable to solution by commercial solvers. Also, there have been advances in optimization solution techniques, such as branch-and-cut, and in heuristic solution of optimization models. All this has created a growing interest in global optimization.

In location design, the trend toward global optimization is quite evolved, especially exploiting discrete location modeling. Optimal location-allocation models have long been exploited, where facilities have to be opened or closed through a set of locations and clients assigned to opened locations, in an attempt to minimize overall travel, opening, and closure costs. These have been extended to capacitated versions where each location has a limited capacity when opened and assigned clients use a fraction of this capacity. Revelle and Laporte (1996) propose models for capacitated facility location, including a multi-period version. Several authors have dealt with multiple stages or levels of facilities, such as factories and central distribution centers, as described by Klose (2000). Gradually, this has led to the creation of network design models, such as production–distribution networks and manufacturing and logistic networks, as exemplified by Geoffrion and Graves (1974), Geoffrion and Powers (1995), Slats et al. (1995), Cruz et al. (1999), Dogan and Goetschalckx (1999), Dasci and Verter (2001), Melkote and Daskin (2001), Martel (2005), and Paquet et al. (2007). Gradually, location decisions are considered in the midst of comprehensive supply chain design models, as illustrated by Cohen and Moon (1990), Arntzen et al. (1995), and Chopra (2003). With the globalization of the economy, many such models are incorporating international issues such as differences and fluctuations in labor rates and availability, transport modes and costs, interest rates, currency rates, transfer prices, fiscal issues, and country risks, as well as dealing with the geographical dispersion of markets, suppliers, and potential site locations (e.g., Goetschalckx et al. 2002, Kouvelis et al. 2004, and Martel 2005). The trend toward global optimization in location design has been fueled by the fact that large-scale, ever more realistic and comprehensive models have been solved to optimality or near-optimality using a variety of solution techniques, such as Bender's decomposition, branch-and-cut, lagrangean relaxation, and lagrangean relax-and-cut.

In layout design, global optimization is much more embryonic, slowed by the inherent higher difficulty of solving layout models to optimality or near-optimality. Global layout optimization began in the early 1990s. Montreuil (1991) introduced a modeling framework for integrating layout design and flow network design, leading to the modeling of net layout design. The framework scope presentation was structured in an ever-increasing scope and complexity. First was introduced a mixed integer linear programming model for block layout and I/O station location which minimizes inter-station rectilinear travel. Second, the modeling was extended to take into consideration travel along a spatially fixed travel network. Third, it was again extended to allow the optional use of links along the fixed travel network. Fourth, the spatially fixed travel network constraint was relaxed to rather impose a less constraining logical travel network. Fifth, the modeling switched from designing a block layout with its travel network(s) to designing a net layout explicitly modeling the aisle system. It started doing so by imposing a set of spatially fixed interconnecting aisles. Sixth, the modeling was relaxed to replace the fixed aisle system by a spatially fixed aisle travel network which had to be transposed into an aisle system through the solution of the net layout model. Seventh, it relaxed the physically fixed aisle travel network by a logical aisle travel network. Eighth, it finally relaxed the model to allow optional aisle travel links, ending up with a net layout design modeling which only required as input a potential aisle travel network from which links could be truncated by the model, resulting in a design with fewer aisles while ensuring minimal travel network usage and implementation cost. Heragu and Kusiak (1991) presented a continuous space layout design model that offers an alternative to the first model in the framework. To this day, several variants of the first model are now being solved optimally or near optimally for small cases and some medium-size cases (Montreuil et al. 2002a, Sherali et al. 2003, Anjos and Vannelli 2006). The more encompassing models are yet subject to investigation by the research community to enable the solution of realistic-size cases. The modeling framework has been used as a problem formalization template by the layout research community over the years. Researchers such as Barbosa-Póvoa et al. (2001), Marcotte (2005), and Ioannou (2007) are now embarking on new global optimization modeling avenues.

5.13 Integrated Location and Layout Design Optimization Modeling

Location and layout are tightly interlaced and complementary. This section introduces two design optimization models which should formalize the relationship between layout and location. The models are part of the global optimization trend described earlier. They both deal explicitly with dynamics and uncertainty. The first is a dynamic probabilistic discrete location model, whereas the second is a dynamic probabilistic discrete location and continuous layout model. The exposition of these models aims to counterbalance the design issue orientation of Sections 5.2 to 5.11 and the solution methodology orientation of Section 5.2 by taking a formal mathematical modeling orientation. In no way should these two models be perceived as the models. They must, rather, be understood as two examples of a vast continuum of potential models to formalize the design issues described in Sections 5.2 through 5.11.

5.13.1 Dynamic Probabilistic Discrete Location Model

This model optimizes the dynamic assignment of a set of centers (or facilities) to a set of discrete locations. The model supports a number of predefined future scenarios, each with a number of successive periods covering the planning horizon. The occurrence of a future is probabilistic. For each period in each future, each center has specific space requirements and pairwise unitary flow travel (or proximity relationship) costs are defined. Each center has also a fixed cost for being assigned in a specific location during a period of a future. Each location can be dynamically made available, expanded or contracted through time, with associated costs. Centers can be moved from a period to another, incurring a moving cost. The model recognizes that decisions relative to the first period are the only rigid ones, as all others will be revisable later on based on further information, as future scenarios will either become past, present, or nearer future scenarios. Below are first exposed the objective function and the constraints, followed by definitions for variables, parameters, and sets. Then the model is described in detail.

Minimize

$$\sum_f P_f \left[\sum_{\forall cl, t>1} a_{cltf} A_{cltf} + \sum_{\forall clc'lt'} c_{clc'l'tf}(A_{cltf} A_{c'l'tf}) + \sum_{\forall cll't} m_{cll'tf} A_{cl,t-1,f} A_{cl'tf} \right.$$

$$\left. + \sum_l (e_l^+ E_l^+ + e_l^- E_l^-) + \sum_{cl} l_{cl} L_{cl} + \sum_f P_f \left[\sum_{\forall l, t>1, f} (s_{ltf} S_{ltf} + s_{ltf}^+ S_{ltf}^+ + s_{ltf}^- S_{ltf}^-) \right] \right] \qquad (5.4)$$

Subject to

$$\sum_{l \in L^c} A_{cltf} = 1 \qquad \forall c, t, f \qquad (5.5)$$

$$A_{cl1f} = L_{cl} \qquad \forall c, l, f \qquad (5.6)$$

$$\sum_{c \in C^l} r_{ctf} A_{cltf} \leq S_{ltf} \leq s_l^m \qquad \forall l, t, f \qquad (5.7)$$

$$S_{ltf} = S_{l,t-1,f} + S_{ltf}^+ - S_{ltf}^- \qquad \forall l, t > 1, f \qquad (5.8)$$

$$S_{l1f} = E_l = s_l^0 + E_l^+ - E_l^- \qquad \forall l, f \qquad (5.9)$$

where the variables are

A_{cltf} Binary variable equal to 1 when center c is assigned to location l in period t of future f, or 0 otherwise.

E_l, E_l^+, E_l^- Continuous non-negative variables deciding the space availability, expansion, and contraction of location l in period 1.

L_{cl} Binary variable equal to 1 when center c is assigned in location l in period 1, or 0 otherwise.

$S_{ltf}, S_{ltf}^+, S_{ltf}^-$ Continuous non-negative variable deciding space availability, expansion, and contraction of location l in period t of future f.

while the parameters and sets are

a_{cltf} Cost of assigning center c to location l in period t of future f.

$c_{clc'l'tf}$ Cost of concurrently assigning center c to location l and center c' to location l' in period t of future f.

C^l Set of centers allowed to be located in location l.

e_l, e_l^+, e_l^- Unit space availability, expansion, and contraction costs for location l in period 1.

l_{cl} Cost of assigning center c to location l in period 1.

L^c Set of locations in which center c is allowed to be located.

$m_{cll'tf}$ Cost of moving center c from location l to location l' in period t of future f.

P_f Probability of occurrence of future f.

r_{ctf} Space requirements for center c in period t of future f.

S_l^0 Initial space availability at location l.

$S_{ltf}, S_{ltf}^+, S_{ltf}^-$ Unit space availability, expansion, and contraction costs for location l in period t of future f.

The objective function 5.4 minimizes the overall actualized marginal cost, here described along two lines. The first line includes the sum of three cost components over all probable futures, weighted by the probability of occurrence of each future f. The first sums, over all allowed combinations, the cost of assigning a center c to a location l in period $t > 1$, independent of where other centers are located in this period t and of where center c was located in the previous period. The second sums, over all allowed combinations, the cost associated with concurrently locating center c in location l and center c' in location l' in period t. This is generically the cost associated with relations, interactions, and flows between centers. The third sums over all allowed combinations, the cost of moving center c from its location l in period $t - 1$ to location l' in period t. This is generically the dynamic center relocation cost.

The second line of the objective function includes three cost components. The first two add up all immediate transition costs from the actual state to the proposed state in period one. First is the cost associated with the space of each location as proposed for the first period. It includes the cost of making this space available and the cost of either expanding or contracting the location from its actual state. Second is the cost of implementing each center in its proposed location. These two components do not explicitly refer to specific futures, as they are common to all futures since they are a direct result of the location decisions and will be incurred in all futures. The third component is similarly the space availability, expansion, and contraction cost for all locations in all later periods of all futures.

Constraint set 5.5 makes sure that a center c is located in a single location l in each period t of each future f. Constraint set 5.6 attaches the location decisions made for time period 1 over all probable futures. So in the first period, each center c is assigned to the same location l in all futures. These are the decisions that have to be taken now that will definitely lead to implementation. These decisions cannot be altered afterward. In all later periods, the location decisions are allowed to vary from one future to another. They define a probabilistic plan that will be alterable subsequently, in light of further information availability, until they are associated with the first period in the revised model and become the hard location decision leading to immediate implementation.

Constraint set 5.7 ensures that the space availability constraint of each location l is respected at each period t of each future f, with the constraint that the sum of the required spaces of each center assigned to a location l does not exceed its space availability at that time. This availability is bounded for each location l to a specified maximum. The space availability of a location l can vary from one period to the next. For each future, constraint set 5.8 keeps an account of planned expansions and contractions of each location at all periods except the first. As constraint set 5.6 does for the location assignments of period one, constraint set 5.9 deals with the incumbent expansion or contraction of each location in the forthcoming first period, common to all futures for each location.

When the space requirement and availability parameters are restrained to one, there is a single time period and a single future, and no location expansion or contraction is allowed, then this model reduces to the well known QAP.

5.13.2 Dynamic Probabilistic Discrete Location and Continuous Layout Model

This model generalizes the above model by allowing treatment of each discrete location as a facility within which its assigned centers have to be laid out. The model thus explicitly deals with center shaping and location within facilities and with avoidance of spatial interference between centers. Centers are restricted to rectangular shapes. They are allowed to be moved between facilities and within facility from a period to the next in a future. Below are first exposed the objective function and the constraints, followed by definitions for variables, parameters, and sets. Then the model is described in detail.

Minimize

$$(4) + \sum_f p_f \left(\sum_{csc's't} d_{csc's'tf} D_{csc's'tf} + \sum_{cst} m_{cs} M_{cstf} \right) \tag{5.10}$$

subject to (5.5) to (5.9) and

$$\left(X_{ctf}^u - X_{ctf}^l\right)\left(Y_{ctf}^u - Y_{ctf}^l\right) = r_{ctf} \qquad \forall c,t,f \tag{5.11}$$

$$\left(X_{ltf}^u - X_{ltf}^l\right)\left(Y_{ltf}^u - Y_{ltf}^l\right) = s_{ltf} \qquad \forall l,t,f \tag{5.12}$$

$$\frac{\left(Y_{etf}^u - Y_{etf}^l\right)}{f_{etf}} \le \left(X_{etf}^u - X_{etf}^l\right) \le f_{etf}\left(Y_{etf}^u - Y_{etf}^l\right) \qquad \forall e \in (C \cup L),t,f \tag{5.13}$$

$$X_{ltf}^l - m\left(1 - A_{cltf}\right) \le X_{ctf}^l \le X_{ctf}^u \le X_{ltf}^u + m\left(1 - A_{cltf}\right) \qquad \forall c,l,t,f \tag{5.14}$$

$$Y_{ltf}^l - m\left(1 - A_{cltf}\right) \le Y_{ctf}^l \le Y_{ctf}^u \le Y_{ltf}^u + m\left(1 - A_{cltf}\right) \qquad \forall l,t,f \tag{5.15}$$

$$x_l^l \le X_{ltf}^l \le X_{ltf}^u \le x_l^u \qquad \forall l,t,f \tag{5.16}$$

$$y_l^l \le Y_{ltf}^l \le Y_{ltf}^u \le y_l^u \qquad \forall l,t,f \tag{5.17}$$

$$X_{ctf}^l \le X_{cstf}^s \le X_{ctf}^u \qquad \forall c,s,t,f \tag{5.18}$$

$$Y_{ctf}^l \le Y_{cstf}^s \le Y_{ctf}^u \qquad \forall c,s,t,f \tag{5.19}$$

$$\left(X_{c'tf}^l - X_{ctf}^u\right) \ge m\left(P_{cc'tf}^x - 1\right) \qquad \forall c,c't,f \tag{5.20}$$

$$\left(Y_{c'tf}^l - Y_{ctf}^u\right) \ge m\left(P_{cc'tf}^y - 1\right) \qquad \forall c,c't,f \tag{5.21}$$

$$P_{cc'tf}^x + P_{c'ctf}^x + P_{cc'tf}^y + P_{c'ctf}^y \ge A_{cltf} + A_{c'ltf} - 1 \qquad \forall c < c'l,t,f \tag{5.22}$$

$$X_{e1f}^l = X_e^l \qquad \forall e \in (C \cup L),f \tag{5.23}$$

$$Y_{e1f}^l = Y_e^l \qquad \forall e \in (C \cup L),f \tag{5.24}$$

$$X_{e1f}^u = X_e^u \qquad \forall e \in (C \cup L),f \tag{5.25}$$

$$Y_{e1f}^u = Y_e^u \qquad \forall e \in (C \cup L),f \tag{5.26}$$

$$X_{cstf}^s - X_{c's'tf}^s = D_{csc's'tf}^{x+} - D_{csc's'tf}^{x-} \qquad \forall c,s,t,f \tag{5.27}$$

$$Y_{cstf}^s - Y_{c's'tf}^s = D_{csc's'tf}^{y+} - D_{csc's'tf}^{y-} \qquad \forall c,s,t,f \tag{5.28}$$

$$D_{csc's'tf} \ge D_{csc's'tf}^{x+} + D_{csc's'tf}^{x-} + D_{csc's'tf}^{y+} + D_{csc's'tf}^{y-} - m\left(2 - A_{cltf} - A_{c'ltf}\right) \qquad \forall c,s,l,t,f \tag{5.29}$$

$$0.5\left(\left(X_{ctf}^l + X_{ctf}^u\right) - \left(X_{c,t-1,f}^l + X_{c,t-1,f}^u\right)\right) = M_{ctf}^{x+} - M_{ctf}^{x-} \qquad \forall c,t,f \tag{5.30}$$

$$0.5\left(\left(Y_{ctf}^l + Y_{ctf}^u\right) - \left(Y_{c,t-1,f}^l + Y_{c,t-1,f}^u\right)\right) = M_{ctf}^{y+} - M_{ctf}^{y-} \qquad \forall c,t,f \tag{5.31}$$

$$M_{ctf} \quad M_{ctf}^{x+} + M_{ctf}^{x-} + M_{ctf}^{y+} + M_{ctf}^{y-} - m\left(2 - A_{cltf} - A_{cl,t-1,f}\right) \qquad \forall c,l,t,f \qquad (5.32)$$

where new variables are

$X_e^l, X_e^u, Y_e^l, Y_e^u$	Continuous variables for the coordinates of the lower and upper boundaries of the sides of entity e along the X and Y axes in period 1 for all futures, where an entity is either a center or a location.
$X_{etf}^l, X_{etf}^u, Y_{etf}^l, Y_{etf}^u$	Continuous variables for the coordinates of the lower and upper boundaries of the sides of entity e along the X and Y axes in period t of future f, where an entity is either a center or a location.
X_{cstf}^s, X_{cstf}^s	Continuous variables for the X and Y coordinates of I/O station s of center c in period t of future f.
$D_{csc's'tf}$	Continuous non-negative variables for the rectilinear distance between station s of center c and station s' of center c' in period t of future f.
$D_{csc's'tf}^{x+}, D_{csc's'tf}^{x-}, D_{csc's'tf}^{y+}, D_{csc's'tf}^{y-}$	Continuous non-negative variables for the positive and negative components along the X and Y axes of the rectilinear distance between station s of center c and station s' of center c' in period t of future f.
$M_{ctf}^{x+}, M_{ctf}^{x-}, M_{ctf}^{y+}, M_{ctf}^{y-}$	Continuous non-negative variables for the positive and negative components along the X and Y axes of the rectilinear move of center c in period t of future f from its coordinates in the previous period of future f.
M_{ctf}	Continuous non-negative variables for the rectilinear move of center c in period t of future f from its coordinates in the previous period of the same future, whenever center c is assigned to the same location in periods t and $t-1$.
$P_{cc'tf}^x, P_{cc'tf}^y$	Binary variables stating whether or not center c is to position lower than center c' along axes X and Y whenever both centers are assigned to the same location in period t of future f.

while new parameters are

$d_{csc'tf}$	Unitary positive interaction cost associated with the rectilinear distance between station s of center c and station s' of center c' whenever both centers are assigned to the same location in period t of future f.
f_{etf}	Maximum allowed ratio between the longest and shortest sides of rectangular entity e, which is either a location or a center; this ratio can be distinct for each period of each future except for the first period, when it has to be the same for all futures.
m_{ctf}	Unitary positive move cost associated with the rectilinear displacement of center c in period t of future f from its coordinates in the previous period of the same future, whenever center c is assigned to the same location in periods t and $t-1$.
m	A very large number.
$x_l^l, x_l^u, y_l^l, y_l^u$	Lower and upper limits for location l along the X and Y axes.

The objective function 5.10 minimizes the sum of objective function 5.4 and the overall expected actualized interaction and move costs. These costs result from the summation over all futures, weighted by their probability of occurrence, of their future-specific costs. When laid out in the same location (site, building, etc.), pairs of centers having significant interactions (flows, relationships) incur a cost when their involved I/O stations are positioned a positive distance from each other. For example, if there is flow from the output station of center A to the input station of center B, then a unitary cost is specified for this pair. Then the interaction cost associated with the pair is the product of their unitary interaction cost and their rectilinear distance. The move cost for a center is computed over all periods of a future, multiplying the rectilinear displacement of its centroid from a period to the next by the unitary move cost specified for this center.

The constraint set includes previously defined constraints 5.5 to 5.9. The new constraints 5.11 to 5.32 are associated with the actual layout of centers assigned to the same location, where they have to share space without interfering with each other while satisfying their shape requirements. Each center and location is restricted to have a rectangular shape and to be orthogonally laid out relative to each other. Each is defined through the positioning of its lower X and Y axis corner and its upper X and Y axis corner.

Constraints 5.11 and 5.12, respectively, enforce that each center and location respect its specified area requirements. These quadratic constraints can be linearized using a set of linear approximation variables and constraints (e.g., Sherali et al. 2003). Constraints 5.13 impose a maximal form ratio between the longest and smallest sides of each center and location.

Constraints 5.14 and 5.15 ensure that whenever a center is assigned to a location, then it is to be laid out within the rectangular area of the location. Constraints 5.16 and 5.17 guarantee that each location is itself located within its maximal allowed coordinates. For example, a building cannot be extended beyond its site boundaries. Similarly, constraints 5.18 and 5.19 impose that each I/O station of a center be positioned within the center's rectangular area.

Constraints 5.20 to 5.22 ensure no physical overlap between centers assigned to the same location in a specific period of a future. They do so by imposing that for any two such centers, the former is either lower or upper along the X axis, or lower or upper along the Y axis.

Similar to constraint 5.6, constraints 5.23 to 5.26 recognize that the first layout decisions are imposed to all futures, to be immediately implemented while all other layout decisions can be subsequently altered depending on future information.

Constraints 5.27 to 5.29 compute the rectilinear distance between any two I/O stations of centers having positive interactions. The first two constraints linearize the computation of the rectilinear distance by adding its positive and negative components along the X and Y axes respectively, while the latter adds up all these components to get the overall rectilinear distance. Constraints 5.30 to 5.32 similarly compute the rectilinear displacement of the centroid of each center from its previous position to its current position. Constraints 5.27 to 5.32 assume positive unitary interaction and move costs. When negative unitary costs are involved, such as when one wants two centers to be far from each other, then the constraints have to be altered using binary variables to adequately compute the rectilinear distances and displacements.

When all centers are a priori assigned to the same location and the layout is to be fixed over the entire planning horizon, then the model simplifies to the static continuous block layout model introduced by Montreuil (1991).

5.14 Conclusion

From the offset, the chapter has warned the reader that location and layout design complexity would be addressed straight in the face, in a hard-nosed way, with the objectives of providing the reader with a holistic vision and equipping him or her to be able to deliver designs that address the real issues at hand. This has been a demanding task, as most of the sections end up presenting material rarely or never yet presented in such a way, often starting with levels of elevation normally achieved only in research papers or in the conclusive remarks of textbooks. As much as possible in such a chapter, practical examples have been provided. Several of these examples are highly elaborate to guarantee that the reader can transpose the material for usage in realistic cases. The overall bet is that the reader will be capable of mastering the essence of the material and achieve levels of design performance much higher than with a more traditional approach.

Even though the chapter is quite long, it has been subject to critical editing choice among the huge number of potential topics. Perhaps the most difficult has been the continuous struggle between presenting more location or layout examples and material, aiming to strike the right balance. It should be clear that this chapter could easily be rewritten without ever mentioning the world layout, or similarly the word *location*. However, both domains constitute a continuum where location is present both at the macro level and micro level surrounding layout. Each center must be located in the network, assigned to an existing or new facility. The union over all interacting centers defines the highly strategic and global location design challenge. At each site location, the facilities must be laid out so as to best deploy their assigned centers, hence defining a layout design challenge at each site. Then given that the main location and layout designs have been set, there appears at the micro level the need for locating a variety of resources through the network and the sites. There is much in common between the two interlaced domains. However, there are also differences which are most evidenced when presenting an example. It should be clear that layout examples have taken a dominant position in the chapter, in an effort to use in many contexts the same basic case. This is surely because of the author's background. Hopefully the overall balance does not penalize too strongly the location facets.

One of the purposeful omissions in the core of the chapter has been a section on the global comparative evaluation of design alternatives. The justification is that its application is much wider in scope than location and layout design. However, as this chapter reaches its closure, it becomes important to address it briefly. Whenever possible, all nondominated design alternatives should be evaluated financially. Their expected return on investment should be computed, as well as their economic value added, taking into consideration all impacts on potential revenues, costs, and investments, as well as the inherent identified risks involved. Furthermore, all the nonfinancial criteria should be analyzed, weighted in terms of their relative importance, and each nondominated alternative design should be evaluated relative to each criterion. Then typical multicriteria decision-making techniques should be used to merge the financial and nonfinancial evaluations to end up with relative rankings of alternatives, as well as sensitivity analyses, so as to best feed the decision-makers (Gal et al. 1999). A wide variety of criteria has been listed through the chapter, yet many more can be found on the reference material. Overall, criteria fall in two categories: performance criteria and capability criteria. All criteria should be in line with the strategic intent of the enterprise. Also, all key stakeholders should be taken into consideration when setting the set of criteria. For example, employees will motivate safety, quality-of-life and visibility criteria. Clients will motivate lead time and flexibility capability criteria. Suppliers may motivate vehicle access criteria. Headquarters will motivate financial performance and may motivate agility and personalization capabilities. The regional community may motivate environment criteria. Such lists of stakeholders and associated criteria are highly case dependent and should be carefully investigated.

It has been said that location and layout design has become a mature domain subject to limited room for significant innovation and impact. The chapter has hopefully contributed to challenge this somber assessment and prove that the domain is highly pertinent and challenging, and that there is a lot of room for professional, academic, and technological research and innovation. Overall, the two main keys appear to take a performance and capability development perspective in line with the strategic intents of the organization and to think of location and layout design as being a continuous process rather than a punctual project, always aiming to proactively adjust to relentless dynamics and turbulence in the organization and in the environment.

Acknowledgments

This work has been supported by the Canada Research Chair in Enterprise Engineering, the NSERC/Bell/ Cisco Business Design Research Chair, and the Canadian NSERC Discovery grant Demand and Supply Chain Design for Personalized Manufacturing. In addition, the author would like to thank Edith Brotherton and Caroline Cloutier, both research professionals in the CIRRELT research center at Laval University, for their help in the validation and proofing of this chapter.

References

Abdinnour-Helm S. and S.W. Hadley (2000). *Tabu search based heuristics for multi-floor facility layout*, International Journal of Production Research, v38, no2, 365–383.

Aleisa, E.E. and L. Lin (2005). *For effective facilities planning: layout optimization then simulation, or vice versa?*, Proc. of 2005 Winter Simulation Conference, 1381–1385.

Anjos M.F. and A. Vannelli (2006). *A new mathematical-programming framework for facility-layout design*, INFORMS Journal on Computing, v18, no1, 111–118.

Anstreicher, K.M., N.W. Brixius, J.P. Goux and J. Linderoth (2002). *Solving large quadratic assignment problems on computational grids*, Journal Mathematical Programming, v91, no3, 563–588.

Armour, G.C. and E.S. Buffa (1963). *A heuristic algorithm and simulation approach to the relative location of facilities*, Management Science, v9, no2, 249–309.

Arntzen, B.C., G.G. Brown, T.P. Harrison and L.L. Trafton (1995). *Global supply chain management at Digital Equipment Corporation*, Interfaces, v25, no1, 69–93.

Askin, R.G. (1999). *An empirical evaluation of holonic and fractal layouts*, International Journal of Production Research, v37, no5, 961–978.

Azadivar, F. and J. Wang (2000). *Facility layout optimization using simulation and genetic algorithms*, International Journal of Production Research, v38, no17, 4369–4383.

Banerjee P., Y. Zhou and B. Montreuil (1997). *Genetically assisted optimization of cell layout and material flow path skeleton*, IIE Transactions, v29, no4, 277–292.

Barbosa-Póvoa A.P., R. Mateus and A.Q. Novais (2001). *Optimal two-dimensional layout of industrial facilities*, International Journal of Production Research, v39, no12, 2567–2593.

Benjaafar, S. (2002). *Modeling and analysis of congestion in the design of facility layouts*, Management Science, v48, 679–704.

Benjaafar, S., S.S. Heragu and S. Irani (2002). *Next generation factory layouts: research challenges and recent progress*, Interfaces, v32, no6, 58–76.

Chhajed, D., B. Montreuil and T.J. Lowe (1992). *Flow network design for manufacturing systems layout*, European Journal of Operational Research, v57, 145–161.

Chittratanawat, S. (1999). *An integrated approach for facility layout, P/D location and material handling system design*, International Journal of Production Research, v37, no3, 683–706.

Chopra, S. (2003). *Designing the distribution network in a supply chain*, Transportation Research Part E: Logistics and Transportation Review, v39, no2, 123–140.

Cohen, M.A. and S. Moon (1990). *Impact of production scale economies, manufacturing complexity, and transportation costs on supply chain facility networks,* Journal of Manufacturing and Operation Management, v3, 269–292.

Cortinhal M.J. and M.E. Captivo (2004). *Genetic algorithms for the single source capacitated location problem*, Metaheuristics: Computer Decision-Making, Kluwer, 187–216.

Crainic T.G., M. Toulouse and M. Gendreau (1996). *Parallel asynchronous tabu search for multicommodity location-allocation with balancing requirements*, Annals of Operations Research, v63, 277–299.

Cruz, F.R.B., J. MacGregor Smith and G.R. Mateus (1999). *Algorithms for a multi-level network optimization problem*. European Journal of Operational Research, v118, no1, 164–180.

Dasci, A. and V. Verter (2001). *A continuous model for production-distribution system design*, European Journal of Operational Research, v129, no2, 287–298.

Dogan, K. and M. Goetschalckx (1999). *A primal decomposition method for the integrated design of multiperiod production-distribution systems*, IIE Transactions, v31, no11, 1027–1036.

Donaghey C.E. and V.F. Pire (1990). Solving the Facility Layout Problem with BLOCPLAN, Industrial Engineering Department, University of Houston, TX, 1990.

Drezner, Z. and H.W. Hamacher, eds (2002). Facility Location, Applications and Theory, Springer-Verlag, Berlin, Germany.

Edmonds, J. (1965). *Paths, trees and flowers*, Canadian Journal of Mathematics, v17, 449–467.

Evans, G.W., M.R. Wilhelm and W. Karwowski (1987). *A layout design heuristic employing the theory of fuzzy sets*, International Journal of Production Research, v25, no10, 1431–1450.

Foulds, L.R., P. Giffons and J.Giffin (1985). *Facilities layout adjacency determination: an experimental comparison of three graph theoretic heuristics*, Operations Research, v33, 1091–1106.

Francis R.L., L.F. McGinnis and J.A. White (1992). Facility Layout and Location: An Analytical Approach, 2nd edition, Prentice-Hall, Englewood Cliffs, NJ, USA.

Gal, T., T.J. Stewart and T. Hanne, eds (1999). Multicriteria Decision Making: Advances in MCDM Models, Algorithms, Theory and Applications, Kluwer Academic Publishers, Boston, USA.

Geoffrion, A.M. and G.W. Graves (1974). *Multicommodity Distribution System Design by Benders Decomposition*, Management Science, v20, no5, 822–844.

Geoffrion, A.M. and R.F. Powers (1995). *Twenty years of strategic distribution system design: an evolutionary perspective*, Interfaces, v25, no5, 105–127.

Goetschalckx, M., L.F. McGinnis and K.R. Anderson (1990). *Toward rapid prototyping of manufacturing facilities*, Proceedings of the IEEE First International Workshop on Rapid System Prototyping, Research Triangle Park, NC, USA.

Goetschalckx, M., C.J. Vidal and K. Dogan (2002). *Modeling and design of global logistics systems: a review of integrated strategic and tactical models and design algorithms*, European Journal of Operational Research, v143, no1, 1–18.

Gomory R.E. and T.C. Hu (1961). *Multi-terminal network flows*, Journal SIAM on Computing, v9, no4, 551–570.

Hardin, C.T. and J.S. Usher (2005). *Facility layout using swarm intelligence*, Proceedings of 2005 IEEE Swarm Intelligence Symposium, 424–427.

Heragu S. and A. Kusiak (1991). *Efficient models for the facility layout problem*, European Journal of Operational Research, v53, 1–13.

Heskett, J.L. (1963). *Cube-per-Order Index—A Key to Warehouse Stock Location*, Transportation and Distribution Review, v3, 27–31.

Huq, F., D.A. Hensler and Z.M. Mohamed (2001). *A simulation analysis of factors influencing the flow time and through-put performance of functional and cellular layouts*, Journal of Integrated Manufacturing Systems, v12, no4, 285–295.

Ioannou, G. (2007). *An integrated model and a synthetic solution approach for concurrent layout and material handling system design*, Computers and Industrial Engineering, v52, no4, 459–485.

Kerbache, L. and L.M. Smith (2000). *Multi-objective routing within large scale facilities using open finite queueing networks*, European Journal of Operational Research, v121, 105–123.

Kincaid, R.K. (1992). *Good solutions to discrete noxious location problems via metaheuristics*, Annals of Operations Research, v40, no1, 265–281.

Klose, A. (2000). *A lagrangean relax-and-cut approach for the two-stage capacitated facility location problem*, European Journal of Operational Research, v126, no 2, 408–421.

Kouvelis, P., M.J. Rosenblatt and C.L. Munson (2004). *A mathematical programming model for global plant location problems: analysis and insights*, IIE Transactions, v36, no2, 127–144.

Kuenh A. and M.J. Hamburger (1963). *A heuristic program for locating warehouses*, Management Science, v9, 643–666.

Lamar, M. and S. Benjaafar (2005). *Design of dynamic distributed layouts*, IIE Transactions, v37, 303–318.

Lee, R.C. and J.M. Moore (1967). *CORELAP—computerized relationship layout planning*, Journal of Industrial Engineering, v18, 195–200.

Leung J. (1992). *A new graph-theoretic heuristic for facility layout*, Management Science, v38, no4, 594–605.

Lilly, M.T. and J. Driscoll (1985). *Simulating facility changes in manufacturing plants*, Proceedings of the 1st International Conference on Simulation in Manufacturing, Stanford Upon Avon, UK.

Marcotte, S. (2005). *Optimisation de la conception d'aménagements holographiques* Holographic Layout Design Optimization, Ph.D. Thesis, Administration Sciences, Laval University, Québec, Canada.

Marcotte, S. and B. Montreuil (2004). *Investigating the impact of heuristic options in a metaheuristic for agile holographic factory layout design*, Proceedings of Industrial Engineering Research Conference 2004, Houston, Texas, 2004/05/15–19.

Marcotte, S. and B. Montreuil (2005). *Factory design robustness: an empirical study*, Proceedings of Industrial Engineering Research Conference 2005, Atlanta, Georgia, 2005/05/14–18.

Marcotte, S., B. Montreuil and P. Lefrançois (1995). *Design of holographic layout of agile flow shops*, Proceedings of International Industrial Engineering Conference, Montréal, Canada.

Marcotte S., B. Montreuil and C. Olivier (2002). Factory design fitness for dealing with uncertainty, Proceedings of MIM 2002: 5th International Conference on Managing Innovations in Manufacturing, Milwaukee, Wisconsin, 2002/09/9-11.

Marcoux, N., A. Langevin and D. Riopel (2005). *Models and methods for facilities layout design from an applicability to real-world perspective*, Logistics Systems: Design and Optimization, edited by A. Langevin and D. Riopel, Springer, USA, 123–170.

Martel, A. (2005). *The design of production-distribution networks: a mathematical programming approach*, Supply Chain Optimization, Geunes, J. and Pardalos, P. (eds), Springer, 265–306.

Maxwell, W.L. and J.A. Muckstadt (1982). *Design of automated guided vehicle system*, IIE Transactions, v1, no2, 114–124.

Melkote, S. and M.S. Daskin (2001). *Capacitated facility location/network design problems*, European Journal of Operational Research, v129, no3, 481–495.

Meller, R.D. and Y.A. Bozer (1996a). *A new simulated annealing algorithm for the facility layout problem*, International Journal of Production Research, v34, 1675–1692.

Meller R.D. and Y.K. Gau (1996b). *The facility layout problem: recent and emerging trends and perspectives*, Journal of Manufacturing Systems, v15, 351–366.

Montgomery, D.C., L.A. Johnson and J.S. Gardiner (1990). Forecasting and Time Series Analysis, 2nd edition, McGraw-Hill, NY, USA.

Montreuil, B. (1982). Interactive Optimization Based Facilities Layout, Ph.D. thesis, ISYE School, Georgia Institute of Technology, Atlanta, GA, USA.

Montreuil, B. (1987). *Integrated design of cell layout, input/output station configuration, and flow network of manufacturing systems*, Intelligent and Integrated Manufacturing Analysis and Synthesis, ed. by C.R. Liu, A. Requicha and S. Chandrasekar, ASME PED, v25, 315–326.

Montreuil, B. (1990). *Representation of domain knowledge in intelligent systems layout design environments*, Computer Aided Design, v22, no2, 90/3, 97–108.

Montreuil, B. (1991). *A modelling framework for integrating layout design and flow network design*, Progress in Material Handling and Logistics, v2, ed. by J.A. White and I.W. Pence, Springer-Verlag, 95–116.

Montreuil, B. (2001). *Design of agile factory networks for fast-growing companies*, Progress in Material Handling: 2001, ed. by R. J. Graves et al., Material Handling Institute, Braun-Brumfield Inc., Ann Arbor, Michigan, USA.

Montreuil, B. (2006). Facilities network design: a recursive modular protomodel based approach, Progress in Material Handling Research: 2006, ed. by R. Meller et al., Material Handling Industry of America (MHIA), Charlotte, North Carolina, USA, 287–315.

Montreuil, B. and P. Banerjee (1988). *Object knowledge environment for manufacturing systems layout design*, International Journal of Intelligent Systems, v3, 399–410.

Montreuil B., E. Brotherton, N. Ouazzani and M. Nourelfath (2004). *AntZone layout metaheuristic: coupling zone-based layout optimization, ant colony system and domain knowledge*, Progress in Material Handling Research: 2004, Material Handling Industry of America (MHIA), Charlotte, North-Carolina, USA, 301–331.

Montreuil, B. and H.D. Ratliff (1988a). *Optimizing the location of input/output stations within facilities layout*, Engineering Costs and Production Economics, v14, 177–187.

Montreuil, B. and H.D. Ratliff (1988b). *Utilizing cut trees as design skeletons for facility layout*, IIE Transactions, v21, no2, 88/06, 136–143.

Montreuil B. and U. Venkatadri (1988). *From gross to net layouts: an efficient design model*, Document de travail FSA 88-56, Faculté des sciences de l'administration, Université Laval, Québec, Canada.

Montreuil, B. and U. Venkatadri (1991). *Scattered layout of intelligent job shops operating in a volatile environment*, Proceedings of International Conference on Computer Integrated Manufacturing, Singapore.

Montreuil, B., E. Brotherton and S. Marcotte (2002a). *Zone based layout optimization*, Proceedings of Industrial Engineering Research Conference, Orlando, Florida, 2002/05/19–22.

Montreuil, B., N. Ouazzani and S. Marcotte (2002b). *WebLayout: an open web-based platform for factory design research and training*, Progress in Material Handling: 2002, ed. by R. J. Graves et al., Material Handling Institute, Braun-Brumfield Inc.

Montreuil, B. and P. Lefrançois (1996). *Organizing factories as responsibility networks*, Progress in Material handling: 1996, ed. by R. J. Graves et al., Material Handling Institute, Braun-Brumfield Inc., Ann Arbor, Michigan, USA, 375–411.

Montreuil, B., H.D. Ratliff and M. Goetschalckx (1987). *Matching based interactive facility layout*, IIE Transactions, v19, no3, 271–279.

Montreuil B., Y. Thibault and M. Paquet (1998). *Dynamic network factory planning and design*, Progress in Material Handling: 1999, ed. by R. J. Graves et al., Material Handling Institute, Braun-Brumfield Inc., Ann Arbor, Michigan, USA, 353–379.

Montreuil, B., U. Venkatadri and E. Blanchet (1993a), *Generating a net layout from a block layout with superimposed flow networks*, Document de travail GRGL 93-54, Faculté des sciences de l'administration, Université Laval, Québec, Canada.

Montreuil, B., U. Venkatadri and H.D. Ratliff (1993b). *Generating a layout from a design skeleton*, IIE Transactions, v25, no1, 93/1, 3–15.

Montreuil, B., U. Venkatadri and R.L. Rardin (1999). *The fractal layout organization for job shop environments*, International Journal of Production Research, v37, no3, 501–521.

Murray A.T. and R.L. Church (1996). *Applying simulated annealing to location-planning models*, Journal of Heuristics, v2, no1, 31–53.

Muther, R. (1961). Systematic Layout Planning, Industrial Education Institute, Boston, USA.

Norman B.A., A.E. Smith and R.A. Arapoglu (1998). *Integrated facility design using an evolutionary approach with a subordinate network algorithm*, Lecture Notes in Computer Sciences, v1498, Springer-Verlag, London, UK, 937–946.

Nourelfath M., N. Nahas and B. Montreuil (2007). *Coupling ant colony optimization and the extended great deluge algorithm for the discrete facility layout problem*, Engineering Optimization, forthcoming.

Nugent C.E., T.E. Vollmann and J. Ruml (1968). *An experimental comparison of techniques for the assignment of facilities to location*. Operations Research, v16, 150–173.

Osman I.H., B. Al-Ayoubi and M. Barake (2003). *A greedy random adaptive search procedure for the weighted maximal planar graph problem*, Computers and Industrial Engineering, v45, no4, 635–651.

Owen, S.H. and M.S. Daskin (1998). *Strategic facility location: a review*, European Journal of Operational Research, v111, no3, 423–447.

Paquet, M., A. Martel and B. Montreuil (2007). *A manufacturing network design model based on processor and worker capabilities*, International Journal of Production Research, 27 p., forthcoming.

Reed, R. (1961). Plant Layout: Factors, Principles and Techniques, R.D. Irwin Inc., Homewood, IL, USA.

Revelle, C.S. and G. Laporte (1996). *The plant location problem: new models and research prospects*, Operations Research, v44, no 6, 864–874.

Rosenblatt M.J. (1986). *The dynamics of plant layout*, Management Science, v32, no1, 76–86.

Seehof, J.M. and W.O. Evans (1967). *Automated layout design program*, Journal of Industrial Engineering, v18, 690–695.

Sherali H.D., B.M.P. Fraticelli and R.D. Meller (2003). *Enhanced model formulations for optimal facility layout*, Operations Research, v51, no4, 629–644.

Slats, P.A., B. Bhola, J.J.M. Evers and G. Dijkhuizen (1995). *Logistic chain modelling*, European Journal of Operational Research, v87, no1, 1–20.

Tate D.M. and A.E. Smith (1995). *Unequal-area facility layout by genetic search*, IIE Transactions, v27, 465–472.

Tompkins, J.A., White, J.A., Bozer, Y.A. and Tanchoco, J.M.A. (2003). Facilities Planning, 3rd edition, John Wiley and Sons, Inc., USA.

Venkatadri U., R. L. Rardin and B. Montreuil (1997). *A design methodology for fractal layout organization*, IIE Transactions (Special Issue of Design and Manufacturing on Agile Manufacturing), v29, no10, 911–924.

Warnecke H.J. and W. Dangelmaier (1982). *Progress in computer aided plant layout*, Technical report, Institute of Manufacturing Engineering and Automation, Fraunhofer, Germany.

Welgama P.S. and P.R. Gibson (1995). *Computer-aided facility layout—a status report*, International Journal of Advanced Manufacturing Technology, v10, no 1, 66–77.

Womack, J.P. and D.T. Jones (1998). Lean Thinking, Free Press, USA.

Wu Y. and E. Appleton (2002). *The optimization of block layout and aisle structure by a genetic algorithm*, Computers and Industrial Engineering, v41, no4, 371–387.

6

Inventory Control Theory: Deterministic and Stochastic Models

6.1 Introduction **6**-1
6.2 Deterministic Models **6**-1
     Economic Order Quantity Model • Economic Order
     Quantity Model for a Series of Two Facilities • Economic
     Order Quantity Model for a Multi-Product Assembly System
6.3 Stochastic Models **6**-11
     Newsvendor Problem • Joint Pricing and Inventory Control in
     a Newsvendor Setting • Multiple Period Models
6.4 Case Study **6**-24
     Exercises
References **6**-25

Lap Mui Ann Chan
*Virginia Polytechnic Institute and
State University*

Mustafa Karakul
York University

6.1 Introduction

We encounter an inventory problem whenever physical goods are stocked for anticipated demand. Inventory is often necessary when there is uncertainty in demand. However, even when demand is known for certain, inventory is built up to satisfy large demands when production is time-consuming. Stocking can also be used as a strategy to take advantage of the economies of scale since suppliers often offer discounts to encourage large orders and administrative costs can be saved by combining orders. Another critical reason for keeping high inventory is the loss of customer goodwill when shortages occur.

On the other hand, inventory ties down capital and incurs storage costs and property taxes. Appropriate cost functions are included in inventory models to capture the trade-off between overstocking and shortage. An optimization of the total profit or cost function generates a best ordering policy that specifies the quantities and times of replenishments.

In this chapter, we consider the problem of keeping inventory for different situations. Deterministic models with known demand and then stochastic models that involve uncertainty in demand are discussed in two separate sections.

6.2 Deterministic Models

Inventory models can be classified into two categories according to the review policy. In continuous review models, inventory is tracked continuously and replenishment is possible at any moment. The second category is periodic review models, in which inventory is checked at prespecified regular epochs, such as the end of a week, and replenishment can be done only at these check points. For continuous review models, we start our discussion with a single facility for a single product and continue to more complicated multi-facility and multi-product systems. On the other hand, for periodic review models, we focus on deriving an optimal inventory policy for stocking a single product for a single facility only.

6.2.1 Economic Order Quantity Model

A classical continuous review inventory problem is the economic order quantity (EOQ) model. This basic model considers a single product that has a known continuous demand of d units per unit time. The cost of replenishment is a fixed setup cost k plus a per unit variable cost of c per unit ordered. The cost of holding each unit of the product is h per unit time. Replenishments are instantaneous. The objective is to find a replenishment policy that satisfies the demand without delay so as to minimize the average replenishment and holding cost over the infinite horizon.

There exists an optimal replenishment policy that has a couple of nice properties. If replenishment is made when there is a positive inventory of the product, we can adjust the order quantities to make sure that replenishment occurs only when inventory is down to zero. Specifically, for an inventory policy that orders at time t_r for $r = 0, 1, 2, \dots$ with $t_0 = 0$, the adjusted order quantity is $d(t_{r+1} - t_r)$ at t_r. Note that for a feasible policy, the inventory level after replenishment at t_r is no less than $d(t_{r+1} - t_r)$.

Thus, this adjustment does not affect the replenishment cost but reduces the inventory holding cost, as illustrated in Figure 6.1a. Hence, to find an optimal replenishment policy, we only need to consider policies in which replenishments are made only when inventory is down to zero. This is called the zero-inventory-ordering (ZIO) property.

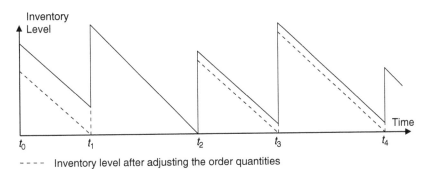

(a) Adjusting a Feasible Policy to a Zero-Inventory-Ordering Policy

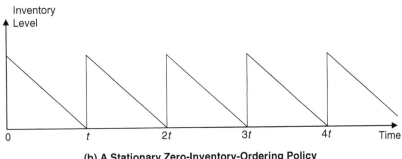

(b) A Stationary Zero-Inventory-Ordering Policy

FIGURE 6.1   A feasible policy and a stationary zero-inventory-ordering policy.

For the second property of an optimal replenishment policy, consider a fixed order quantity of Q in a ZIO policy. The replenishment cost of this order is $k + cQ$, the time till the next replenishment is Q/d and the average inventory until the next replenishment is $Q/2$. Together, we have the average cost of this replenishment:

$$AC(Q) = (k + cQ)/(Q/d) + hQ/2 = kd/Q + hQ/2 + cd \qquad (6.1)$$

Suppose replenishments of different quantities are made in a ZIO policy; then there must be one among all these replenishments that is associated with the smallest average cost. Thus, a ZIO replenishment policy in which every order is for a quantity that is the same as this smallest average cost replenishment has a smaller average cost than the one with different order quantities. Specifically, the average cost of the inventory policy shown in Figure 6.1b is no more than that of the policies shown in Figure 6.1a when $t = t_{k+1} - t_k$ with $AC(d[t_{k+1} - t_k]) \leq AC(d[t_{r+1} - t_r])$ for $r = 1, 2, 3, \ldots$. Hence, to find an optimal policy, we only need to consider ZIO policies that always order the same quantity. This is called the stationary property.

To find an optimal stationary ZIO policy, it remains to obtain the best order quantity that minimizes the average cost provided in Equation 6.1. Since the first derivative of $AC(Q)$ with respect to Q is $AC'(Q) = -kd/Q^2 + h/2 = 0$ when $Q = (2kd/h)^{1/2}$ and the second derivative of $AC(Q)$ with respect to Q is $A''(Q) = 2kd/Q^3 > 0$, $AC(Q)$ is a convex function that attains its minimum possible value of $(2kdh)^{1/2} + cd$ at $Q^* = (2kd/h)^{1/2}$. In summary, an optimal policy is to order $(2kd/h)^{1/2}$ unit of the product, when inventory is down to zero, every $(2kd/h)^{1/2}/d = [2k/(dh)]^{1/2}$ units of time with an average cost of $(2kdh)^{1/2} + cd$ per unit time. Note that in the existence of a constant replenishment lead time l, orders are placed a lead time ahead to make sure that they arrive when the inventory level is down to zero in a ZIO policy. Hence, constant lead time has no effect on the optimal order quantity Q^* or the optimal reorder interval Q^*/d.

Various efforts have been made by researchers to extend the EOQ model from the stocking of a single product for a single facility to more complicated systems. However, optimality results are elusive. Roundy (1985) introduces the class of near optimal power-of-two policies, which are stationary ZIO policies with reorder intervals that are power-of-two multiples of each other. We discuss the derivation of near optimal power-of-two policies for a two-facility in-series and then a multi-product assembly system in the following.

6.2.2 Economic Order Quantity Model for a Series of Two Facilities

Consider a retailer who faces the demand of a product that occurs at a constant rate of d per unit time. The retailer obtains the product from a warehouse at a cost of k_1 per order and holds inventory at a cost of H_1 per unit product per unit time. The warehouse in turn obtains the product from a supplier at a cost of k_0 per order and holds inventory at a cost of H_0 per unit product per unit time. Orders are satisfied instantaneously for both the warehouse and the retailer. The objective is to obtain an ordering policy that satisfies the demands at the warehouse and retailer without delay so as to minimize the long-run average ordering and holding cost for both the warehouse and retailer over the infinite horizon.

The retailer faces an EOQ problem. However, since the warehouse receives discrete orders from the retailer, he does not face an EOQ problem. On the other hand, if the warehouse does not consider the retailer as a separate facility and considers the product held at the retailer as part of its own inventory, then the warehouse is facing an EOQ problem with fixed order cost k_0, per unit holding cost rate of $h_0 = H_0$ and demand that occurs at a constant rate of d. As the warehouse accounts for a per unit holding cost rate of h_0 for the inventory at the retailer, the retailer has to pay a holding cost of only $h_1 = H_1 - h_0$ per unit product per unit time for its inventory. h_j, $j = 0,1$ are referred to as echelon holding costs. The EOQ problem faced by the retailer is modified to one with fixed order cost k_1, per unit holding cost rate of h_1 and demand that occurs at a constant rate of d. From the analysis of the EOQ model, the optimal

ZIO policy has reorder intervals of $T_j^* = [2k_j/(dh_j)]^{1/2}$ with an average cost of $(2k_j dh_j)^{1/2}$, $j = 0,1$ for these two EOQ problems.

If $T_0^* = T_1^*$, then the warehouse and retailer can synchronize with each other by ordering dT_0^* units of the product simultaneously when inventory is down to zero, every T_0^* units of time. Note that in ordering simultaneously, the product is delivered to the retailer through the warehouse, but is never stored there. In implementing an optimal policy for each one of the two EOQ models, system-wide average cost is minimized.

Since inventory is kept at the retailer but not at the warehouse, the average cost of $(2k_0 dh_0)^{1/2} + (2k_1 dh_1)^{1/2}$ for this optimal policy should be a function of H_1 but not of H_0. To rewrite the average cost in terms of H_1 only, note that since

$$[2k_0/(dh_0)]^{1/2} = T_0^* = T_1^* = [2k_1/(dh_1)]^{1/2},$$

$T_0^* = \{2(k_0 + k_1)/[d(h_0+h_1)]\}^{1/2} = [2(k_0 + k_1)/(dH_1)]^{1/2}$ and the optimal average cost

$$(2k_0 dh_0)^{1/2} + (2k_1 dh_1)^{1/2} = dT_0^*(h_0 + h_1) = dT_0^* H_1 = [2(k_0 + k_1)dH_1]^{1/2}. \qquad (6.2)$$

We use this observation to help determine an optimal policy for the case when $T_0^* < T_1^*$.

If $T_0^* < T_1^*$, then the warehouse and retailer cannot synchronize with each other to implement the optimal ZIO policies for the two EOQ models simultaneously. The problem is that $T_0^* = [2k_0/(dh_0)]^{1/2}$, a decreasing function of $h_0 = H_0$, is too small. In other words, the per unit holding cost rate h_0 at the warehouse is too large. Consider a duplicate system with the same H_1 but a smaller holding cost rate $h'_0 = H'_0$ at the warehouse so that $[2k_0/(dh_0)]^{1/2} < \{2(k_0 + k_1)/[d(h_0 + h_1)]\}^{1/2} = [2(k_0 + k_1)/(dH_1)]^{1/2} = [2k_0/(dh'_0)]^{1/2}$. As discussed earlier, an optimal policy for this duplicate system is for the warehouse and the retailer to order simultaneously every $[2(k_0 + k_1)/(dH_1)]^{1/2}$ units of time when inventory at the retailer is down to zero for an average cost of $[2(k_0 + k_1)dH_1]^{1/2}$. Since H_1 is the same for both systems, in following the same policy for the original system, inventory is kept only at the retailer and the average cost is:

$$\{(k_0 + k_1) + h_1 d[2(k_0 + k_1)/(dH_1)]/2\}/[2(k_0 + k_1)/(dH_1)]^{1/2} = [2(k_0 + k_1)dH_1]^{1/2}.$$

Note that the result is the same as Equation 6.2. Since the costs for the original system are no less than the duplicate system and the optimal policy for the duplicate system results in the same average cost for the original system, it is an optimal policy for the original system as well.

Note that in keeping H_1 constant and reducing the holding cost H_0 by δ, h_0 is reduced by δ while h_1 is increased by δ. In other words, reducing the holding cost by δ at the warehouse is equivalent to redistributing δ units of the echelon holding cost at the warehouse to the retailer. Furthermore, if the warehouse and the retailer have the same reorder interval, then inventory is kept only at the retailer and the average cost is not affected by this redistribution of the echelon holding cost from the warehouse to the retailer. These observations are used in the discussion of the multi-product systems.

If $T_0^* > T_1^*$ and the retailer orders every T_1^* units of time, then the warehouse can synchronize with the retailer and place an order every T_0^* units of time only if $T_0^* = rT_1^*$ for some positive integer r. In that case, optimality is achieved by placing every order from the warehouse simultaneously with an order from the retailer, since optimal ZIO policies are implemented for the two EOQ models. On the other hand, in the case T_0^* is not an integer multiple of T_1^*, Roundy (1985) suggests a heuristic from the class of power-of-two policies, which satisfy the ZIO and stationary property with the reorder interval for the warehouse equals to a power-of-two multiple of the reorder interval for the retailer. In particular, let $2^m T_1^* \leq T_0^* < 2^{m+1} T_1^*$ for some non-negative integer m. If $T_0^*/(2^m T_1^*) \leq 2^{m+1} T_1^*/T_0^*$, then the warehouse places an order every $T_0 = 2^m T_1^*$ units of time. Otherwise, the warehouse places an order every $T_0 = 2^{m+1} T_1^*$ units of time. In either case, every order from the warehouse is placed simultaneously with one from the retailer to make sure that ZIO policies are implemented for the two EOQ models. However, optimality is achieved for only one of the two EOQ models to attain an average cost of $(2k_1 dh_1)^{1/2}$. For the other EOQ model, the reorder interval is T_0 and the corresponding average cost is $k_0/T_0 + h_0 dT_0/2$. For the effectiveness of this power-of-two policy, note that if $T_0^*/(2^m T_1^*) \le 2^{m+1} T_1^*/T_0^*$, then $T_0^*/T_0 = T_0^*/(2^m T_1^*) \le 2^{m+1} T_1^*/T_0^* = 2T_0/T_0^*$ and hence $1 \le T_0^*/(2^m T_1^*) = T_0^*/T_0 \le 2^{1/2}$. Otherwise, $T_0/T_0^* = 2^{m+1} T_1^*/T_0^* \ge T_0^*/(2^m T_1^*) = 2T_0^*/T_0$ and hence $1 \le 2^{m+1} T_1^*/T_0^* = T_0/T_0^* \le 2^{1/2}$. Together, we have $2^{-1/2} \le T_0/T_0^* \le 2^{1/2}$. Since $(k_0/T_0 + h_0 dT_0/2)/(2k_0 dh_0)^{1/2} = (T_0^*/T_0 + T_0/T_0^*)/2$ is a convex function of T_0/T_0^* that attains its minimum value at $T_0/T_0^* = 1$, we have

$$(k_0/T_0 + h_0 dT_0/2)/(2k_0 dh_0)^{1/2} \le (2^{-1/2} + 2^{1/2})/2 \sim 1/0.94.$$

This implies that the average cost of an optimal policy is at least 94% of that of the power-of-two policy. In other words, this power-of-two policy is 94% optimal.

By adjusting the reorder interval for the warehouse only, an optimal policy is used for one EOQ model, while 94% optimality is achieved for the other one. Roundy (1985) suggests another power-of-two policy that is obtained by adjusting the reorder intervals for both the warehouse and retailer in order to minimize the total cost of optimality for the two EOQ models. This more complicated power-of-two policy is 98% optimal.

The results for power-of-two policies can be extended to systems with facilities that form an acyclic network. We illustrate this by considering a multi-product assembly system.

6.2.3 Economic Order Quantity Model for a Multi-Product Assembly System

Consider a manufacturer of n products. Demand of each product occurs at a constant rate. By scaling, we can assume without loss of generality that the demand rate of each product is 2 units per unit time. Each product i is manufactured by a number of assemblies of parts specified by an assembly directed network $T_i = (N_i, A_i)$. N_i represents the set of parts involved in the production of product i. We will refer to product i also as a part. Hence, i is in N_i. Node i has no successor, while each one of the other nodes in N_i has exactly one immediate successor in T_i. Each part j in N_i is produced by assembling the parts in the set P_j^i of its immediate predecessors in T_i. Figure 6.2 illustrates the production assembly networks for products 1 and 2. The holding cost rate of each part is linear. For each part j required for the production of part i, let H_j^i be the holding cost of part j per unit production of product i per unit time. For example, if the demand rate of product 2 is 6 lb per day, the holding cost of part 4 is \$2/lb per day, and ¼ lb of part 4 is required per pound production of product 2; then in using each day as a time unit, product 2 is measured in units of 3 lbs, and part 4 for the production of product 2 is measured in units of $(3)(¼) = ¾$ lbs with $H_4^2 = \$2(¾) = \$3/2$. Independent of the amount of part j to produce, each assembly is instantaneous and incurs a setup cost of k_j. The objective is to obtain a production policy that satisfies the demands without delay so as to minimize the long-run average setup and holding cost over the infinite horizon.

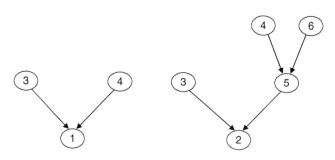

FIGURE 6.2 The assembly networks for products 1 and 2.

Similar to the earlier discussion of the two facilities in-series system, to obtain a 94% optimal EOQ production policy for this model, we first transform the problem into EOQ models. Then, the assembly policies are synchronized by redistributing the echelon holding costs and adjusting the inter-setup intervals to power-of-two multiples of each other.

6.2.3.1 Constructing the EOQ Models Network

To transform the problem into EOQ models, each node j in N_i considers the part j designated to the production of product i still in the system, either as part j or assembled inside other parts already, as its inventory. As an illustration, the echelon inventory of part 4 for the production of part 2 includes the quantity of part 4 that is designated for the production of part 2, the quantity of part 4 inside part 5 that is designated for the production of part 2, and the quantity of part 4 inside part 2 that is still in the system. Since the holding cost of each part l in P^i_j required for the production of part j is already accounted for by its predecessors in T_i, the echelon holding cost for node j in N_i is

$$h^i_j = H^i_j - \sum_{l \in p^i_j} H^i_l.$$

Thus, each node j in N_i corresponds to an EOQ model with a setup cost of k_j, a holding cost of h^i_j per unit product per unit time, and a demand rate of 2 units per unit time for the product.

However, a part j might be required by different products for production. To avoid multiple counts of setup cost for an assembly, each part j that is required for the production of multiple products is identified with an EOQ model with a setup cost of k_j, no holding cost, and a demand rate of 2 units per unit time for the product. At the same time, for each N_i that includes j, the setup cost is removed from the EOQ model corresponding to node j in N_i. That is, node j in N_i corresponds to an EOQ model with zero setup cost, a holding cost of h^i_j per unit product per unit time, and a demand rate of 2 units per unit time for the product.

These EOQ models are presented in the EOQ models network for the system. The EOQ models network for the system is a directed network $G^E = (N^E, A^E)$ with

$$N^E = U_i\{i_j: j \in N_i\}U\{j: j \in N_i \text{ for at least 2 different } i\} \text{ and}$$

$$A^E = U_i\{(i_j, i_l): (j, l) \in A_i\}U\{(i_j, j): j \in N_i \text{ and } j \in N^E\}.$$

Associated with each node x in N^E is an ordered pair $(k^E(x), h^E(x))$ that represents the EOQ model, with a setup cost of $k^E(x)$, a holding cost of $h^E(x)$ per unit product per unit time and a demand rate of 2 units per unit time for the product, associated with node x. In particular,

$$k^E(x) = k_j \text{ if } x = j \in N_i \quad \text{for some } i, \quad \text{or} \quad x = i_j \text{ for some } j \in N_i \text{ and } j \notin N^E;$$

otherwise,

$$k^E(x) = 0$$

and

$$h^E(x) = h^i_j \quad \text{if } x = i_j \in N_i \text{ for some } i;$$

otherwise,

$$h^E(x) = 0.$$

To illustrate this with an example, the EOQ models network for the two products with the assembly networks in Figure 6.2 is shown in Figure 6.3. In Figure 6.3, $x:k^E(x),h^E(x)$ is shown inside each node x.

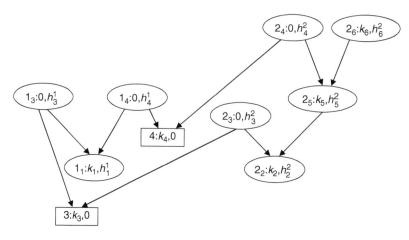

FIGURE 6.3 The EOQ models network for products 1 and 2.

Redistributing the Echelon Holding Costs

The optimal reorder interval for an EOQ model with a setup cost of $k^E(x)$, a holding cost of $h^E(x)$ per unit product per unit time and a demand rate of 2 units per unit time for the product is $[k^E(x)/h^E(x)]^{1/2} \cdot [k^E(x)/h^E(x)]^{1/2}$ is infinite if $h^E(x) = 0$. For any (x, y) in A^E, the optimal assembly policies corresponding to nodes x and y cannot be synchronized if $[k^E(x)/h^E(x)]^{1/2} < [k^E(y)/h^E(y)]^{1/2}$. As discussed earlier for the two-facilities in-series system, the problem that $h^E(x)$ is too large can be rectified by redistributing some of the echelon holding cost $h^E(x)$ from node x to node y. Note that while redistributing some of the echelon holding cost $h^E(i_j)$ from node i_j to node i_l for some product i and $(j,l) \in A_i$ is equivalent to reducing the holding cost of part j, that is, designated for the production of product i, redistributing some of the echelon holding cost $h^E(i_j)$ from node i_j to node j for some product i and $j \in N_i$ is equivalent to not changing that part of the holding cost of part j that is designated for the production of product i.

For any subset N of N^E and the corresponding subnetwork $G = (N, A)$ of G^E with

$$A = \{(x,y): x,y \in N \text{ and } (x,y) \in A^E\}.$$

It is optimal to assemble the parts corresponding to the nodes in N simultaneously, if the echelon holding costs can be redistributed from predecessors to successors in G until the resulting echelon holding costs $h(x)$ satisfies $[k^E(x)/h(x)]^{1/2}$ is a constant for all the nodes in G with $h(x) = 0$ in the case $k^E(x) = 0$. That is, for each $x \in N$ with $k^E(x) > 0$,

$$h(x)/k^E(x) = \sum_{x \in N} h(x)/\sum_{x \in N} k^E(x) = \sum_{x \in N} h^E(x)/\sum_{x \in N} k^E(x).$$

Such an even redistribution of the echelon holding costs is possible if and only if we can flow the amount of excess echelon holding cost, $h^E(x) - h(x)$, from the source nodes x with $h^E(x) > h(x)$ to cover the lack of echelon holding cost, $h(y) - h^E(y)$, at the sink nodes y with $h^E(y) > h(y)$ through the network G. In particular, the maximum flow network is $G^F = (N^F, A^F)$ with

$$N^F = N \cup \{s,t\},$$

$$A^F = A \cup \{(s,x): h(x) > h^E(x)\} \cup \{(y,t): h(y) < h^E(y)\}.$$

In addition, the capacity $c(x, y)$ associated with each arc in A^F is infinite if $(x, y) \in A$. Each $(s, x) \in A^F$ has a capacity $c(s, x) = h^E(x) - h(x)$, while each $(y, t) \in A^F$ has a capacity $c(y, t) = h(x) - h^E(x)$. The objective is to maximize the flow from node s to t through the network G^F, where there is a capacity $c(x, y)$ on the flow to send along arc (x, y). The maximum s-t flow problem is a typical application of linear programming (LP). In solving the LP for the maximum s-t flow problem, either the optimal objective flow value = $\Sigma_{(s,x) \in A^F} c(s,x)$, then an even redistribution of the echelon holding cost is possible. Otherwise, the dual minimum s-t cut (X, X') with $s \in X$ and $t \in X'$ partitions N into two sets $N^1 = X \setminus \{s\}$ and $N^2 = X' \setminus \{t\}$. Since excess echelon holding costs, that cannot flow to cover the lack of echelon holding cost at the nodes in N^2, are still available at the nodes in N^1; nodes in N^2 are predecessors of the nodes in N^1. That is, predecessors do not have enough while successors have too many echelon holding costs. In other words, there is no problem of a predecessor having a smaller optimal inter-setup interval than its successor between the nodes in N^2 and N^1, and redistribution of the echelon holding cost can be considered separately for the two sets of nodes.

Start with $N = N^E$. Solve the maximum flow problem for the network subnetwork G, and if an even redistribution of the echelon holding cost is possible for G, then set $h^F(x) = h(x)$ for each $x \in N$. Otherwise, partition N into two sets N^1 and N^2 according to the optimal dual minimum cut and repeat the process for $N = N^1$ and $N = N^2$ until $h^F(x)$ is determined for each $x \in N^E$. As indicated by the earlier discussion, this redistribution of echelon holding costs results in $h^F(x)$, $x \in N^E$ that satisfy $h^F(x) = 0$ in the case $k^E(x) = 0$, and $[k^E(x)/h^F(x)]^{1/2} \geq [k^E(y)/h^F(y)]^{1/2}$ for each $(x, y) \in A^E$, with $k^E(x) > 0$ and $k^E(y) > 0$.

Adjusting the Inter-setup Intervals of the Assemblies

Since $[k^E(x)/h^F(x)]^{1/2} \geq [k^E(y)/h^F(y)]^{1/2}$ for each $(x, y) \in A^E$ with $k^E(x) > 0$ and $k^E(y) > 0$, $\min\{[k^E(x)/h^F(x)]^{1/2}: k^E(x) > 0$ and $x \in N^E\} = [k^E(z_z)/h^F(z_z)]^{1/2}$ for some product $z = 1, 2, \ldots, n$. Assemble product z every $T_z = T^*_z = [k^E(z_z)/h^F(z_z)]^{1/2}$ units of time.

For any $x = i_j \in N^E$ with $k^E(i_j) > 0$ for some part j and product i, or $x = j \in N^E$ for some part j, let $T^*_j = [k^E(x)/h^F(x)]^{1/2}$ and $2^{m(j)}T^*_z \leq T^* < 2^{m(j)+1}T^*_z$ for some positive integer $m(j)$. If $T^*_j/(2^{m(j)}T^*_z) \leq 2^{m(j)+1}T^*_z/T^*_j$, then part j is assembled every $T_j = 2^{m(j)}T^*_z$ units of time. Otherwise, the part j is assembled every $T_j = 2^{m(j)+1}T^*_z$ units of time. For any $i_j \in N^E$ with $k^E(i_j) > 0$ for some part j and product i, let $T^i_j = T_j$.

For any $x = i_j \in N^E$ with $k^E(i_j) = 0$ for some part j and product i, assembly inter-setup time is set backward for successors first and then predecessors up the assembly network for product i. Let l be the unique immediate successor of j in the assembly network for product i; then part j designated for the production of product i is assembled every $T^i_j = \max\{T_j, T^i_l\}$ units of time.

The assemblies are synchronized by assembling $2T^i_j$ units of part j designated for the production of product i simultaneously. Then, $2T^i_j$ units of part j designated for the production of product i are assembled every T^i_j units of time. Since $[k^E(x)/h^F(x)]^{1/2} \geq [k^E(y)/h^F(y)]^{1/2}$ for each $(x, y) \in A^E$ with $k^E(x) > 0$ and $k^E(y) > 0$ implies that $T^i_j \geq T^i_l$ for any product i and $(j, l) \in A_i$, inventory is down to zero at every assembly of part j designated for the production of product i.

Since echelon holding cost is redistribution from a node x to a node y in N^E only when they have the same corresponding assembly inter-setup time, accounting for the redistributed part of the holding cost at the assembly corresponding to node x or that at node y makes no difference to the average cost of the assembly policy. Hence, the average setup and holding cost of the power-of-two policy is $\Sigma\{k^E(x)/T_j + h^E(x) T_j: x = i_j \in N^E$ with $k^E(i_j) > 0$ for some part j and product i, or $x = j \in N^E$ for some part $j\}$.

For any $x = i_j \in N^E$ with $k^E(i_j) > 0$ for some part j and product i, or $x = j \in N^E$ for some part j, since $2^{-1/2} \leq T_j/T^*_j \leq 2^{1/2}$ by the choice of T_j, $\Sigma\{k^E(x)/T_j + h^E(x)T_j: x = i_j \in N^E$ with $k^E(i_j) > 0$ for some part j and product i, or $x = j \in N^E$ for some part $j\} \leq [(2^{-1/2} + 2^{1/2})/2]\Sigma\{2[k^E(x)h^E(x)]^{1/2}: x = i_j \in N^E$ with $k^E(i_j) > 0$ for some part j and product i, or $x = j \in N^E$ for some part $j\}$.

In other words, it is a 94% optimal policy. A 98% optimal power-of-two policy can be obtained using a more complicated adjustment of the assembly inter-setup intervals.

6.2.3.2 Multi-Period Inventory Model

A general periodic review inventory model considers the problem of satisfying the demand of a single product without delay for T periods of time. Replenishment can be made at the beginning of each period and used to satisfy demand in that and later periods. Holding cost of a period is charged against inventory left at the end of the period. For each period $t = 1, 2, \ldots, T$, the demand is d_t, the cost of ordering Q_t units is $C_t(Q_t)$, and the cost of holding I_t units of inventory is $H_t(I_t)$. It is assumed that d_t is a nonnegative integer whereas $C_t(Q_t)$ and $H_t(I_t)$ are nondeceasing functions for $t = 1, 2, \ldots, T$, as is often true in practice. The objective is to find a replenishment policy that satisfies the demand without delay so as to minimize the total replenishment and holding costs over the T periods.

Typically, a multi-period inventory problem is formulated as a dynamic program.

The optimal value function: Let $F_t(I_{t-1})$ be the minimum cost of satisfying the demand from period t to T starting with an inventory of I_{t-1} at the beginning of period t.

The boundary condition: Since the replenishment and holding costs are nondecreasing, holding inventory at the end of period T will not lower the cost of a replenishment policy. Hence, we only need to consider replenishment policy that does not hold inventory at the end of period T to find an optimal one and set $F_{T+1}(0) = 0$.

The recursive formula: Since we only consider policies that end with no inventory at period T, the starting inventory, I_{t-1}, at the beginning of period t is no more than t to T. The total demand for periods starting with an inventory I_{t-1} at the beginning of period t, the decision is on how much to order. A quantity of at least $d_t - I_{t-1}$ must be ordered to satisfy the demand at period t without delay. Furthermore, since we only consider policies that end with no inventory at period T, at most the total demand from period t to T minus I_{t-1} units of the product will be ordered in period t. In ordering Q_t units of the product, the replenishment cost at period t is $C_t(Q_t)$, whereas the inventory at the end of period t is $I_t = I_{t-1} + Q_t - d_t$. Hence, the holding cost at period t is $H_t(I_{t-1} + Q_t - d_t)$, while the minimum cost for periods $t + 1$ to T is $F_{t+1}(I_{t-1} + Q_t - d_t)$. $F_t(I_{t-1})$ is obtained by selecting the order quantity, Q_t, that minimizes the total cost at period t, $C_t(Q_t) + H_t(I_{t-1} + Q_t - d_t)$, and the remaining periods $t + 1$ to T, $F_{t+1}(I_{t-1} + Q - d_t)$. That is, for $I_{t-1} = 0, 1, \ldots, \Sigma_{t \le i \le T} d_i$,

$$F_t(I_{t-1}) = \text{Min}\{C_t(Q_t) + H_t(I_{t-1} + Q_t - d_t) + F_{t+1}(I_{t-1} + Q_t - d_t):$$

$$\text{Max}\{0, d_t - I_{t-1}\} \le Q_t \le \sum_{t \le i \le T} \{d_i - I_{t-1}\} \qquad (6.3)$$

An optimal policy: To obtain an optimal replenishment policy, we start with setting $F_{T+1}(0) = 0$. Using the recursive formula, we calculate backwards the function $F_t(I_{t-1})$ and store the corresponding optimal order quantity choice $Q_t(I_{t-1})$ for $t = T, T - 1, \ldots, 2$ and $I_{t-1} = 0, 1, \ldots, \Sigma_{t < i < T} d_i$. For an initial inventory level of I_0, we can then find $F_1(I_0)$ and the corresponding optimal order quantity $Q_1^* = Q_1(I_0)$ for period 1 using Equation 6.3. In ordering Q_1^* units of the product in period 1, the inventory at the end of period 1 is

$$I_1^* = I_0 + Q_1^* - d_0$$

Hence, the optimal order quantity at period 2 is $Q_2^* = Q_1(I_1^*)$. We then continue for $t = 3, \ldots, T$ in using

$$I_{t-1}^* = I_{t-2}^* + Q_{t-1}^* - d_{t-1}$$

and

$$Q_t^* = Q_t(I_{t-1}^*)$$

to obtain an optimal replenishment policy Q_t^*, $t = 1, 2, 3, \ldots, T$.

6.2.3.3 Multi-Period Inventory Model with Concave Costs

Economies of scale often exist for large quantities. Incremental discount is a popular model that reflects this phenomenon. An incremental discount cost model $C(Q)$ is associated with $B + 1$ quantities $0 = Q_0 < Q_1 < \cdots < Q_B$. The cost for the jth unit of product is c_b if $Q_{b-1} \leq j < Q_b$ for some $b = 1, 2, 3, \ldots, B$ and c_{b+1} if $Q_B \leq j$ with $c_1 > c_2 > \cdots > c_{b+1}$. An incremental discount model with $B = 3$ is illustrated in Figure 6.4. Since the incremental discount model has a nonincreasing marginal cost, it is a concave function. Concave cost functions are very popular and have many nice properties that a lot of research has been focused on.

A nice property of the concave function is that the linear combination of a set of concave functions results in a concave function. Another nice property is that it induces consolidation. To illustrate this, consider buying a product from two different sources that offer concave cost models. The cost of buying Q_j units of the product from source j is $C_j(Q_j)$ for $j = 1, 2$. Suppose a nonnegative quantity Q_j of product is bought from source j for $j = 1, 2$. Since $C_j(Q_j), j = 1, 2$ have nonincreasing marginal values, $C_1(Q_1) + C_2(Q_2) > C_1(Q_1 + Q_2)$ if $C_1(Q_1) - C_1(Q_1 - 1) < C_2(Q_2) - C_2(Q_2 - 1)$; otherwise, $C_1(Q_1) + C_2(Q_2) \geq C_2(Q_1 + Q_2)$. Hence,

$$C_1(Q_1) + C_2(Q_2) \geq \min\{C_1(Q_1 + Q_2), C_2(Q_1 + Q_2)\}.$$

In other words, multiple sourcing does not result in lower cost than single sourcing, and we only need to consider a single sourcing policy to obtain a minimum cost policy.

For a multi-period inventory model with concave functions C_t and H_t for $1, 2, \ldots, T$, the first property implies that the cost of having the product available at period t from an ordering in each period $j = 1, 2, \ldots, t$ is a concave function. In addition, the second property implies that the product available at a period can be consolidated to come from a single order. In other words, there exists an optimal replenishment policy that satisfies the ZIO property.

Thus, to obtain an optimal replenishment policy for the multi-period inventory model with concave costs, we can use the following dynamic program which determines an optimal ZIO replenishment policy.

The optimal value function: Since an order will be made only when there is no inventory at the beginning of a period, we only need to identify the periods with positive orders to fully determine a ZIO policy. Let F_t be the minimum cost of satisfying the demand from period t to T starting with no inventory at the beginning of period t.

The boundary condition: Since the replenishment and holding costs are nondecreasing, holding inventory at the end of period T will not lower the cost of a replenishment policy. Hence, we only need to consider the replenishment policies that do not hold inventory at the end of period T to find an optimal one and set $F_{T+1} = 0$.

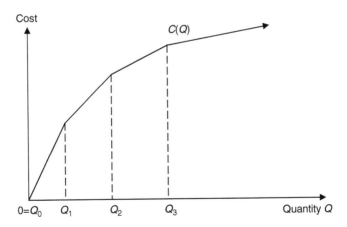

FIGURE 6.4 An incremental discount model.

The recursive formula: The next replenishment can be at any one of the future periods $t + 1$ to T. If the next replenishment is made at period i, then the order placed at period t is for a quantity that equals the total demand from period t to i-1.

$$F_t = \text{Min}_{t < i \leq T+1} \left\{ C_t \left(\sum_{t \leq j \leq i-1} d_j \right) + \sum_{t \leq l \leq i-2} H_l \left(\sum_{l+1 \leq j \leq i-1} d_j \right) + F_i \right\} \tag{6.4}$$

An optimal policy: To obtain an optimal ZIO policy, we start with setting $F_{T+1} = 0$. Using the recursive formula (6.4), we calculate backwards the function F_t and store the corresponding optimal next order period $P(t)$ for $t = T, T - 1, \ldots, 1$. To obtain the optimal order periods through the function P, start with $t_1^* = 1$ and for $j = 1, 2, 3, \ldots$, set $t_{j+1}^* = P(t_j^*)$ until $t_{j+1}^* = T + 1$. Let $t_{p+1}^* = T + 1$. Then the optimal policy is to place an order at period t_i^* for a quantity equal to the total demand from period t_i^* to $t_{i+1}^* - 1$.

6.3   Stochastic Models

In Section 6.2 we considered inventory models which assume that demand is known. Without uncertainty in demand, these models focus on balancing the trade-offs between setup and inventory holding costs. However, in many reallife situations, demand is forecasted with quite a lot of uncertainty as reflected by the main principles of forecasting: (*i*) forecasts are always wrong and (*ii*) forecasts weaken as the length forecast horizon increases. When demand is uncertain, besides the trade-offs among setup and inventory holding costs, one has to consider the costs related to possible shortages. In this section, we consider several inventory control models that incorporate demand uncertainty. Instead of assuming that demand is known, it is assumed that demand is a random variable with known probability distribution.

We start with a discussion of the classical newsvendor problem (a.k.a. newsboy problem), the simplest yet possibly the most celebrated and powerful of all the single-period stochastic inventory control models, in Section 6.3.1. Then, we extend this model in several ways. Section 6.3.2 discusses the scenarios where price is also a decision variable. In Section 6.3.3, we focus our attention on the multiple-period stochastic inventory control of a single product.

6.3.1   Newsvendor Problem

Consider any retailer who needs to make a single procurement decision for a perishable product that is sold over a single period during which demand is uncertain. There are several examples of such businesses. A newsvendor sells newspapers in a day and weekly magazines over a week. A retailer sells summer clothing over a summer season, or T-shirts and hats for the Super Bowl football event, and a manufacturer may design, produce, and sell winter fashion items such as ski jackets and coats over a winter season. The main characteristics of such businesses are: first, the products are perishable. That is, at the end of the selling period the excess inventory is not of any use in the current market; a day-old newspaper cannot be sold as newspaper anymore but can be disposed of as recycled paper or possibly sold to rural areas where paper is not delivered daily; summer clothing is not generally for sale in winter unless the excess stock is shipped to other parts of the world; T-shirts and hats for the 2006 Super Bowl are not in demand after the event. Similarly, winter fashion items are not generally sold after the season is over; they are either shipped off to discount stores or cleared through sales. Second, the procurement lead time is assumed to be too long to make secondary procurements. Hence, there is only one procurement opportunity before the sales season and the retailer has to commit himself to a certain procurement quantity well in advance.

Based on realized demand from past sales, current economic conditions, and expert judgment, randomness in the demand, D, is assumed to follow a known product-specific demand distribution $F(\cdot)$. Our discussion in this section will assume continuous distributions unless otherwise stated. Products are procured at a per unit cost of c, sold at a per unit price of r. Due to the randomness in demand there could be excess inventory or demand at the end of the sales season. Excess inventory is assumed to be returned to the supplier or salvaged at a per unit price of v, which is less than c, and excess demand is assumed to be lost causing not only a loss of the possible profit , $r - c$, but also a possible shortage cost of s dollars per unit that represents the loss of goodwill. Note that $r > c > v$; otherwise, the problem can trivially be solved by either ordering as much as necessary if $v > c$, or not ordering at all if $c > r$.

Since demand is random, the procurement decisions are very much dependent on the risk averseness of the retailer. In this chapter, we only consider the risk-neutral decision-makers. Hence, our risk-neutral retailer needs to determine a procurement quantity Q such that the single period expected total inventory ordering, holding, and shortage cost is minimized, or, equivalently, the single-period expected profit is maximized. That is, the retailer needs to solve

$$\max_{Q \geq 0} \Pi(Q)'$$

where the expected profit $\Pi(Q)$ can be expressed as

$$\Pi(Q) = rE[\min(Q,D)] - cQ + vE[\max(0,Q-D)] - sE[\max(0,D-Q)]$$

The operator $E(\cdot)$ denotes the expectation. Each expectation, respectively, represents the expected sales, excess inventory, and excess demand for any given Q. This model is known as the newsvendor model (or, more commonly, newsboy problem). Note that, for any procurement quantity Q and any realization d of the random demand D,

$$\min(Q, d) = d - \max(0,d-Q) \text{ and } Q = d + \max(0,Q-d) - \max(0,d-Q).$$

Hence,

$$E[\min(Q,D)] = E[D] - E[\max(0,D-Q)]$$

and

$$Q = E[D] + E[\max(0,Q-D)] - E[\max(0,D-Q)] \cdot$$

Substituting these identities in the given equation, the expected profit function can be rewritten as

$$\Pi(Q) = E[D](r-c) - (c-v)E[\max(0,Q-D)] - (r+s-c)E[\max(0,D-Q)] \cdot \qquad (6.5)$$

Interpretation of this function is interesting by itself. The first term is the riskless profit for the equivalent certainty problem that experiences a known demand of $E[D]$. The second term represents the total expected holding cost, which is the per unit holding (overage) cost of $c_0 = c - v$ charged against every unit of excess inventory $E[\max(0, Q - D)]$. And finally, the third term is the total expected shortage (underage) cost, which is the per unit shortage cost of $c_u = r + s - c$ (where $r - c$ is the lost sales profit) charged against each unit of the excess demand $E[\max(0, D - Q)]$. In the literature (see Silver and Peterson 1985) total expected cost

$$L(Q) = (c-v)E[\max(0,Q-D)] + (r+s-c)E[\max(0,D-Q)] \qquad (6.6)$$

is known as the single-period *loss function*. Since riskless profit $E[D](r - c)$, which would occur in the absence of uncertainty, is independent of Q, maximizing $\Pi(Q)$ is equivalent to minimizing $L(Q)$. Before finding the optimal procurement policy, let us write $L(Q)$ explicitly as

$$L(Q) = (c-v)\int_{x=0}^{Q}(Q-x)dF(x) + (r+s-c)\int_{x=Q}^{\infty}(x-Q)dF(x).$$

Taking the derivative of $L(Q)$ with respect to Q and applying Leibnitz' Rule, the first-order optimality condition can be written as:

$$(c-v)\Pr(D\leq Q)-(r+s-c)\Pr(D\geq Q)=0 .$$

This condition suggests that the optimal procurement quantity S is such that the marginal cost of overage, which is the probability of a shortage multiplied by the unit overage cost $(c-v)$, is equal to the marginal cost of underage, which is the probability of a shortage multiplied by the unit cost of a shortage $(r+s-c)$.

Solving this equation for Q, the optimal procurement quantity S is found from the fractile formula

$$F(S) = \frac{c_u}{c_u + c_o}, \text{ that is, } F(S) = \Pr(D \leq S) = \frac{r+s-c}{r+s-v}.$$

The assumption $r > c > v$ implies that the right-hand side of the formula is greater than 0 and less than 1, $F(\cdot)$ is a continuous nondecreasing function, and hence a finite positive S always exists. Furthermore, the second derivative of $L(Q)$, $(c-v)f(Q)+(r+s-c)f(Q)\geq 0$ for all $Q \geq 0$, implies the convexity of $L(Q)$. In addition, $L(Q)$ has a negative slope at $Q=0$, $-(r+s-c)$, and a positive slope, $c-v$, as Q tends to ∞, implying that $L(Q)$ has a finite minimizer S over $(0, \infty)$.

Sometimes customers order in bulk. In such cases, the number of customers might be low and their demand structure might not assume a continuous distribution. Also, some products such as planes, trains, and so on cannot be ordered in fractions. An airline can order an integral number of jumbo jets, but it does not quite make sense to order 0.11 planes! Hence, the assumption of a continuous demand distribution might not make sense for all cases. Luckily, for the newsvendor model, this is not a problem. If demand distribution F is actually discrete, the above analysis follows similarly with a small adjustment. The expectation terms in the loss function have to be explicitly represented by summations rather than the integrals. That is, let F be a discrete distribution with probability density function (*pdf*)

$$f(d_j)= q_j, j=1,2,...N, \text{ and } \sum_{j=1}^{N} q_j = 1.$$

Without loss of generality, one can assume that $d_1 = 0$ and $d_1 < d_2 < \cdots < d_N < \infty$. Then, the loss function for $Q \in [d_j, d_{j+1}]$ for any $j = 1, 2, \ldots, N$ is

$$L(Q) = (c-v)\sum_{i=1}^{j}(Q-d_i)q_i + (r+s-c)\sum_{i=j+1}^{N}(d_i - Q)q_i,$$

which is a piece-wise linear convex function of Q. Analyzing the first derivative of $L(Q)$, this property can be easily observed:

$$L'(Q) = (c-v)\sum_{i=1}^{j}q_i - (r+s-c)\sum_{i=j+1}^{N}q_i$$

$$=(c-v)\Pr(D\leq Q)-(r+s-c)\Pr(D>Q)$$

which is constant for all $Q \in [d_j, d_{j+1}]$, meaning that $L(Q)$ is linear over this range. For $j = 1$, that is, for all $Q < d_1$, the derivative is a negative constant, $-(r+s-c) < 0$. Hence, $L(Q)$ is a decreasing function at

$Q = 0$. As j increases, $L'(Q)$ is nondecreasing (increasing if all $q_j > 0$) because $\Pr(D \leq Q)$, which multiplies the positive quantity $(c - v)$, increases or stays the same, and $\Pr(D > Q)$, which multiplies the positive quantity $(r + s - c)$, decreases or stays the same. Hence, $L(Q)$ has a nondecreasing first derivative, and thus is a convex function. Since $L(Q)$ is decreasing at $Q = 0$ and increasing at $Q = d_N$, a minimizer of this function exists.

Finally, realize that when the demand distribution is discrete, the optimal quantity is equal to a possible demand point d_j. Furthermore, this demand point is easily found by finding the smallest index such that $L'(Q) > 0$. Note that, as j increases, $L'(Q)$ increases from a negative value $-(r + s - c)$ to a positive value $(c - v)$. Hence, the optimal procurement quantity $S = d_z$ where z is the smallest j such that

$$(c-v)\sum_{i=1}^{j}q_i - (r+s-c)\sum_{i=j+1}^{N}q_i > 0$$

There are several tacit assumptions in the earlier analysis: first, there is no initial inventory; second, there is no fixed ordering cost; third, the excess demand is lost; fourth, price is exogenous; and fifth, salvage value is guaranteed to be achieved. The first three of these assumptions can easily be dealt with by making some observations in the earlier analysis, but we will discuss the other two assumptions in more detail in the coming subsections.

Let us assume that before the retailer places an order, which costs her a setup cost of k dollars per order (paper work, labor etc.), she realizes that there are I units of the product in her warehouse. If the retailer would like to increase the inventory level to Q, the expected cost of procuring $(Q - I)$ units is $k - cI + L(Q)$, which is still minimized by S if we actually decide to procure any units at all. Setup cost k is only incurred if we decide to procure any item at all, and hence if we do not procure any units on top of I, k is not incurred. Under what conditions should the retailer decide to procure on top of the initial inventory I? There are two cases: (*i*) if $I > S$, no units should be procured, and (*ii*) if $I < S$, then the retailer needs to compare the cost of procuring the extra $S\text{-}I$ units, that is, $k - cI + L(S)$, with the cost of not procuring any extra units at all, that is $-cI + L(I)$. If $k + L(S) < L(I)$, $S\text{-}I$ units should be procured; otherwise, none should be procured.

If we let s be a value such that $k + L(S) = L(s)$, the earlier discussion suggests that the optimal procurement policy is an (s, S) policy. That is, procure $S\text{-}I$ if the initial inventory I is less than or equal to s; otherwise. do not procure. Quantity S is known as the order-up-to level, and s is known as the reorder point. Note that, if $k = 0$, $s = S$, this kind of a procurement policy is known as the base-stock policy. That is, if the initial inventory level I is less than S, procure $S\text{-}I$; otherwise, do not procure at all.

Let us now consider the case where the excess demand is not lost, but backordered, and the shortage cost not only reflects the loss of goodwill but also the emergency shipment costs. In this case, the single-period loss function is

$$L(Q) = (c - v)\int_{x=0}^{Q}(Q - x)dF(x) + (s - c)\int_{x=Q}^{\infty}(x - Q)dF(x),$$

which is almost identical to the lost sales case except that the shortage cost, $s - c$, does not include the lost revenue anymore. Hence, the optimal procurement quantity is found from

$$F(S) = \frac{s - c}{s - v}.$$

Example 1

A hot dog stand at Toronto SkyDome, home of the Blue Jays baseball club, sells hot dogs for $3.50 each on game days. Considering the labor, gas, rent, and material, each hot dog costs the vendor $2.00 each. During any game day, based on the past sales history, the daily demand at SkyDome is found to be normally distributed with mean 40 and standard deviation 10. If there are any hot dogs left at the end of the day, they can be sold at the entertainment district for $1.50 each. If the vendor sells out at SkyDome, she closes shop and calls it a day (lost sales).

(a) If the vendor buys the hot dogs daily, how many should she buy to maximize her profit?

The optimal procurement level S satisfies

$$F(S) = \frac{r + s - c}{r + s - v}$$

where $r = 3.50$, $c = 2.00$, $s = 0$, $v = 1.50$, and $F(\cdot)$ is normally distributed. That is, S satisfies $P(D \leq S) = 1.5/2.0 = 0.75$. Standardizing the normal distribution, we have $P(Z < (S - 40)/6) = 0.75$. From the normal table or Microsoft Excel, $z = 0.675$ and $S = 40 + 10(0.675) = 46.75$. Rounding up, the vendor should procure 47 hot dogs with an expected profit of $53.64.

(b) If she buys 55 hot dogs on a given day, what is the probability that she will meet all day's demand at SkyDome?

She needs to determine the probability that demand is going to be less than or equal to 55. This is easily done by calculating $Pr(D \leq 55) = Pr(Z \leq (55 - 40)/10) = Pr(Z \leq 1.5) = 0.9332$. Hence, she has a 93.32% chance that she will satisfy all the demand at SkyDome and have an expected profit of $51.92.

(c) If we assume that the vendor can purchase hot dogs from the next hot dog stand for $2.50 each in case she sells out her own stock (backorder case), how many hot dogs should she buy?

In the backorder case, the critical fractile is found as $(s - c)/(s - v)$, where $s = 2.50$. Hence, $Pr(D \leq S) = (2.5 - 2)/(2.5 - 1.5) = 0.5$. Standardizing the normal distribution, $P(Z < (S - 40)/10) = 0.50$. From the normal table or Microsoft Excel, $z = 0.0$ and $S = 40 + 10(0.0) = 40$. The vendor should procure 40 hot dogs with an expected profit of $65.98.

6.3.2 Joint Pricing and Inventory Control in a Newsvendor Setting

In this section, we assume that the retailer has the capability of setting the price as well as the procurement quantity of the product. For now, we consider the lost sales case with zero setup cost and initial inventory. This can very well be the case for many innovative companies who introduce the product first to the market and have some patent rights to charge the price they would like. Even though they might charge any price they wish, companies still need to consider the effect of the price on demand. Companies need to jointly determine the optimum price and procurement quantity with respect to the demand–price relationship that they assume in order to maximize their expected profit.

In the operations management and economics literature, demand is often modeled in an additive or a multiplicative fashion and the randomness in demand is assumed to be price independent. Specifically, demand is defined as $D(r, e) = m(r) + \varepsilon$ in the additive case and $D(r, \varepsilon) = m(r)\varepsilon$ in the multiplicative case, where $m(r)$ is a decreasing function that captures the price–demand relationship and ε is a random variable defined over $[0, \Delta]$ with mean m. Note that, if ε is a random variable defined over $[A, B]$, it can easily be converted to another random variable defined over $[0, B - A]$. We will consider $m(r) = a - br$ $(a > 0, b > 0)$ in the additive case and $m(r) = ar^b$ $(a > 0, b > 1)$ in the multiplicative case. Both representations of $m(r)$ are popular in the economics literature, with the former representing a linear demand curve and the second representing an iso-elastic demand curve. Due to several reasons, there might be bounds on the price charged, that is, $r_L \leq r \leq r_U$. Note that any realization of the demand needs to be non-negative, there might be profit margin requirements from upper management, and finally, competitive or government forces might not allow one to charge any price desired. Hence, the retailer needs to solve the nonlinear program

$$\max \ \Pi(Q,r)$$
$$st \quad r_U \geq r \geq r_L$$
$$Q \geq 0.$$

The expected single-period profit very much depends on the demand–price relationship. Each demand–price relationship scenario needs independent treatment in the lost sales case. However, a unified approach is possible in the backorder case.

6.3.2.1 Lost Sales Models

Additive Demand–Price Relationship

In the joint pricing and procurement problem, minimizing the single-period loss function $L(Q)$ is not equivalent to maximizing the single-period expected profit. Hence, the retailer needs to maximize her profit which is identical to the newsvendor profit in (6.5) except that demand D is replaced by $D(r, \varepsilon)$, which is equal to $a - br + \varepsilon$.

$$\Pi(Q,r) = E[D(r,\varepsilon)](r - c) - (c - v)\int_{x=0}^{Q-m(r)}(Q - x - m(r))dF(x) - (r + s - c)\int_{x=Q-m(r)}^{\Delta}(x + m(r) - Q)dF(x)$$

As opposed to the exogenous price case, this expected profit function is not necessarily concave for all possible values of the parameters. However, it is shown by Karakul (2007) that if demand distribution satisfies a weak condition, it is still a well-behaved function and it has a unique stationary point in the feasible region which is also the unique local maximum. That is, it is a unimodal function. To see this, we first introduce a change of variable $u = Q - m(r)$ which is interpreted as a safety stock factor representing the type 1 service level, that is, the probability of not stocking out. For given u, the service level is $F(u)$, but the procurement quantity Q does not have this one-to-one correspondence with the service level: for given Q, the service level is $F(Q - m(r))$ and is dependent on the price. Carrying out this change of variable, the expected profit in terms of u and r is:

$$\Pi(u,r) = E[D(r,\varepsilon)](r - c) - (c - v)\int_{x=0}^{u}(u - x)dF(x) - (r + s - c)\int_{x=u}^{\Delta}(x - u)dF(x) ,$$

where expectations are taken over the random variable ε. Now, consider the first-order conditions of this function with respect to r and u:

$$\frac{\partial \Pi(u,r)}{\partial r} = -2br + a + bc + -\int_{x=u}^{\Delta}(x - u)dF(x) = 0 \qquad (6.7)$$

$$\frac{\partial \Pi(u,r)}{\partial u} = (r + s - v)(1 - F(u)) - (c - v) = 0. \qquad (6.8)$$

From Equation 6.7, optimal price r as a function of u is found as:

$$r(u) = \frac{a + bc + -\int_{x=u}^{\Delta}(x - u)dF(x)}{2b}.$$

Substituting this in (6.8) and assuming that the demand distribution $F(\cdot)$ has a hazard rate $z(\cdot) = f(\cdot)/(1-F(\cdot))$ such that $2z(x)^2 + dz(x)/dx > 0$ for all $x \in (0, \Delta)$,[*] Karakul (2007) shows that there is a unique solution that satisfies the first-order conditions and it corresponds to a local maximum.

Define $\Pi_u = d\Pi(u, r(u))/du$ and consider its first and second derivatives

$$d\Pi_u / du = \frac{-f(u)}{2b}\left[2b(r(u)+s-v) - \frac{(1-F(u))}{z(u)}\right],$$

$$d^2\Pi_u / du^2 = \frac{df(u)/d(u)}{2b}\left[2b(r(u)+s-v) - \frac{(1-F(u))}{z(u)}\right] - \frac{f(u)(1-F(u))}{2bz(u)^2}\left[2z(u)^2 + dz(u)/du\right].$$

Note that any stationary point of Π_u (not any stationary point of Π) needs to satisfy the first-order condition $d\Pi_u/du = 0$, and hence

$$d^2\Pi_u / du^2 \big|_{d\Pi_u/du=0} = -\frac{f(u)(1-F(u))}{2bz(u)^2}\left[2z(u)^2 + dz(u)/du\right] < 0$$

if $2z(u)^2 + dz(u)/du > 0$ for all $u \in (0, \Delta)$. This suggests that all stationary points of Π_u (the total derivative of Π) are local maxima, which means that it actually has a unique stationary point and it is a local maximum. This implies that Π_u can vanish at most twice over $[0, \Delta]$ and consequently, Π might have two stationary points, with the larger one being the local maximum over this range. However, $\Pi_u(0) = r(0) + s - c > 0$ and hence Π_u equals zero at most once in $(0, \Delta)$, proving the unimodality of $\Pi(u, r)$.

The optimal stocking factor and price (u^*, r^*) can be found by first solving the nonlinear equation

$$\frac{\partial\Pi(u,r(u))}{\partial u} = (r(u) + s - v)(1 - F(u)) - (c - v) = 0$$

with respect to u to obtain u^* and then substituting u^* in $r(u)$ to obtain r^*. The optimal procurement quantity is calculated as $S = a - br^* + u^*$.

Example 2

Consider the hot dog stand example. Assuming that excess demand is lost and there is not any competition, the vendor would like to determine the best price and procurement level. Luckily, the vendor has an operations research background and she was able to figure out that the demand is a linear function of the price $100 - 10r + \varepsilon$, where ε is a random variable with a normal distribution 40 and standard deviation 10. What is her best price and procurement quantity?

Remember that $c = 2$, $s = 0$, and $v = 1.5$. Solving

$$\frac{\partial\Pi(u,r(u))}{\partial u} = (r(u) + 2 - 1.5)(1 - F(u)) - (2 - 1.5) = 0$$

---

[*] Note that all log-concave distribution functions, that is, the distribution functions whose logarithms are concave, satisfy this condition (see An 1995 for a discussion of log-concave distributions which include normal, gamma, Erlang and many other well-known distributions).

for u, we find $u^* = 54.25$. Note that F represents the normal distribution, and it is necessary to use a package like Maple or Matlab to solve this nonlinear equation. The optimal price is $r(54.25) = \$7.98$ and the optimal order quantity is $S = 100 - 10 * 7.98 + 54.25 = 74.45$. The closest integer value is 74, and hence the vendor should order 74 hot dogs and charge $7.98 each for a total profit of $350.62.

The rounding of the order quantity is not necessarily always up or down. Since, in this case, a continuous distribution is used to approximate a discrete one, the integer number that is closest to S is more likely to bring the highest profit. Note that the hot dogs would be quite expensive if there were not competition and the demand–price relationship were given by $100 - 10r$. (What would the price be if the demand–price relationship was $100 - 15r$?)

Multiplicative Demand–Price Relationship

In the case the demand and price have a multiplicative relationship, the change of variable is somewhat different. We define $u = Q/m(r)$. Substituting $D(r,\varepsilon) = m(r)\varepsilon$ for D and $u = Q/m(r)$ in the objective function of the newsvendor problem in Equation 6.5, the single-period expected profit function is

$$\Pi(u,r) = E[D(r,\varepsilon)](r-c) - m(r)\{(c-v)\int_{x=0}^{u}(u-x)dF(x) + (r+s-c)\int_{x=u}^{\Delta}(x-u)dF(x)\}$$

$$= m(r)\left\{E[\varepsilon](r-c) - (c-v)\int_{x=0}^{u}(u-x)dF(x) - (r+s-c)\int_{x=u}^{\Delta}(x-u)dF(x)\right\}.$$

As in the additive case, this expected profit function is not necessarily concave or convex for all parameter values, but is unimodal for the demand distributions considered earlier. From the first-order condition that $\partial\Pi(u,r)/\partial r = 0$, the optimal price r as a function of u can be obtained as:

$$r(u) = \frac{bc}{(b-1)} + \frac{b\left[(c-v)\int_{x=0}^{u}(u-x)dF(x) + s\int_{x=u}^{\Delta}(x-u)dF(x)\right]}{(b-1)\left[\quad - \int_{x=u}^{\Delta}(x-u)dF(x)\right]},$$

Substituting this into

$$\frac{\partial\Pi(u,r)}{\partial u} = m(r)[(r+s-v)(1-F(u)) - (c-v)] = 0,$$

assuming that the demand distribution $F(\cdot)$ has a hazard rate $z(\cdot) = f(\cdot)/(1-F(\cdot))$ such that $2z(x)^2 + dz(x)/dx > 0$ for all $x \in (0, \Delta)$, and following similar ideas as in the proof for the additive case, one can show that $d\Pi(u, r(u))/du$ is increasing at $u = 0$, decreasing at $u = \Delta$, and is itself a unimodal function over $[0, \Delta]$ for $b > 2$. This proves that there is a unique solution that satisfies the first-order conditions and it corresponds to a local maximum (see Petruzzi and Dada 1999 for a proof). Hence, the optimal stocking factor and price (u^*, r^*) can be found by first solving the nonlinear equation:

$$\frac{d\Pi(u, r(u))}{du} = m(r(u))[(r(u) + s - c)(1 - F(u)) - (c - v)F(u)] = 0$$

with respect to u to obtain u^* and then substituting u^* in $r(u)$ to obtain r^* and in $u = Q/m(r(u))$ to obtain the optimal procurement quantity $S = u^* a r^{*-b}$.

6.3.2.2 Backorder Models

The analysis of the joint pricing and procurement problem of a single product with random demand follows a different approach when it is assumed that the excess demand is backlogged rather than lost.

As in the discussion of the backorder case in the newsvendor problem, the per unit shortage cost is now represented by s and it does not consider the loss of profit $(r - c)$. Note that s does not only represent the loss of goodwill but also the cost of fulfilling the unmet demand with an emergency order and $s > c$ is a reasonable assumption. Furthermore, by defining $h = h^+ - v$ as the per unit adjusted holding cost (which can be a negative value because it is defined as the real holding cost h^+ minus the salvage v) and realizing that the expected sales is equal to the expected demand, the single-period profit is:

$$\Pi(Q, r) = E[rD(r, \varepsilon)] - cQ - E[h \max(0, Q - D(r, \varepsilon)) + s \max(0, D(r, \varepsilon) - Q)].$$

For some specific demand–price relationships, further analysis is possible. Let the demand function satisfy $D(r, \varepsilon) = \alpha m(r) + \beta$, where $\varepsilon = (\alpha, \beta)$, α is a non-negative random variable with $E[\alpha] = 1$ and β is a random variable with $E[\beta] = 0$. By scaling and shifting, the assumptions $E[\alpha] = 1$ and $E[\beta] = 0$ can be made without loss of generality. Furthermore, assume that $m(r)$ is continuous and strictly decreasing, and the expected revenue $R(d) = dm^{-1}(d)$ is a concave function of the expected demand d. Note that $D(r, \varepsilon) = a - br + \beta$ $(a > 0, \ b > 0)$ and $D(r, \varepsilon) = \alpha \, ar^{-b}$ $(a > 0, \ b > 1)$ are special cases that satisfy these conditions.

Since there is a one-to-one correspondence between the selling price r and the expected demand d, the single-period expected profit function can be equivalently expressed as:

$$\Pi(Q, d) = R(d) - cQ - E[h \max(0, Q - \alpha d - \beta) + s \max(0, \alpha d + \beta - Q)]$$

Observing that $h \max(0, y) + s \max(0, -y)$ is a convex function of y, one can see that $h \max(0, Q - \alpha d - \beta) + s \max(0, \alpha d + \beta - Q)$ is a convex function of (Q, d) for any realization of α, β (see Bazaraa et al. 1993, page 80). Furthermore, taking the expectation over α and β preserves convexity and hence, $H(Q) = E[h \max(0, Q - \alpha d - \beta) + s \max(0, \alpha d + \beta - Q)]$ is convex in (Q, d). This proves that $\Pi(Q, d)$ is a concave function and the optimal expected demand, d^*, and procurement quantity, Q^*, can be obtained from the first-order conditions. Optimal price is determined as $r^* = m^{-1}(d^*)$. In the existence of initial inventory, it is shown by Simchi-Levi et al. (2005) that the optimal procurement quantity is determined by a base-stock policy. That is, if the initial inventory I is less than the optimal procurement level S, then we replenish our stock to bring the inventory level up to S; otherwise, we do not order. The optimal price is determined as a nonincreasing function of the initial inventory level.

There are several extensions to the given single-period joint pricing and inventory control problems with stochastic demand. Karakul and Chan (2004) and Karakul (2007) consider the case in which the excess inventory is not salvaged for certain, but they are sold at a known discounted price to a group of clients who exhibit a discrete demand distribution for this excess stock. This case is known as the newsvendor problem with pricing and clearance markets. Cachon and Kok (2007) analyze the importance of estimating the salvage price correctly. Karakul and Chan (2007) consider the product introduction problem of a company which already has a similar but inferior product in the market. Authors consider a single-period model that maximizes the expected profit from the optimal procurement of these two products and the optimal pricing of the new product. A detailed review of the inventory control of substitutable products that include the seats in flights, hotel rooms, technologically improved new products, fashion goods, etc. can be found there.

6.3.3 Multiple Period Models

In this section we extend the newsvendor model such that the retailer needs to make procurement decisions for a specific product over the next N periods. At the beginning of each period t, for example, day, week, month, the inventory amount of the product is counted and noted as I_t. Then, an order of size Q_t may be placed or not depending on the quantity on hand. Initially, we assume that the orders are filled instantly, that is, the lead time is zero. A discussion of the nonzero lead times will be provided at the end of this section.

Although the analysis can be carried out for time-varying demand distributions, for the sake of simplicity, we assume that demand at each period D is independent and identically distributed following the continuous distribution $F(\cdot)$ defined over a bounded non-negative region $(0, \Delta)$. We focus on backorder models in this section.

Although the costs involved in this model are very similar to the newsvendor model, they might have a different interpretation. Initially, let us assume that setup cost k is zero. There is a non-negative holding cost h for each unit of the excess inventory at the end of each period; this can be thought of as the capital, insurance, and handling cost per unit carried in inventory. For each unit of demand that is not met at the end of the period, the retailer incurs s dollars of backorder penalty cost.

Since the price is exogenous, the retailer needs to determine the optimal procurement quantities Q_t for $t = 0, 1, 2 \ldots, N-1$ that minimize the total expected cost

$$TC(\vec{Q}) = \sum_{t=0}^{N-1} \left\{ cQ_t + hE\left[\max(0, I_t + Q_t - D)\right] + sE\left[\max(0, D - I_t - Q_t)\right]\right\}.$$

The most natural and appropriate technique to solve this problem is Dynamic Programming (DP). The appropriate DP algorithm has the following cost-to-go function:

$$J_t(I_t) = \min_{Q_t \geq 0} \left\{ cQ_t + H(I_t + Q_t) + E\left[J_{t+1}(I_t + Q_t - D)\right]\right\}, \tag{6.9}$$

where

$$H(y) = hE\left[\max(0, y - D)\right] + sE\left[\max(0, D - y)\right].$$

The cost-to-go function represents the minimum expected cost from periods $t, t+1, \ldots, N-1$ for an initial inventory of I_t at the beginning of period t and optimal procurement quantities Q_j, $j = t, t+1, \ldots, N-1$. Note that the inventory at the beginning of period $t+1$ is found as $I_{t+1} = I_t + Q_t - d$, where d is a realization of the demand variable D. Assuming that any excess inventory at the end of period N is worth nothing, the DP algorithm has the boundary condition:

$$J_N(I_N) = 0.$$

A change of variables is useful in analyzing (6.9). We introduce the variable $y_t = I_t + Q_t$ that represents the inventory level immediately after the order in period t is placed. With this change of variable, the right-hand side of Equation 6.9 can be rewritten as:

$$\min_{y_t \geq x_t} \left\{ cy_t + H(y_t) + E[J_{t+1}(y_t - D)]\right\} - cI_t.$$

The function H is easily seen to be convex because, for each realization of D, $\max(0, y - D)$ and $\max(0, D - y)$ are convex in y and taking the expectation over D preserves convexity. If we can prove that J_{t+1} is convex, the function in the curly brackets, call it $G_t(y_t)$, is convex as well. Then the only result that remains to be shown is $\lim_{|y| \to \infty} G_t(y) = \infty$ which proves the existence of an unconstrained minimum S_t. If these properties are proven, which we will do shortly, then a base-stock policy is optimal. That is, if S_t is the unconstrained minimum of $G_t(y_t)$ with respect to y_t, then considering the constraint $y_t = I_t$, a minimizing y_t equals S_t if $I_t \leq S_t$ and equals I_t otherwise. Using the reverse transformation $Q_t = y_t - I_t$, the minimum of the DP Equation 6.9 is attained at $Q_t = S_t - I_t$ if $I_t \leq S_t$, and at $Q_t = 0$ otherwise. Hence, an optimal policy is determined by a sequence of scalars $\{S_0, S_1, \ldots, S_{N-1}\}$ and has the form

$$Q_t^*(I_t) = \begin{cases} S_t - I_t, & \text{if } I_t < S_t \\ 0, & \text{if } I_t \geq S_t \end{cases} \tag{6.10}$$

where each S_t, $t = 0, 1, \ldots, N - 1$ solves

$$G_t(y) = cy + H(y) + E[J_{t+1}(y - D)].$$

The earlier-discussed convexity and existence proofs are done inductively. We have $J_N = 0$, so it is convex. Since $s > c$ and the derivative of $H(y)$ tends to $-s$ as $y \to -\infty$, $G_{N-1}(y) = cy + H(y)$ has a negative derivative as $y \to -\infty$ and a positive derivative as $y \to \infty$. Therefore, $\lim_{|y| \to \infty} G_{N-1}(y) = \infty$ and the optimal policy for period $N - 1$ is given as:

$$Q^*_{N-1}(I_{N-1}) = \begin{cases} S_{N-1} - I_{N-1}, & \text{if } I_{N-1} < S_{N-1} \\ 0, & \text{if } I_{N-1} \geq S_{N-1} \end{cases},$$

where S_{N-1} minimizes $G_{N-1}(y)$. From the DP Equation 6.9 we have

$$J_{N-1}(I_{N-1}) = \begin{cases} c(S_{N-1} - I_{N-1}) + H(S_{N-1}), & \text{if } I_{N-1} < S_{N-1} \\ H(I_{N-1}), & \text{if } I_{N-1} \geq S_{N-1} \end{cases},$$

which is a convex function because: first, both $H(I_{N-1})$ and $c(S_{N-1} - I_{N-1}) + H(S_{N-1})$ are convex; second, it is continuous; and, finally, at $I_{N-1} = S_{N-1}$ its left and right derivatives are both equal to $-c$. For $I_{N-1} < S_{N-1}$, J_{N-1} is a linear function with slope $-c$ and, as I_{N-1} approaches S_{N-1} from the right-hand-side, its derivative is $-c$ because S_{N-1} minimizes the convex function $cy + H(y)$ whose derivative $c + H'(y)$ vanishes at $y = S_{N-1}$ (see Fig. 6.5).

Note that if the initial inventory at the beginning of period $N - 1$ is greater than the unconstrained minimizer S_{N-1}, we do not order any more and hence do not incur any extra procurement cost but, rather, face the possible holding or shortage cost H. On the contrary, if the initial inventory is less than the unconstrained minimizer S_{N-1}, then we procure enough to increase the on-hand inventory level to S_{N-1}. Hence, we incur not only the procurement cost $c(S_{N-1} - x_{N-1})$ but also the possible holding or shortage cost $H(S_{N-1})$.

Hence, given the convexity of J_N, we proved that J_{N-1} is convex and $\lim_{|y| \to \infty} J_{N-1}(y) = \infty$. This argument can be repeated to show that if J_{t+1} is convex for $t = N - 2, N - 3, \ldots, 0$, $\lim_{|y| \to \infty} J_{t+1}(y) = \infty$, and $\lim_{|y| \to \infty} G_t(y) = \infty$, then

$$J_t(I_t) = \begin{cases} c(S_t - I_t) + H(S_t) + E\left[J_{t+1}(S_t - D)\right], & \text{if } I_t < S_t \\ H(I_t) + E\left[J_{t+1}(I_t - D)\right], & \text{if } I_t \geq S_t, \end{cases}$$

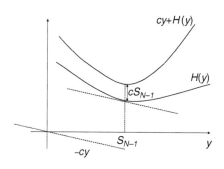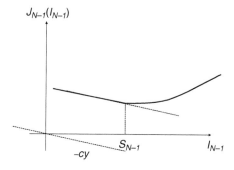

FIGURE 6.5 Structure of the cost-to-go function.

where S_t minimizes $cy + H(y) + E[J_{t+1}(y - D)]$. Furthermore, J_t is convex, $\lim_{|y| \to \infty} J_t(y) = \infty$, and $\lim_{|y| \to \infty} G_{t-1}(y) = \infty$. This completes the proof that J_t is convex for all $t = 0, 1, \ldots, N = 1$ and a base-stock policy is optimal.

Analysis is more complicated in the existence of a positive setup cost k.

6.3.3.1 Positive Setup Cost

If there is a setup cost for any non-negative procurement quantity Q_t, then the procurement cost is:

$$C(Q) = \begin{cases} k + cQ, & \text{if } Q > 0 \\ 0, & \text{if } Q = 0. \end{cases}$$

The DP algorithm has the following cost-to-go function:

$$J_t\left(I_t\right) = \min_{Q_t \geq 0}\left\{C(Q_t) + H\left(I_t + Q_t\right) + E\left[J_{t+1}\left(I_t + Q_t - D\right)\right]\right\}, \tag{6.11}$$

with the boundary condition $J_N(I_N) = 0$.

Considering the functions $G_t(y) = cy + H(y) + E[J_{t+1}(y - D)]$ and the piecewise linear procurement cost function $C(Q)$,

$$J_t(I_t) = \min\{G_t(I_t), \min_{Q_t > 0}[k + G_t(I_t + Q_t)]\} - cI_t,$$

or by the change of variable $y_t = I_t + Q_t$,

$$J_t(I_t) = \min\{G_t(I_t), \min_{y_1 > I_t}[k + G_t(y_t)]\} - cI_t.$$

If G_t can be shown to be convex for all $t = 0, 1, \ldots, N - 1$, then it can be easily seen that an (s, S) policy will be optimal. That is,

$$Q_t^*(I_t) = \begin{cases} S_t - I_t, & \text{if } I_t < s_t \\ 0, & \text{if } I_t \geq s_t \end{cases}$$

would be optimal, where S_t minimizes $G_t(y)$ and s_t is the smallest y value such that $G_t(y) = k + G_t(S_t)$. Unfortunately, if $k > 0$, it is not necessarily true that G_t is convex. However, it can be shown that G_t is still a well-behaved function that satisfies the property

$$k + G_t(z + y) \geq G_t(y) + z\left(\frac{G_t(y) - G_t(y - b)}{b}\right), \qquad \text{for all } z \geq 0, b > 0, y.$$

Since the proof is mathematically involved, we skip the proof and refer the interested readers to Bertsekas (2000). Functions that satisfy the stated property are known as K-convex functions. There are several properties of K-convex functions, which we provide in the next lemma without its proof [for proofs, see Bertsekas (2000), pp. 159–160], that help us show that the (s, S) policy is still optimal in the existence of a non-negative fixed ordering cost.

Lemma 1: Properties of K-convex functions

(a) *A real-valued convex function g is also 0-convex and hence also K-convex for all $K \geq 0$.*
(b) *If $g_1(y)$ and $g_2(y)$ are K-convex and L-convex $(K \geq 0, L \geq 0)$, respectively, then $ag_1(y) + bg_2(y)$ is $(aK + bL)$-convex for all $a > 0$ and $b > 0$.*
(c) *If $g(y)$ is K-convex and w is a random variable, then $E\{g(y - w)\}$ is also K-convex, provided $E\{|g(y - w)|\} < \infty$ for all y.*

(d) *If g is a continuous K-convex function and g(y) − > ∞, then there exist scalars s and S with s < S such that*

    i.   *$g(S) \leq g(y)$, for all y.*
    ii.   *$g(S) + K = g(s) < g(y)$, for all $y < s$.*
    iii.   *$g(y)$ is a decreasing function on $(-\infty, s)$.*
    iv.   *$g(y) \leq g(z) + K$ for all y, z with $s \leq y \leq z$.*

Part (a) is a technical result showing the relationship between convex and K-convex functions. Part (b) extends a result that holds for the convex functions to K-convex ones, that is, affine combination of K-convex functions is still K-convex (with a different K). Part (c) states that the expectation operator preserves K-convexity. Finally, part (d) gives the results that are necessary to see that an (s, S) policy is optimal if J_t for all $t = 0, 1, \ldots, N − 1$ are K-convex.

Following similar lines of the proof in the zero setup cost case and using the K-convexity properties, optimality of the (s, S) policy can be shown inductively. Since $J_N = 0$, it is convex. As in the previous case, $G_{N-1}(y) = cy + H(y)$ is convex [hence K-convex from Lemma 1(a)] and $\lim_{|y| \to \infty} G_{N-1}(y) = \infty$. Since we have

$$J_{N-1}(I_{N-1}) = \min\left\{ G_{N-1}(I_{N-1}), \min_{y \geq x_{N-1}}\left[k + G_t(y)\right]\right\} - cI_{N-1},$$

it can be seen that

$$J_{N-1}(I_{N-1}) = \begin{cases} k + G_{N-1}(S_{N-1}) - cI_{N-1}, & \text{if } I_{N-1} < s_{N-1} \\ G_{N-1}(I_{N-1}) - cI_{N-1}, & \text{if } I_{N-1} \geq s_{N-1}, \end{cases}$$

where S_{N-1} minimizes $G_{N-1}(y)$ and s_{N-1} is the smallest value of y such that $G_{N-1}(y) = k + G_{N-1}(S_{N-1})$. Note that for $k > 0$, $s_{N-1} < S_{N-1}$. Furthermore, the derivative of G_{N-1} at s_{N-1} is negative and hence the left derivative of J_{N-1} at s_{N-1}, $−c$, is greater than the right derivative, $−c + G'_{N-1}(s_{N-1})$, which implies that J_{N-1} is not convex (but it is continuous; see Fig. 6.6). However, based on the K-convexity of G_{N-1}, it can be shown that J_{N-1} is also K-convex. Using part (c) of the lemma, G_{N-2} is a K-convex function whose limit is infinity as $|y|$ approaches infinity. Repeating the earlier-stated arguments, J_{N-2} is K-convex. Continuing in this manner, one can show that for all t, G_t is a K-convex and continuous function which approaches ∞ as $|y|$ approaches ∞. Hence, by using part (d) of the lemma, an (s, S) policy is optimal.

So far we assume that demands are independent and identically distributed; cost parameters c, h, s are time-invariant; excess demand is backordered; total expected holding and shortage cost is convex; there are no capacity constraints; the time horizon is finite; the decision-maker is risk-neutral; and the price is exogenous. All these assumptions can be relaxed, and it can actually be proven that an (s, S) type policy is

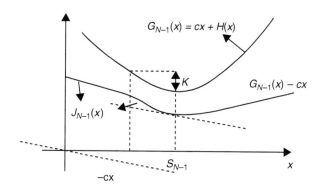

FIGURE 6.6 Structure of the cost-to-go function with positive setup cost.

still optimal. But due to the similarity of the proofs and conciseness concerns, we leave them as an extra reading to the reader. For models with time-invariant parameters, capacity constraints, and exogenous pricing assumptions, see Simchi-Levi et al. (2005); lost sales and correlated demand, see Bertsekas (2000); quasiconvex loss functions, see Veinott (1966) or a slightly simplified version of it in Simchi-Levi et al. (2005); infinite horizon, see Zheng (1991); and risk-averse decision making, see Agrawal and Seshadri (2000) and Chen et al. (2004).

6.4 Case Study

AMS is a growing fashion house. It started as a small family business in selling novelty T-shirts a couple of decades ago. Nowadays, it is a recognized forerunner in the global casual apparel industry. Its products are divided into two categories: novel and basic. Novel products are designed to be put in the market for one season only, while basic products are offered for at least two seasons. Unlike the basic products that might have inventory left over from previous years, all the excess novel products are salvaged at the end of their selling season. Furthermore, all the novel products are produced before their selling season.

Five new novelty T-shirts are designed for the next season. The cost of each T-shirt is $3. Traditionally, the cost is 30% of the selling price and the quantity of production is the average of the modes of the forecast. According to a $10 selling price, the expert forecasts of their independent demand are shown in Table 6.1.

Steven James is a product manager just hired to work under the Director of Novel Products and asked to report on the sales plans of these T-shirts. As a top graduate from an industrial engineering department who has a keen interest in inventory and pricing models, Steven is very enthusiastic in his new job and is confident that he will contribute significantly to AMS. After checking the current sales plans for the new novelty T-shirts, he wants to improve the current production plan and also try to convince his boss that a better pricing scheme should be implemented. In order to achieve these objectives, he needs to answer the following questions in his report:

- What is the expected profit for the current sales plan?
- What is the optimal production plan for the current pricing scheme? What is the corresponding expected profit?
- What is the potential increase in expected profit in deploying a different pricing scheme?

Discussion with the sales department reveals that excess novelty T-shirts have a salvage value of $.50. Furthermore, for a selling price from $5 to $15, each independent demand can be approximated by an additive model with a 10% drop in the $10 low-demand estimate per dollar increase in selling price. That is, demand = $a - br + \varepsilon$ for $5 \leq r \leq 15$, with a, b and ε as shown in Table 6.2.

6.4.1 Exercises

1. Suppose that the demand for a product is 20 units per month and the items are withdrawn at a constant rate. The setup cost each time a production run is undertaken to replenish inventory is $10. The production cost is $1 per item, and inventory holding cost is $0.20 per item per month. Assuming

TABLE 6.1 Forecasts of Low, Medium, and High Demand for the Novelty T-Shirts

T-Shirts	Demand (Probability)		
Swirl	10,000 (.2)	40,000 (.5)	80,000 (.3)
Strip	5000 (.25)	10,000 (.25)	50,000 (.5)
Sea	4000 (.1)	7000 (.5)	15,000 (.4)
Stone	3000 (.3)	9000 (.4)	20,000 (.3)
Star	8000 (.4)	10,000 (.4)	12,000 (.2)

TABLE 6.2   Parameters for the Additive Demand-Price Models

T-Shirts	a	b		ε (Probability)	
Swirl	20,000	1000	0 (.2)	30,000 (.5)	70,000 (.3)
Strip	10,000	500	0 (.25)	5000 (.25)	45,000 (.5)
Sea	8000	400	0 (.1)	3000 (.5)	11,000 (.4)
Stone	6000	300	0 (.3)	6000 (.4)	17,000 (.3)
Star	16,000	800	0 (.4)	2000 (.4)	4000 (.2)

TABLE 6.3   Requirement and Production Information

Month	Requirement	Setup Cost ($)	Production Cost ($)
1	2	500	800
2	4	700	900
3	3	400	900

shortages are not allowed, determine the optimal production quantity in a production run. What are the corresponding time between consecutive production runs and average cost per month?

2. Consider a situation in which a particular product is produced and placed in in-process inventory until it is needed in a subsequent production process. The number of units required in each of the next three months, the setup cost, and the production cost that would be incurred in each month are shown in Table 6.3. There is no inventory of the product, but 1 unit of inventory is needed at the end of the three months. The holding cost is $200 per unit for each extra month the product is stored. Use dynamic programming to determine how many units should be produced in each month to minimize the total cost.

3. In Example 2 (Section 6.3.2.1), if the demand–price relationship were $(100r^{-3})\varepsilon$, what would the optimal price and procurement level be?

4. Consider the hot dog stand example (Example 1 in Section 6.3.1). Now suppose that we would like to determine the optimal procurement policy over the next week (assume four games a week and we are only concerned about the game days). Each order costs the vendor $10.00 for gas and parking. Assume that any hot dog left at the end of the day is stored for the next game day and is not sold at the entertainment district. Each excess hot dog costs us $0.50 for handling and proper refrigeration. Also, let us assume that there are other vendors next door. In case of a shortage, extra hot dogs can be purchased from the neighboring hot dog vendors for $2.50 each and, hence, no demand is lost. Find the optimal procurement policy for the vendor over the next four sales periods.

5. Following the outline given in Section 6.3.2, prove that J_t for $t = 0, 1, \ldots, N - 1$ is K-convex when order setup cost k is positive.

References

Agrawal, V. and S. Seshadri (2000). Impact of uncertainty and risk aversion on price and order quantity in the newsvendor problem. *M&SOM*, 2(4):410–423.

An, M. (1995). Log-concave probability distributions: theory and statistical testing. Technical Report NC 27708-0097, Department of Economics, Duke University.

Bazaraa, M.S., H.D. Sherali, and C.M. Shetty (1993). *Nonlinear Programming: Theory and Algorithms*. 2nd Edition, Wiley, New York, NY.

Bertsekas, D. (2000). *Dynamic Programming and Optimal Control*. 2nd Edition, Athena Scientific, Belmont, MA.

Cachon, G.P. and A.G. Kok (2007). Implementation of the newsvendor model with clearance pricing: How to (and how not to) estimate a salvage value, *M&SOM*, 9(3):276–290.

Chen, X., M. Sim, D. Simchi-Levi, and P. Sun (2004). Risk aversion in inventory management. Working paper, Massachusetts Institute of Technology.

Karakul, M. (2007). Joint pricing and procurement of fashion products in the existence of a clearance market. *International Journal of Production Economics*, forthcoming.

Karakul, M. and L.M.A. Chan (2004). Newsvendor problem of a monopolist with clearance markets. *Proceedings of YA/EM 2004 on CD*.

Karakul, M. and L.M.A. Chan (2007). Analytical and managerial implications of integrating product substitutability in the joint pricing and procurement problem. *European Journal of Operational Research*, doi: 10.1016/j.ejor.2007.06.026.

Petruzzi, N. and M. Dada (1999). Pricing and the newsvendor problem: a review with extensions. *Operations Research*, 47:183–194.

Roundy, R. (1985). 98%-effective integer-ratio lot-sizing for one-warehouse multi-retailer systems. *Management Science*, 31:1416–1430.

Silver, E.A. and R. Peterson (1985). *Decision Systems for Inventory Management and Production Planning*. John Wiley, New York.

Simchi-Levi, D., X. Chen, and J. Bramel (2005). *Logic of Logistics: Theory, Algorithms, and Applications for Logistics and Supply Chain Management*. 2nd Edition, Springer Verlag, New York, NY.

Veinott, A. (1966). On the optimality of (s, S) inventory policies: new condition and a new proof. *SIAM Journal of Applied Mathematics*, 14:1067–1083.

Zheng, Y.S. (1991). A simple proof for the optimality of (s, S) policies for infinite horizon inventory problems. *Journal of Applied Probability*, 28:802–810.

7

Material Handling System

7.1	Introduction	**7-1**
7.2	Ten Principles of Material Handling	**7-2**
	Planning • Standardization • Work • Ergonomics • Unit Load • Space Utilization • System • Automation • Environmental • Life Cycle	
7.3	Material Handling Equipment	7-9
	Types of Equipment	
7.4	How to Choose the "Right" Equipment	7-15
7.5	Analytical Model for Material Handling Equipment Selection	7-15
7.6	Warehousing	7-17
7.7	Warehouse Functions	7-18
7.8	Material-Handling System Case Study	7-19
7.9	Automated Storage and Retrieval Systems Case Study	7-19
7.10	Summary	7-26
	References	7-26

Sunderesh S. Heragu
University of Louisville

7.1 Introduction

Material handling systems are hardware systems that move material through various stages of processing, manufacture, assembly, and distribution within a facility [1]. Material movement occurs everywhere in a factory or warehouse—before, during, and after processing. The cost of material movement is estimated to be anywhere from 5% to 90% of overall factory cost with an average around 25% [2]. Material movement typically does not add value in the manufacturing process. However, this step is necessary to make a product.

The increasing demand for high product variety and short response times in today's manufacturing industry emphasizes the importance of highly flexible and efficient material handling systems. The operation of the material handling system is determined by product routings, factory layout, and material flow control strategies. Most existing textbooks cover just parts of these aspects. In this chapter, we try to introduce the material handling system from an integrated system point of view and include most factors related to the material handling system. In Section 7.2, 10 principles of the material handling system are discussed. They provide some general guidelines while selecting equipment, designing layout, and standardizing, managing, and controlling the material as well as the handling system. Section 7.3

discusses the material handling equipment topic. The multiple types of equipment and how to select this equipment are discussed in this section. Section 7.4 discusses the material handling equipment selection problem. An analytical model for the material handling selection is presented in Section 7.5. Warehousing and its functions are presented in Sections 7.6 and 7.7. Case studies illustrating applications in material handling and warehousing are presented in Sections 7.8 and 7.9.

7.2   Ten Principles of Material Handling

The 10 principles of material handling developed by the Material Handling Industry of America are: planning, standardization, work, ergonomics, unit load, space utilization, system, automation, environmental, and life cycle. A multimedia education CD explaining various aspects of the 10 principles is available upon request (see [3]).

7.2.1   Planning

A material handling plan is a prescribed course of action that specifies the material, moves, and the method of handling in advance of implementation. Four key aspects need to be considered in developing a sound material handling plan.

1. The communication between designers and users is very important in developing the plans for operations and equipments. For large-scale material handling projects, a team including all stakeholders is required.
2. The material handling plan should incorporate the organization's long-term goals and short-term requirements.
3. The plan must be based on existing methods and problems, subject to current physical and economic constraints, and meet organizational requirements and goals.
4. The plan should build in flexibility so that sudden changes in the process can be assimilated.

7.2.2   Standardization

Standardization is a way of achieving uniformity in the material handling methods, equipment, controls and software without sacrificing needed flexibility, modularity, and throughput. Standardization of material handling methods and equipment reduces variety and customization. This is a benefit so long as overall performance objectives can be achieved. The key aspects of achieving standardization are as follows:

1. The planner needs to select methods and equipment that can perform a variety of tasks under a variety of operating conditions and anticipate changing future requirements. Therefore, the methods and equipment can be standardized at the same time ensuring flexibility. For example, the conveyor system in Figure 7.1 can carry different sizes of parcels.
2. Standardization can be applied widely in material handling methods, such as the sizes of containers and other characteristics, as well as operating procedures and equipment.
3. Standardization, flexibility, and modularity need to complement each other, providing compatibility.

7.2.3   Work

Material handling work is equal to the product of material handling flow (volume, weight, or count per unit of time) and distance moved. It should be minimized without sacrificing productivity or the level of service required of the operation. The work can be optimized from three aspects:

FIGURE 7.1   Conveyor system. (Courtesy of Vanderlande Industries, The Netherlands. With permission.)

1. Combine, shorten, or eliminate unnecessary moves to reduce work. For example, in dual command storage and retrieval cycles, two commands, storage or retrieval, are executed in one trip so it has less work than single storage and retrieval cycles.
2. Consider each pick-up and set-down or placing material in and out of storage as distinct moves and components of distance moved.
3. Material handling work can be simplified and reduced by efficient layouts and methods (Fig. 7.2). Gravitational force is used to reduce material handling work.

FIGURE 7.2   Gravity roller conveyor. (Courtesy of Sunderesh S. Heragu, 10 Principles of Materials Handling, CD. With permission.)

7.2.4 Ergonomics

Ergonomics is the science that seeks to adapt work and working conditions to suit the abilities of the worker. It is important to design safe and effective material handling operations by recognizing human capabilities and limitations.

1. Select equipment that eliminates repetitive and strenuous manual labor and that the user can operate effectively. Equipment specially designed for material handling is usually more expensive than standard equipment. But using standard equipment will result in fatigue, hurt particular parts of the worker's body, and result in error and low-operating efficiency. Therefore, it may be necessary to select specialized equipment to minimize long-term costs and injury.
2. In material handling systems, ergonomic workplace design and layout modification, it is important to pay more attention to the human physical characteristics. For example, in Figure 7.3 the work-place design on the left does not provide toe space for the worker, requiring him or her to bend forward. Maintaining this posture will produce fatigue and injury. The modified workplace with toe space is more comfortable for the worker because his or her body is in an erect position (see right side in Fig. 7.3).
3. The ergonomics principle embraces both physical and mental tasks. For example, when a printed label or message must be read quickly and easily, the plain and simple type font should be chosen preferentially. Less familiar designs and complex font may result in errors, especially when read in haste. Aesthetic fonts are poor choices. Obviously, extremes like Old English should never be used. In one word, keep it simple.
4. Safety is the priority in workplace and equipment design.

7.2.5 Unit Load

A unit load is one that can be stored or moved as a single entity at one time, regardless of the number of individual items that make up the load. When unit load is used in material flow, the following key aspects deserve attention:

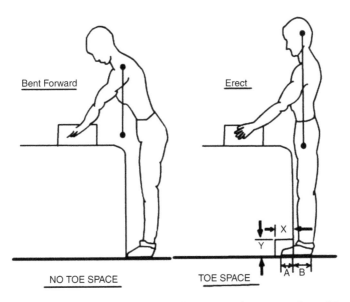

FIGURE 7.3 Modified work place. (From DeLaura, D. and Kons, D., *Advances in Industrial Ergonomics and Safety II*, Taylor & Francis, 1990. With permission.)

1. Less effort and work are required to collect and move a unit load than to move many items one at a time. But this does not mean bigger unit load size is always better. As the unit load size increases, the total transportation cost decreases. This decrease is offset by the increase in the inventory cost. Figure 7.4 shows the relationship between the two.
2. Load size and composition may change as material and product move through various stages of manufacturing and the resulting distribution channels.
3. Large unit loads of raw material are common before manufacturing and also after manufacturing when they constitute finished goods.
4. During manufacturing, smaller unit loads, sometimes just one item, yield less in process inventory and shorter item throughput times. From Little's law [4], when a system has reached steady state, the average number of parts in the system is equal to the product of the average time per part in the system and its arrival rate.
5. Smaller unit loads are consistent with manufacturing strategies that embrace operational objectives such as flexibility, continuous flow, and just-in-time delivery.

7.2.6 Space Utilization

A good material handling system should try to improve the effectiveness and efficiency of all the available space. There are three key points for this principle.

1. In work areas, eliminate cluttered, unorganized spaces and blocked aisles. For example, blocked aisles will add more material flow work. In Figure 7.5, the product on the floor will force the forklift to pick the product on the shelf using a longer material flow path, while the storage in Figure 7.6 will result in inefficient use of vertical storage space (called honeycombing loss).
2. In storage areas, the objective of maximizing storage density must be balanced against accessibility and selectivity. If items are going to be in the warehouse for a long time, storage density is an important consideration. If items enter and leave the warehouse frequently, their accessibility and selectivity are important. If the storage density is too high to access or select the stored product, high storage density may not be beneficial.
3. A Cube per Order Index (COI) storage policy is often used in a warehouse. COI is a storage policy in which each item is allocated warehouse space based on the ratio of its storage space requirements (its cube) to the number of storage/retrieval transactions for that item. Items are listed in a

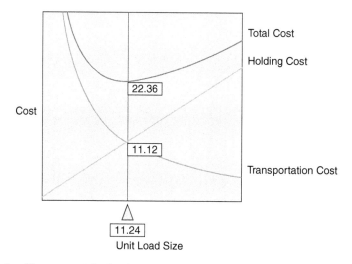

FIGURE 7.4 Trade-off between unit load and inventory costs. (Courtesy of Sunderesh S. Heragu, 10 Principles of Materials Handling, CD. With permission.)

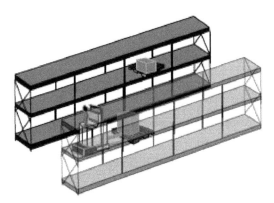

FIGURE 7.5 Retrieving material in blocked aisles. (Courtesy of Sunderesh S. Heragu, 10 Principles of Materials Handling, CD. With permission.)

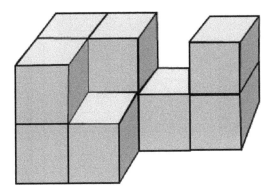

FIGURE 7.6 Honeycombing loss. (Courtesy of Sunderesh S. Heragu, 10 Principles of Materials Handling, CD. With permission.)

nondecreasing order of their COI ratios. The first item on the list is allocated to the required number of storage spaces that are closest to the input/output (I/O) point; the second item is allocated to the required number of storage spaces that are next closest to the I/O point, and so on. Figure 7.7 shows an interactive "playspace" in the "10 Principles of Materials Handling" CD that allows a learner to understand the fundamental concepts of the COI policy.

7.2.7 System

A system is a collection of interdependent entities that interact with each other. The main components of the supply chain are suppliers, manufacturers, distributions, and customers. The activities to support material handling both within and outside a facility need to be integrated into a unified material handling system. The key aspects of the system principle are:

1. At all stages of production and distribution, minimize inventory levels as much as possible.
2. Even though high inventory allows a company to provide a higher customer service level, it can also conceal the production problems which, from a long-term point of view, will hurt the company's operations. These problems can eventually result in low production efficiency and high product cost.
3. Information flow and physical material flow should be integrated and treated as concurrent activities. The information flow typically follows material flow.

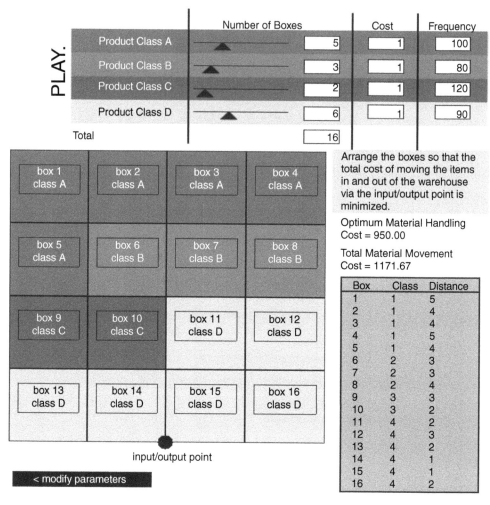

FIGURE 7.7 Example of COI policy. (Courtesy of Sunderesh S. Heragu, 10 Principles of Materials Handling, CD. With permission.)

4. Materials must be easily identified in order to control their movement throughout the supply chain. For example, bar coding is the traditional method used for product identification. Radio frequency identification (RFID) uses radio waves to automatically identify people or objects as they move through the supply chain. Due to two unique product identification mandates, one from the private sector (Wal-Mart) and another from the public sector (Department of Defense), RFID has become very popular in recent years. The big difference between the two automatic data capture technologies is that bar codes is a line-of-sight technology. In other words, a scanner has to "see" the bar code to read it, which means people usually have to orient the bar code toward a scanner for it to be read. RFID tags can be read as long as they are within the range of a reader even if there is no line of sight. Bar codes have other shortcomings as well. If a label is ripped, soiled, or falls off, there is no way to scan the item. Also, standard bar codes identify only the manufacturer and product, not the unique item. The bar code on one milk carton is the same as every other, making it impossible to identify which one might pass its expiration date first. RFID can identify items individually.

5. Meet customer requirements regarding quantity, quality, and on-time delivery and fill orders accurately.

7.2.8 Automation

Automation in a material flow system means using electro-mechanical devices, electronics and computer-based systems with the result of linking multiple operations to operate and control production and service activities. These automated devices and systems are usually controlled by programmed instructions. Automation enables equipment or systems to run with little or no operator intervention. It improves safety, operational efficiency, consistency, and predictability, while increasing system responsiveness. Automation also decreases operating costs. In order to make the automation serve the material flow system properly, the following key aspects should be considered:

1. Simplify pre-existing processes and methods before installing mechanized or automated systems.
2. Consider computerized material handling systems where appropriate for effective integration of material flow and information management.
3. In order to automate handling, items must have features that accommodate mechanization.
4. Treat all interface issues in the situation as critical to successful automation.

7.2.9 Environmental

The environmental principle in material handling involves designing material handling methods and selecting and operating equipment in a way that preserves natural resources and minimizes adverse effects on the environment coming from material handling activities. The following three key aspects need to be considered:

1. Design containers, pallets, and other products used in material handling so they are reusable and/or biodegradable. For example, use recyclable pallets.
2. By-products of material handling should be considered in the system design.
3. Give special handling considerations to hazardous material handling.

7.2.10 Life Cycle

Life cycle costs include all cash flows that occur between the time the first dollar is spent on the material handling equipment or method until its disposal or replacement. Its key aspects are:

1. Life cycle costs in material handling system include: capital investment; installation, setup, and equipment programming; training, system testing, and acceptance; operating, maintenance, and repair; and recycle, resale, and disposal.
2. Plan for preventive, predictive, and periodic maintenance of equipment. Include the estimated cost of maintenance and spare parts in the economic analysis. There are three types of equipment failures that occur over the equipment's useful life—early failures when the product is being debugged, constant failures associated with the normal use of equipment, and increasing failure rate during the wear-out stage, when products fail due to aging and fatigue. A sound maintenance program will postpone the wear-out period and extend the useful life of equipment. Maintenance cost should be considered in the life cycle.
3. Prepare a long-range plan for equipment replacement.
4. In addition to measurable cost, other factors of a strategic or competitive nature should be quantified when possible.

The 10 principles are vital to material handling system design and operation. Most are qualitative in nature and require the industrial engineer to employ these principles when designing, analyzing, and operating material handling systems.

7.3 Material Handling Equipment

In this section we list the various equipment that actually transfers materials between different stages of processing. In manufacturing companies, various material handling devices (MHDs) are used and together they constitute a material handling system (MHS). If we regard materials as the blood of a manufacturing company, then MHSs are the vessels that transport blood to the necessary parts of the body. The major function of an MHS is to transport parts and materials; this type of activity does not add any value to products and can be regarded as a sort of "necessary waste." However, in some cases, MHSs perform value-added activities. The MHS is an important subsystem of the entire manufacturing system; it interacts with the other subsystems. Thus, when we try to design or run an MHS, we should look at it from a system perspective. If we isolate an MHS from other subsystems, we might get an optimal solution for the MHS itself, but one that is suboptimal for the entire system.

In the following sections, we will first introduce seven basic types of MHDs. We then discuss how to choose the "right" equipment and how to operate equipment in the "right" way.

7.3.1 Types of Equipment

Several different types of MHDs are available for manufacturing companies to choose. These companies need to consider a number of factors including size, volume of loads, shape, weight, cost, and speed. As mentioned in the introduction, we need to consider the entire system when we try to make our choices. Of course, in order to make good decisions, we need to have an overview of different MHDs. There are seven basic types of MHDs [1]: conveyors; palletizers; trucks; robots; automated guided vehicles; hoists, cranes, and jibs; and warehouse material-handling devices. We will introduce these types one by one briefly.

7.3.1.1 Conveyors

Conveyors are fixed path MHDs. They are only used when the volume of material to be transported is large and relatively uniform in size and shape. Depending upon the application, many types of conveyors are possible, including: accumulation conveyor, belt conveyor, bucket conveyor, can conveyor, chain conveyor, chute conveyor, gravity conveyor, power and free conveyor, pneumatic or vacuum conveyor, roller conveyor, screw conveyor, slat conveyor, tow line conveyor, trolley conveyor, and wheel conveyor. Pictures of a few conveyors are shown in Figure 7.8. The above list is not complete. Readers can refer to www.mhia.org for additional information on conveyors (and other types of MHDs).

FIGURE 7.8 Various conveyor types and their applications in material movement and sorting (a–d). (Courtesy of FKI Logistex, Dematic Corporation. With permission.)

(b)

(c)

(d)

FIGURE 7.8   (continued)

7.3.1.2 Palletizers

Palletizers are used to palletize items coming out of a production or assembly line so that unit loads can be formed directly on a pallet. Palletizers are typically automated, high-speed MHDs with a user-friendly interface so that operators can easily control them. Another type of equipment that is related to a pallet is a pallet lifting device. This MHD is used to lift and/or tilt pallets and raise or lower heavy cases to desired heights so that operators can pick directly from the pallets. A palletizer is shown in Figure 7.9.

7.3.1.3 Trucks

Trucks are particularly useful when the material moved varies frequently in size, shape, and weight; when the volume of the parts/material moved is low; and when the number of trips required for each part is relatively few. There are many different types of trucks on the market with different weight, cost, functionality, and other features. A sample is shown in Figure 7.10.

7.3.1.4 Robots

Robots are programmable devices that mimic the behavior of human beings. With the development of artificial intelligence technology, robots can do a number of tasks not suitable for human operators. However, robots are relatively expensive. But they can perform complex or repetitive tasks automatically. They can work in environments that are unsafe or uncomfortable to the human operator, work under extreme circumstance including very high or low temperature, and handle hazardous material.

7.3.1.5 Automated Guided Vehicles

Automated guided vehicles (AGVs) have been very popular since they were introduced about 30 years ago and will continue to be an important MHD in the future. AGVs can be regarded as a type of specially designed robot. Their paths can be controlled in a number of different ways. They can be fully automated or semiautomated. They can also be embedded into other MHDs. A sample of AGVs and their applications is illustrated in Figure 7.11.

FIGURE 7.9   Palletizer. (Courtesy of FKI Logistex, Dematic Corporation. With permission.)

(a)

(b)

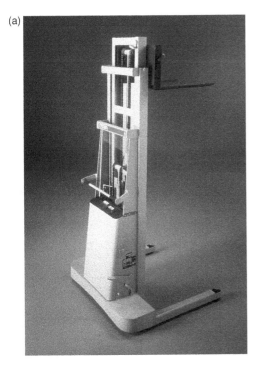

FIGURE 7.10   Order-picking trucks. (Courtesy of Crown Corporation. With permission.)

7.3.1.6   Hoists, Cranes, and Jibs

These MHDs use the overhead space. The movement of material in the overhead space will not affect the production process and workers in a factory. Typically, these MHDs are expensive and time consuming to install. They are preferred when the parts to be moved are bulky and require more space for transportation (Fig. 7.12).

7.3.1.7   Warehouse Material-Handling Devices

Warehouse material-handling devices are also referred to as storage and retrieval systems. If they are highly automated, they are referred to as automated storage and retrieval systems (AS/RSs). The primary functions of warehouse material-handling devices are to store and retrieve materials as well as transport them between the pick/deposit (P/D) stations and the storage locations of the materials. Some AS/RSs are shown in Figure 7.13.

(a)

FIGURE 7.11   Application of AGVs (a and b).

(b)

FIGURE 7.11 (continued)

X_{ijkl} number of units of part type i to be transported from machine j to k using MHD l

(a)

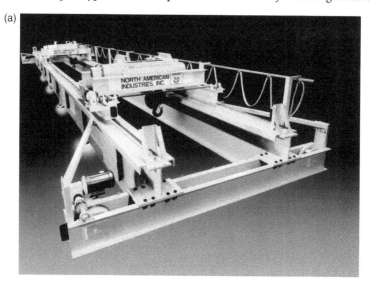

FIGURE 7.12 Gantry crane and hoist (a and b). (Courtesy of North American Industries and Wallace Products Corporation. With permission.)

FIGURE 7.12   (continued)

FIGURE 7.13   Automated storage and retrieval systems (AS/RSs). (Courtesy of Jervis B. Webb Company. With permission.)

7.4   How to Choose the "Right" Equipment

Apple[5] has suggested the use of the "material handling equation" in arriving at a material handling solution. As shown in Figure 7.14, it involves seeking thorough answers to six major questions—why (select material handling equipment), what (is the material to be moved), where and when (is the move to be made), how (will the move be made), and who (will make the move). It should be emphasized that all the six questions are extremely important and should be answered satisfactorily. Otherwise, we may end up with an inferior material handling solution. In fact, it has been suggested that analysts come up with poor solutions because they jump from the what to the how question [5].

The material handling equation can be specified as: *Material + Move = Method*, as shown in Figure 7.14. Very often, when the *material* and *move* aspects are analyzed thoroughly, it automatically uncovers the appropriate material handling *method*. For example, analysis of the type and characteristics of *material* may reveal that the material is a large unit load on wooden pallets. Further analysis of the logistics, characteristics and type of *move* may indicate that a 20-foot load/unload lift is required, distance traveled is 100 feet, and some maneuvering is required while transporting the unit load. This suggests that a forklift truck would be a suitable material handling device. Even further analysis of the method may tell us more about the specific features of the forklift truck; for example, a narrow aisle fork lift truck with a floor load capacity of 200 pounds.

7.5   Analytical Model for Material Handling Equipment Selection

Several analytic approaches have been proposed to select the required number and type of MHDs and to assign them to material-handling moves so that different objectives are achieved optimally. These models fall into three catalogs: deterministic approach, probabilistic approach, and knowledge-based approach. A deterministic model is presented below.

The objective of the model for simultaneously selecting the required number and type of MHDs and assigning them to material-handling moves is to minimize the operating and annualized investment costs of the MHDs. A material-handling move or simply a move is the physical move that an MHD has to execute in order to transport a load between a pair of machines. The number of moves depends upon not only the volume and transfer batch size of each part type manufactured, but also the number of machines it visits. All candidate MHD types that can perform the moves are evaluated and an optimal selection and assignment are determined by this model. If necessary, we can modify the objective function of the model to incorporate equipment idle time in conjunction with capital and operating costs. Before presenting the model, we define its variables and parameters.

i part type index, $i = 1, 2, ..., p$

j machine type index, $j = 1, 2, ..., m$

l MHD type index, $l = 1, 2, ..., n$

L_i set of MHDs that can be used to transport part type i

H length of planning period

D_i number of units of part type i required to be produced

K_{ij} set of machines *to* which part type i can be sent from machine j for the next processing step

M_{ij} set of machines *from* which part type i can be sent to machine j for the next processing step

A_i set of machine types required for the first operation on part type i

B_i set of machine types required for the last operation on part type i

V_l purchase cost of MHD H_l

T_{ijkl} time required to move one unit of part type i from machine type j to k using MHD l

C_{ijkl} unit transportation cost to move part type i from machine j to k using MHD l

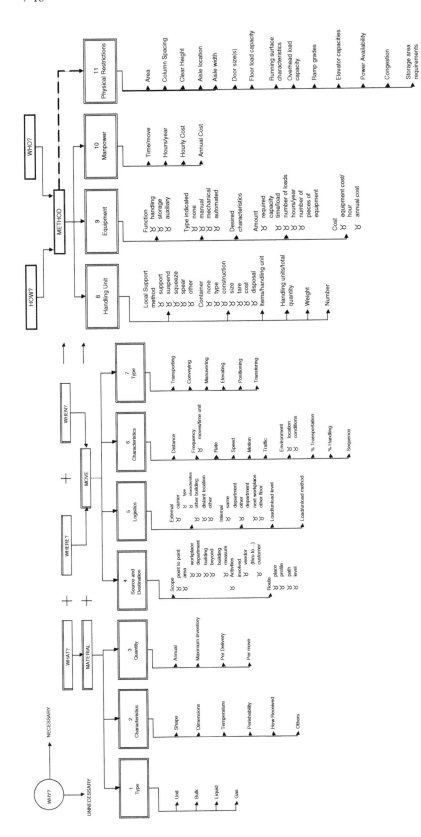

FIGURE 7.14   Material handling "equation." (From J.M. Apple, *Plant Layout and Modeling*, Wiley, New York, 1977. With permission.)

Y_l number of units of MHD type l selected

Model

$$\text{Minimize} \quad \sum_{l=1}^{r} V_l Y_l + \sum_{i=1}^{p} \sum_{j=1}^{m} \sum_{k \in K_{ij}} \sum_{l \in L_i} C_{ijkl} X_{ijkl} \tag{7.1}$$

$$\text{subject to} \quad \sum_{j \in A_i} \sum_{k \in K_{ij}} \sum_{l \in L_i} X_{ijkl} = D_i \tag{7.2}$$

$$\sum_{k \in M_{ij}} \sum_{l \in L_i} X_{ijkl} - \sum_{k \in K_{ij}} \sum_{l \in L_i} X_{ijkl} = 0 \qquad i = 1, 2, ..., p; \; j: j \notin A_i \cup B_i \tag{7.3}$$

$$\sum_{j \in B_i} \sum_{k \in M_{ij}} \sum_{l \in L_i} X_{ijkl} = D_i \quad i = 1, 2, ..., p \tag{7.4}$$

$$\sum_{i=1}^{p} \sum_{j=1}^{m} \sum_{l \in K_i} T_{ijkl} X_{ijkl} \leq H Y_l \quad l = 1, 2, ..., n \tag{7.5}$$

$$X_{ijkl} \geq 0 \quad i = 1, 2, ..., p; \; j = 1, 2, ..., m; \; k = 1, 2, ..., n; \; l = 1, 2, ..., n \tag{7.6}$$

$$Y_l \geq 0 \text{ and integer} \quad l = 1, 2, ..., n \tag{7.7}$$

The objective function of the above model minimizes not only the operating costs (measured as a function of the move transportation costs), but also the MHD purchase costs. When only one part type is being considered, the above model is a fixed charge network flow problem in which the number of nodes depends upon the number of machines and MHDs capable of processing and transporting the part type. In the network flow context, Constraint 7.2 ensures that the "flow" generated at the first (supply) node is equal to the number of parts to be processed. In other words, the number of units of each part type leaving their respective machines after the first operation should be equal to the number of units of that part type to be produced. These part types are "absorbed" at the last (demand) node as enforced by Constraint 7.4. In other words, the number of units of each part type coming to their respective machines for the last operation should be equal to the number of units of that part type produced. Constraint 7.3 is a material balance expression and ensures that for each intermediate transhipment node corresponding to the machines required for the in-between operations, that is, other than first and last, all the units received are passed on to node(s) at the next stage. Thus, all the parts received at each intermediate machine are sent to the appropriate machine(s) for the next processing step. Constraint 7.5 imposes that the MHD capacity not be exceeded. Because Constraint 7.6 is an integer constraint, the required number of each type of MHD necessary for transporting material between machines will be selected. It can be shown that X_{ijkl} variables will automatically be integers in the optimal solution. Hence, no additional integer restrictions for these variables are necessary. When there are load limits on the MHDs, these can be enforced by introducing capacities on the appropriate arcs. This means that the corresponding X_{ijkl} variables will have an upper bound as well.

7.6 Warehousing

Many manufacturing and distribution companies maintain large warehouses to store in-process inventories or components received from an external supplier. Businesses that lease storage space to other companies for temporary storage of material also own and maintain a warehouse. In the former case, it has been argued that warehousing is a time-consuming and nonvalue-adding activity. Because additional paperwork and time are required to store items in storage spaces and retrieve them later when needed, the JIT manufacturing philosophy suggests that one should do away with any kind of temporary storage and maintain a pull strategy in which items are produced only as and when they are required; that is, items should be produced at a certain stage of manufacturing, only if they are required at the next stage. Moreover, the quantity produced should directly correspond to the amount demanded at the next stage of manufacturing. JIT philosophy requires that the same approach be taken toward components received from suppliers. The supplier is considered as another (previous) stage in manufacturing. However, in practice, because of a variety of reasons, including the need to maintain a sufficient inventory of items because of the unreliability of suppliers, and to improve customer service and respond to their needs quickly, it is not possible or, at least, not desirable to completely do away with temporary storage.

Consider the following situation at Nike, a company that makes athletic wear. Nike has recently built a large distribution warehouse in Belgium because one of their main business objectives is to serve 75% of their customers within 24 hours. Without appropriate warehousing facilities, it is impossible for Nike to achieve this objective because many of their manufacturing plants and suppliers are overseas—in the Far East! Members club stores such as Sam's Club, Costco, and B.J.'s Warehouse Club have found a niche in the consumer retailing business in the past decade. These stores provide memberships to businesses and their employees or friends and allow only members to shop in their stores. They generally sell merchandise in bulk and directly out of their warehouse, eliminating the need to build and maintain costly retail stores. While this significantly reduces overhead costs for the warehouse, for the consumer, it typically costs less to shop in such stores than in traditional malls because he or she buys in bulk. The primary function in such warehouse stores is not warehousing but retailing!

The above two examples amply demonstrate the need for establishing warehouses to satisfactorily service end customers despite the lack of value-added services in many of them. This section is devoted to warehouse and storage design and planning.

7.7 Warehouse Functions

As seen in the previous section, there are several reasons for building and operating warehouses. In many cases, the need to provide better service to customers and be responsive to their needs appears to be the primary reason. While it may seem that the only function of a warehouse is warehousing, that is, temporary storage of goods, in reality, many other functions are performed. Some of the more important ones are listed and briefly discussed below [6].

Temporary storage of goods: To achieve economies of scale in production, transportation and handling of goods, it is often necessary to store goods in warehouses and release them to customers as and when the demand occurs.

Put together customer orders: Warehouses, for example, the Nike distribution center in Laakdal, Belgium, receives shipments in bulk from overseas and, using an automated or manual sortation system, puts together individual customer orders and ships them directly to the stores.

Serve as a customer service facility: Because warehouses ship goods to customers and therefore are in direct contact with them, a warehouse can serve as a customer service facility and handle replacement of damaged or faulty goods, conduct market surveys and even provide after-sales service. For example, many Japanese electronic goods manufacturers let warehouses handle repair and after-sales service in North America.

Protect goods: Because warehouses are typically equipped with sophisticated security and safety systems, it is logical to store manufactured goods in warehouses to protect against theft, fire, floods, and weather elements.

Segregate hazardous or contaminated materials: Safety codes may not allow storage of hazardous materials near the manufacturing plant. Because no manufacturing takes place in a warehouse, this may be an ideal place to segregate and store hazardous and contaminated materials.

Perform value-added services: Many warehouses routinely perform several value-added services such as packaging goods, preparing customer orders according to specific customer requirements, inspecting arriving materials or products, testing products not only to make sure they function properly but also to comply with federal or local laws, and even assemble products. Clearly, inspection and testing do not add value to the product. However, we have included them here because they may be a necessary function because of company policy or federal regulations.

Inventory: Because it is difficult to forecast product demand accurately, in many businesses, it may be extremely important to carry inventory and safety stocks to allow them to meet unexpected surges in demand. In such businesses, not being able to satisfy a demand when it occurs may lead to a loss in revenues, or worse yet, may severely impact customer loyalty toward the company. Also, companies that produce seasonal products, for example, lawn mowers and snow throwers, may have excess inventory left over at the end of the season and have to store the unsold items in a warehouse.

A typical warehouse consists of two main elements: the storage medium and the material handling system. Of course, there is a building that encloses the storage medium, goods and the storage/retrieval (S/R) system. Because the main purpose of the building is to protect its contents from theft and weather elements, it is made of strong, lightweight material. Warehouses come in different shapes, sizes, and heights, depending upon a number of factors, including the kind of goods stored inside, volume, and type of S/R systems used. The Nike warehouse in Laakdal, Belgium, covers a total area of 1 million square feet. Its high-bay storage is almost 100 feet in height, occupies roughly half of the total warehouse space and is served by a total of 26 man-aboard stacker cranes. On the other hand, a "members club" store may have a total warehouse space of 200,000 square feet with a building height of 35 feet.

7.8   Material-Handling System Case Study

The European Combined Terminals (ECT) in Rotterdam, the Netherlands, is the largest container terminal in the world. Goods to and from Europe are transported to the outside world primarily via two types of containers—large and small. The newer docks have been built on reclaimed land in the North Sea (Fig. 7.15a). Trucks arriving from Belgium, Germany, France, the Netherlands, and other countries wait their turn in a designated spot for their load, that is, container, to be picked up by a straddle carrier (Fig. 7.15b). The straddle carrier holds the load under the operator and moves it (Fig. 7.15c and 7.5d) to a temporary hold area from where it is loaded onto ships (Fig. 7.15e). Containers are usually held for two days in this area. When they are ready to be loaded onto ships, mobile overhead gantry cranes that move on tracks and have special container-holding attachments lift the containers from above and take them to another location where AGVs are waiting to receive the load (Fig 7.15d and 7.5f). A fleet of AGVs then transports the containers to tower cranes (Fig. 7.15g). The tower cranes are positioned very close to the loading area of the ships. Moreover, one of their arms can be tilted upward at a 90° angle to allow for tall ships to pass under them (Fig. 7.15f). Using overhead cranes, the containers are picked up from the AGV and transported one by one to the ship deck (Fig. 7.15g). While the figures illustrate how ships are loaded, unloading is done in a similar manner—only the steps are reversed. Effective use of AGVs, cranes and trucks allows ECT to load or unload a ship in about one day.

7.9   Automated Storage and Retrieval Systems Case Study

Phoenix Pharmaceuticals, a German pharmaceutical company established in 1994, has a 150,000-square-foot warehouse in Herne, Germany. This warehouse, which has an annual turnover of $400 million,

(a)

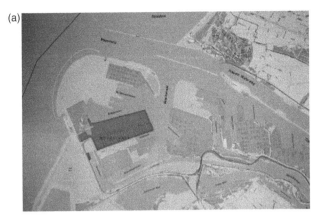

(b)

(c)

FIGURE 7.15 MHSs in action (a–g). (From Sunderesh S. Heragu, *Facilities Design*, iUniverse, Lincoln, NE. With permission.)

(d)

(e)

(f)

FIGURE 7.15 (continued)

(g)

FIGURE 7.15 (continued)

receives pharmaceutical supplies from 19 plants all across Germany and distributes them to area drugstores. Phoenix has a 30% market share and is a leader in the pharmaceutical business. Due to competitive and other business reasons, the company must fill each order from drugstores and ship it in less than 30 minutes. There are roughly 87,000 items stored in the warehouse, of which 61% are pharmaceutical and the remainder are cosmetic supplies. The number of picks ranges anywhere from 150 to 10,000 in any given month. If Phoenix did not have warehouses located at strategic locations, it would obviously not be able to respond to its customers, that is, fill and ship orders, accurately and adequately. Not only it is very costly for the company to ship the pharmaceutical supplies from the plant to each drugstore directly, but it is also not possible to do so because of the distances.

Order picking in Phoenix is done using three levels of automation:

1. Manual order picking using flow racks
2. Semiautomated order picking using an automatic dispensing system
3. Full automation using a robotic order picker

Incoming customer orders are printed on high-speed printers and the orders are attached manually to totes and sent via conveyors (Fig. 7.16a) to manual order picking areas. Here, operators pick items specified in an order from flow racks, fill the container and send it to shipping areas from where it is sent to the customer (i.e., drug stores). Order picking in Phoenix is done manually for bulky items that are not suitable for the AS/RS.

Semiautomated order picking is used for small items (e.g., a box containing a few dozen aspirin tablets, nasal spray, or medicine), which are stacked up on the outside of automatic vertical dispensers (Fig. 7.16b) in their respective columns. The dispenser has several columns—one for each brand of medicine picked. The dispensers are inclined over a conveyor forming an A shape and a computer controlled mechanism kicks items specified in an order from their respective columns on to the moving conveyor belt (Fig. 7.16c). The items then proceed to the end of the conveyor line, where they are dropped into a waiting tote. Each tote corresponds to a specific order. The tote (similar to those shown in Fig. 7.16a) are at a lower level than the conveyor line. Hence, there is no need for manual handling of the picked items. A light signal (Fig. 7.16b) tells the operators when items need to be replenished—typically when the item has reached or gone below its safety stock level. The automatic dispensing mechanism is very effective for picking a large variety of items for which the picking frequency is medium. The automation level with the dispenser mechanism is medium. It is relatively inexpensive and the order picking is done at a much faster rate than manual order picking. The degree of accuracy is also very high. However, it usually can be used only for handling relatively small items.

FIGURE 7.16 Order picking in Phoenix Pharmaceuticals warehouse, Germany (a–i). (From Sunderesh S. Heragu, *Facilities Design*, iUniverse, Lincoln, NE. With permission.)

(e)

(f)

(g)

FIGURE 7.16   (continued)

(h)

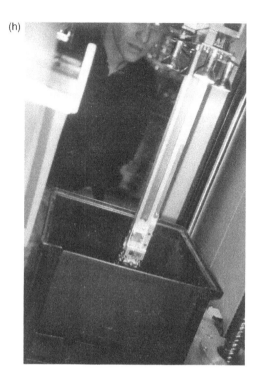

(i)

FIGURE 7.16 (continued)

The third level of order picking in Phoenix is done via expensive robotic order picker. Phoenix has two sets of robots—one for storage and another for retrieval. The retrieval robots (Fig. 7.16d) pick items from narrow aisles whose width is just a little more than that of the robot (Fig. 7.16e). Equipped with computers (Fig. 7.16f) and optical scanners, the robot retrieves items specified in an order and puts them into one of several compartments (see the circular compartmentalized drum in the middle of Fig. 7.16f). Each compartment corresponds to a customer order. The required items are picked from their respective locations, loaded onto the compartments, and taken to a conveyor line (see the right side of Fig. 7.16g), where they are dropped into waiting totes. Each compartment in the circular drum has a metal flap at the bottom that automatically opens and allows all the items in an order to be dropped into its specified tote.

The storage robots (Fig. 7.16g) have a deck that can hold large bins. Items to be stored in racks are put into these bins, which are then loaded on the robot deck one at a time. A robot arm plunges into the bin and picks items using vacuum suction cups (again one at a time—see Fig. 7.16h). The items are then put into their respective storage bins using robot arms equipped with optical scanners (Fig. 7.16i). The bins are then transported and stored by the robot.

7.10 Summary

A material handling system is a complex system that provides a vital link between successive workstations. In this chapter, we tried to introduce it from the 10 principles point of view, different types of available material handling equipment, a model for selection of MHDs, warehousing, and its functions, and we presented two case studies to illustrate MHSs in action.

References

1. S. S. Heragu, (2006), *Facilities Design*, 2nd Edition, iUniverse, Lincoln, NE.
2. J. A. Tompkins, J. A. White, Y. A. Bozer, and E. H. Frazeller, (2003), *Facilities Planning*, 3rd Edition, Wiley, New York.
3. S. S. Heragu, "10 Principles of Materials Handling," Educational CD available upon request.
4. J. D. C. Little (1961), A proof for the queuing formula $L = IW$, *Operations Research*, Vol. 9, pp. 383–385.
5. J. M. Apple, (1977), *Plant Layout and Material Handling*, 3rd Edition, Wiley, New York.
6. R. A. Kulwiec, (1980), *Advanced Material Handling*, The Material Handling Institute, Charlotte, NC.
7. DeLaura, D. and D. Kons (1990), *Advances in Industrial Ergonomics and Safety II*, Taylor & Francis, Boca Raton, FL.

8

Warehousing

8.1	Introduction	**8**-1
	Role of the Warehouse in the Supply Chain • Product and Order Descriptions	
8.2	Functional Departments and Flows	**8**-3
	Receiving and Stowing • Piece Pick Operations • Carton Pick Operations • Pallet Pick Operations • Sorting, Packing, Staging, Shipping Operations • Support Operations, Rewarehousing, Returns Processing	
8.3	Storage Department Descriptions and Operations	**8**-9
	Bin Shelving • Carton Flow Rack • Pallet Rack • Pallet Floor Storage	
8.4	Sorting, Packing, Consolidation, and Staging Descriptions	**8**-13
8.5	Warehouse Management	**8**-15
8.6	Facility Layout and Flows	**8**-16
	Translation of Abstract Flow Diagram to Layout • General Building Description • Flows and Circulation	
8.7	Performance and Cost Analyses	**8**-19
	Turnover Analysis • Productivity Analysis • Cost Analysis	
8.8	Summary	**8**-22
	Acknowledgments	**8**-22
	References	**8**-22

Gunter P. Sharp
Georgia Institute of Technology

8.1 Introduction

This chapter presents a description of a small, fictitious warehouse that distributes office supplies and some office furniture to small retail stores and individual mail-order customers. The facility was purchased from another company, and it is larger than required for the immediate operation. The operation, currently housed in an older facility, will move in a few months. The owners foresee substantial growth in their high-quality product lines, so the extra space will accommodate the growth for the next few years. The description of the warehouse is of the planned operation after moving into the facility.

The purpose of this chapter is to introduce the reader to the operations of warehouses. Basic functions are described, typical equipment types are illustrated, and operations within departments are presented in some detail so that the reader can understand the relationships among products, orders, order lines, storage space, and labor requirements. Storage assignment and retrieval strategies are briefly discussed. Evaluation of the planned operation includes turnover, performance, and cost analyses. Additional information can be found in other chapters of this volume and in the reference material.

8.1.1   Role of the Warehouse in the Supply Chain

Warehouses can serve different roles within the larger organization. For example, a stock room serving a manufacturing facility must provide a fast response time. The major activities would be piece (item) picking, carton picking, and preparation of assembly kits (kitting). A mail-order retailer usually must provide a great variety of products in small quantities at low cost to many customers. A factory warehouse usually handles a limited number of products in large quantities. A large discount chain warehouse typically "pushes" some products out to its retailers based on marketing campaigns, with other products being "pulled" by the store managers. Shipments are often full and half truckloads. The warehouse described here is a small chain warehouse that carries a limited product line for distribution to its retailers and independent customers.

The purpose of the warehouse is to provide the utility of time and place to its customers, both retail and individual. Manufacturers of office supplies and furniture are usually not willing to supply products in the quantities requested by small retailers and individual customers. Production schedules often result in long runs and large lot sizes. Thus, manufacturers usually are not able to meet the delivery dates of small retailers and individuals. The warehouse bridges the gap and enables both parties, manufacturer and customer, to operate within their own spheres.

8.1.2   Product and Order Descriptions

8.1.2.1   Product Descriptions

The products handled include paper products, pens, staplers, small storage units, other desktop products, low-priced media like CD and DVD blanks, book and electronic titles, and office furniture. High-value electronic products are delivered directly from other distributors and not handled by the warehouse. One would say that the warehouse handles relatively low-value products from the viewpoint of manufacturing cost.

Products are sold by the warehouse as pieces, cartons, and on pallets. Figure 8.1 shows the relationships among these load types. Individuals usually request pieces; retailers may also request pieces of slow movers, products that are not in high demand. Retailers usually request fast movers, products that are in high demand, in carton quantities. Bulky products like large desktop storage units may be in high enough demand that they are sold by the warehouse in pallets. Furniture units are also sold on pallets for ease of movement in the warehouse and in the delivery trucks. Table 8.1 shows the number of products to be stored and the number of storage locations needed. The latter issue is discussed in Section 8.3.

The typical dimensions of a piece is $10 \times 25 \times 3.5$ cm, with a typical volume of 0.875 liters. A carton has typical dimensions of $33 \times 43 \times 30$ cm, with a typical volume of 42.6 liters. Thus, a typical carton contains 48.7 pieces. The typical dimension of a pallet is $80 \times 120 \times 140$ cm, with the last dimension being the height. The pallet base is about 10 cm high, so the typical product volume is 1.25 m^3, corresponding

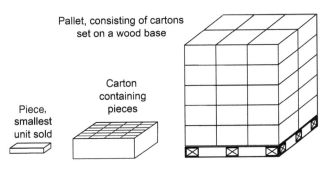

FIGURE 8.1   Load types.

TABLE 8.1 Product Storage Requirements Summary

	Piece Pick, Slow Movers	Piece Pick, Fast Movers	Carton Pick	Pallet Pick	Total
Number of products	1000	500	500	140	2140
Number of pick locations	1000	500	540	208	2248
Number of total locations	1050	550	1620	1560	4780

TABLE 8.2 Product Dimensions and Conversion Factors

Unit	Width, cm	Length, cm	Height, cm	Volume, Liters	Units in Next-Larger Unit
Piece	10	25	3.5	0.875	48.7
Carton	33	43	30	42.6	29.3
Pallet	80	120	130	1250	—

Note: About 10 cm needs to be added to pallet height for the base.

TABLE 8.3 Order Characteristics

	Order Size	From Piece Pick, Slow Movers	From Piece Pick, Fast Movers	From Carton Pick	From Pallet Pick	Total
Lines/order	Small	2	6	2	0.1	10.1
	Large	3	9	30	1	43
Quantity/line	Small	2	2	1.5	1	6.5
	Large	6	6	2.5	1	15.5

to 29.3 cartons. The pallet base allows for pickup by forklift truck from any of the four sides. Table 8.2 summarizes these values. Different products, of course, have different dimensions and relationships. The conversion factors can vary depending on whether the product is sold mainly in piece, carton, or pallet quantities. We will not introduce further complexity here and use the values given here for determining storage and labor requirements.

8.1.2.2 Order Descriptions

There are two types of orders processed at the warehouse. Large orders are placed by the retailers who belong to the same corporation; these are delivered by less-than-truckload (LTL) carrier. Small orders are placed by individuals, and these are delivered by package courier service like United States Postal Service (USPS), United Parcel Service (UPS), and Federal Express (FedEx). Large orders contain more products and the quantity per product is greater than for small orders, as shown in Table 8.3.

8.2 Functional Departments and Flows

An overall view of the functions that represent the distribution center is shown in Figure 8.2, the function flow map of the operations in the facility. This diagram shows the logical flow of products all the way from receiving through storage and retrieval to shipping. Solid arrows represent main flows, and dashed arrows show minor and occasional flows. We maintain a distinction between functional departments and physical areas. A functional department, although it may be affected by a physical area boundary, is not restricted by the ordinary physical boundaries that might appear on a layout plan.

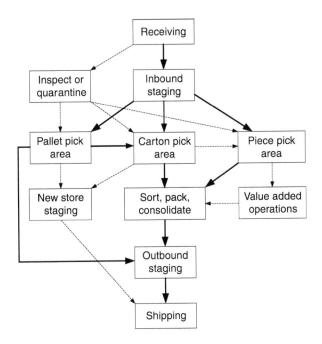

FIGURE 8.2   Function flow map of operations.

8.2.1   Receiving and Stowing

Products enter the facility at the receiving and/or shipping dock after being unloaded from trucks with the use of forklift trucks. Figure 8.3 shows a typical vehicle that can be used to load or unload trucks and store pallets in storage racks up to about 4 m high (load support height). Products are inspected using vehicle-mounted and handheld barcode scanners that contain integrated radio-frequency (RF) communication devices (see Fig. 8.4). If the product does not match an incoming purchase order, or if inspection and/or quarantine are needed, the product is moved to the inspect or quarantine area. This happens infrequently, and most products are either moved to the inbound staging area or staged in the dock area.

From inbound staging, products are moved to storage locations and stowed. Pallets are moved by forklift truck to pallet reserve storage areas. Exceptions may occur if a corresponding product pick location in either the carton pick area or the piece pick area is empty. In that situation, the pick area is replenished first, using one or more cartons from the incoming pallet, and the remainder is sent to a pallet storage area. Products that are received in carton quantities are moved by either pallet jack or cart to a piece pick area. Table 8.4 shows the daily quantities of receipts, number of trips, and labor hours needed.

Products in the pallet reserve storage area (see Fig. 8.5) are assigned locations using a shared storage concept, with the more active products located closer to the receiving or shipping dock. The storage area is divided into three areas, (A, B, and C), corresponding to (fast, medium, slow). An incoming lot of pallets of identical product is classified as (A, B, or C) on the basis of adjusted turnover of the lot:

$$\text{Adjusted turnover} = \text{pallet sales per period/number of pallets in lot} \qquad (8.1)$$

The incoming lot is assigned to the first available space in its area (A, B, or C). This method of storage assignment is called class-based storage. It combines the advantages of shared and dedicated

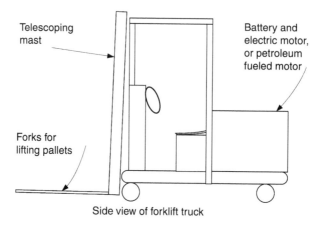

FIGURE 8.3 Typical forklift truck.

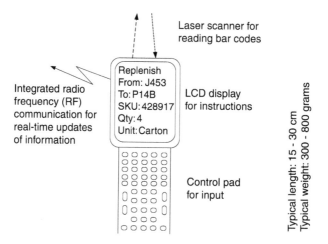

FIGURE 8.4 Handheld barcode scanner with integrated RF device.

TABLE 8.4 Receiving Operations, Daily Summary

Storage Area	Receive Units	Equivalent Pick Units	cu.m.	Method	Capacity per Trip	Number of Trips	Time per Trip	Labor Hours
Piece pick	40.4 cartons	1968 pieces	1.72	Cart, batch if possible	5 cartons	8.08	10 min.	1.35
Carton pick	50.4 pallets	1476 cartons	62.9	Forklift and pallet jack	1 pallet	50.4	4 min.	3.36
Pallet pick	22 pallets	22 pallets	27.5	Forklift	1 pallet	22	3 min.	1.10
							Total	5.81

Note: Approximately the same labor is needed for loading outbound LTL carriers.

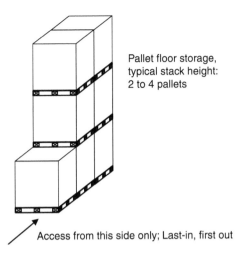

Pallet floor storage,
typical stack height:
2 to 4 pallets

Access from this side only; Last-in, first out

FIGURE 8.5   Pallet reserve storage area, floor stacking.

storage: lower overall space needs due to sharing and faster cycle times because A products are in better locations (Tompkins et al. 1996).

The pallet rack area (see Fig. 8.6) is a hybrid area, with picking (retrieval) by carton from the lower level, and the upper levels used for full pallet reserve storage. The reserve pallets for a product are stored, to the extent possible, in the same aisle and near the lower-level location where the product is picked. The lower-level positions are dedicated: each product is assigned a fixed location, with a few fast movers being assigned two locations. The upper-level positions are shared. The classification of products is based on the number of access trips per period.

The piece pick area is divided into fast and slow movers: carton flow rack for fast movers and bin shelving for slow movers. The products are given dedicated assignments, using one of the two indices. More details on these methods of storage assignment are in Goetschalckx and Ratliff (1990), Sharp (2001), and Bartholdi and Hackman (2006).

$$Cube\,per\,order\,index = \frac{Access\,trips\,per\,period}{Maximum\,storage\,space\,needed} \tag{8.2}$$

$$Viscosity\,index = \frac{Retrieval\,visits\,per\,period}{(Cubic\,volume\,of\,product\,retrieved\,per\,period)^{0.5}} \tag{8.3}$$

8.2.2   Piece Pick Operations

In the carton flow (see Fig. 8.7) and bin shelving (see Fig. 8.8) areas, order pickers move along the product locations and select items in response to customer orders. The carton flow area is for relatively fast-moving and bulkier products, according to one of the methods given above, and the bin shelving area is for slower-moving and smaller products. Most of the products in the carton flow area are not stored anywhere else in the warehouse. The products are received as cartons and brought to the replenishment (back) sides of the flow racks, inserted, and selectively picked from the front end. Some fast moving products may be picked as pieces in this area and as cartons in the pallet rack area. For these products, there is a replenishment movement from the pallet rack area to the carton flow rack, instead of from the receiving dock. Nearly all of the products in the bin shelving area are received as cartons and moved directly from receiving to the storage area.

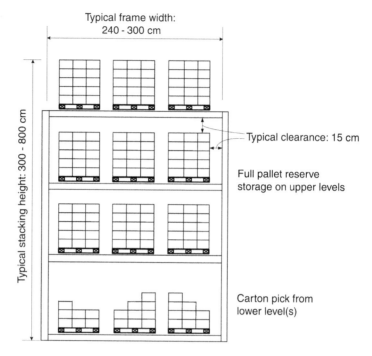

FIGURE 8.6 Pallet rack storage medium.

The main purpose of the piece pick operation is to enable the transformation of carton quantities of product into piece quantities. Some pieces are picked by order using a cart, and these move to packing and consolidation. Others are picked using batch picking, where requests from several small orders are combined into one pick list to minimize total travel during the picking process. If the items are not kept separate on the cart, they must first undergo sorting before going to packing and then consolidation.

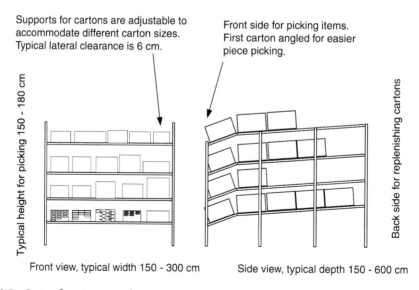

FIGURE 8.7 Carton flow storage medium.

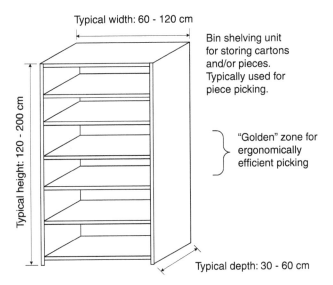

FIGURE 8.8   Bin shelving storage medium.

8.2.3   Carton Pick Operations

The lower level of the pallet rack area is used for selective retrieval (picking) of cartons in response to customer orders. The purpose of the carton pick operation is to enable the transformation of pallet quantities into carton quantities. Some cartons are picked by order using a pallet jack, and these move directly to consolidation and outbound staging. Other cartons are picked using batch picking and these must first go to sorting before going to consolidation. If the total volume of activity is small, sorting and consolidation can be combined.

When the pick location at the lower level of the pallet rack becomes empty, a replenishment operation moves a full pallet from an upper level to the lower one and removes the empty pallet base. These operations are anticipated based on the orders to be filled during the next time window.

8.2.4   Pallet Pick Operations

Full pallet picking is done primarily in the floor storage area and occasionally in the pallet rack area. These pallets move directly to outbound staging. A forklift truck has the capacity to transport one pallet at a time. Travel within the pallet floor storage area follows the rectilinear distance metric (Francis et al. 1992).

8.2.5   Sorting, Packing, Staging, Shipping Operations

Pieces and cartons that are picked using batch picking must first be sorted by order before further processing. The method of batch picking, described in the following, is designed to facilitate this process without requiring extensive conveyor equipment. In addition, all pieces must be packed into overpack cartons, and these are then consolidated with regular (single product) cartons by order. Some cartons and overpacks move to outbound staging for package courier services like USPS, UPS, and FedEx. Others move to outbound staging for LTL carrier service. The package courier services load their vehicles manually, and the LTL carriers are loaded by warehouse personnel using either forklift trucks or pallet jacks.

8.2.6 Support Operations, Rewarehousing, Returns Processing

At irregular times, the warehouse staff must perform additional functions that are not part of the normal process. Whenever a new store is being prepared for opening, a large quantity of product, for the full product line, must be picked and staged. There is a separate area set aside for this staging.

Occasionally, some products need to be repackaged and/or labeled for retail stores. This value-added processing is performed between picking and packing. Returned merchandise must be inspected, possibly repackaged, and then returned to storage locations. The volume is not significant, and it is handled in the value-added area. Periodically, product locations must be changed to reflect changing demand. This rewarehousing is performed during slack periods so as not to require additional labor.

In addition, the warehouse contains an office for management and sales personnel, toilets for both staff and truck drivers, and a break room with space for vending machines and dining. There is a battery-charging room for the electric batteries used by forklifts and pallet jacks, and a small maintenance room.

8.3 Storage Department Descriptions and Operations

This section presents details on the individual storage departments and their operations. Here we determine the storage space requirements, and we describe the pick methods and obtain labor requirements.

8.3.1 Bin Shelving

The bin shelving area contains 1,000 slow-moving products that are picked as pieces. They are housed in shelving units that are 40 cm deep, 180 cm high, and 100 cm wide, for a cubic volume of 0.72 m³. Using a cubic space utilization factor of 0.6 to allow for clearances and mismatches of carton dimensions with the shelves, each shelving unit can accommodate on average $0.72 \times 0.6/0.0426 = 10.14$ cartons. If each product requires at most one carton, then we need $1000/10.14 = 98.6$ or 99 shelving units. Rounding this to 100 units implies a pick line of $100/2 = 50$ m. One way to implement this is to establish two pick aisles, each 25 m long, as shown in Figure 8.9. In the final layout, the system is expanded to a length of 30 m. In addition, space is provided for two future aisles. Although all the products stored here are considered slow movers, with some exceptions for products with small total required inventory measured in cubic volume, the principle of activity-based storage is extended further to identify the faster-moving products (among the slow movers). These are placed in the ergonomically desirable golden zone (see Fig. 8.8).

The small number of requests per order for slow-moving products (see Table 8.3) makes it appropriate to use a sort-while-pick (SWP) method for retrieval. An order picker uses a cart with multiple compartments (see Fig. 8.10) to pick items for several orders on one trip past the shelves. The compartments prevent items for different orders from being mixed. Later, when the cart is moved to sorting, consolidation, and packing, there is actually little sorting work to do, but mainly consolidation and packing.

8.3.1.1 Time Windows

The warehouse operates one shift per day, with two time windows: an A.M. (morning) and a P.M. (afternoon) window. This reflects a balance between having a short response time at the warehouse and some

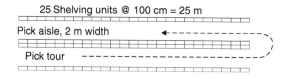

FIGURE 8.9 Bin shelving area layout.

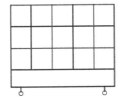 Sort-while-pick (SWP)
cart for piece picking items for
more than one order at a time.
Approximate dimensions: 150 cm
long, 60 cm wide, 120 cm high.
Cart has 15 compartments on
each side, for a total of 30.

FIGURE 8.10   Sort-while-pick (SWP) cart.

TABLE 8.5   Window Characteristics, Number of Orders

Window	Small Orders	Large Orders	Total for Window
AM, W1	20	8	28
PM, W2	22	10	32
Total for day	42	18	60

fixed truck departure times, especially the LTL carriers. Most orders that are received before 6:00 A.M. are processed during the morning window; those that cannot be processed and those that arrive during the morning are processed in the afternoon window. Table 8.5 shows how the 60 orders per day are split between A.M. and P.M., and between large and small orders.

8.3.1.2   Operations Analysis

The small orders during the A.M. window represent 20 orders, 40 order lines (lines), and 80 total pieces. These are picked on one U-shaped tour, using the SWP method, with a cart. At a rate of 30 lines per hour, this translates into 1.3 labor hours. In a similar manner, the large orders during the A.M. window are picked on one U-shaped tour. The cart is similar to that used for small orders, but it has fewer and larger compartments. In the P.M. window, the process is repeated, with the result that four pick tours per day, using SWP, are made in the slow-moving, bin-shelving area. These results are summarized in Table 8.6. The employees who work in this area move to the sorting, consolidation, and packing area and continue with the same orders, to the extent possible. This allows for easier tracking of quality problems, such as errors in selecting the wrong item, the wrong quantity, or errors in consolidation.

8.3.2   Carton Flow Rack

The carton flow area contains 500 fast-moving products, housed in carton flow rack frames that are 250 cm deep, 180 cm high, and 200 cm wide. Each frame is 4 levels high, and on average 5 lanes wide, thus containing 20 lanes. The staggering of the levels means that each lane is less than 250 cm deep, but closer to 220 cm, and thus able to accommodate 5 cartons. For the 500 products, 540 lanes are needed since some products need more than one lane. Thus, 540/20 = 27 frames are needed. This is rounded up to 30 frames that are arranged in a single aisle 30 m long, as shown in Figure 8.11. Any future expansion

TABLE 8.6   Piece Pick Operations, Slow Movers, Bin Shelving, Daily Summary

Order Size	Window	Number of Orders	Lines per Order	Total Lines	Lines per Hour	Labor Hours	Pick Method	Qty. per Line	Total Pieces	Number of Trips
Small	W1	20	2	40	30	1.3	SWP	2	80	1
	W2	22	2	44	30	1.5	SWP	2	88	1
Large	W1	8	3	24	30	0.8	SWP	6	144	1
	W2	10	3	30	30	1.0	SWP	6	180	1
Total		60		138		4.6			492	4

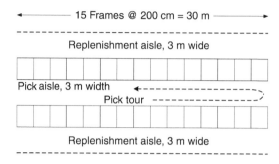

FIGURE 8.11 Carton flow area layout.

would be in the pallet floor storage area, where another 30-m-long aisle could be placed. The adjusted turnover principle of the golden zone is also applied here.

The retrieval process for small orders is similar to that in the bin-shelving area: SWP using a cart with multiple compartments. For the large orders, there is enough volume and the length of the pick line (30 m) is short enough that a single-order-pick (SOP) method with a cart can be used. The results are shown in Table 8.7. The employees who work in this area also move to the sorting, consolidation, and packing area and continue with the same orders, to the extent possible.

In many warehouses, there is the design question of which products to assign to an area naturally suited for piece picking and how much space to allocate to each product. If the product is also stored in a carton pick area, there is always the possibility of retrieving pieces from that area, with some loss of efficiency. If the replenishment of the piece pick area is from carton picking, then this is an example of the forward-reserve problem (Bartholdi and Hackman, 2006). The essence of this problem is how to maximize the gains from improved picker efficiency in the forward area, like bin shelving or carton flow rack, with the number of replenishment trips from the reserve area, like carton pick. Three questions can be posed for such a problem: (*i*) Which products should be assigned to the forward area? (*ii*) How much space should be assigned to each product in the forward area? and (*iii*) How large should the forward area be?

8.3.3 Pallet Rack

The pallet rack area physically has the appearance of one storage area. In fact, it consists of two functional areas, a carton pick area at the first level and a pallet reserve storage area directly above. This is a common arrangement. The requirement is for 540 pick locations for 500 products; some products move faster and require two pick locations to avoid replenishment delays. The second, third, and fourth levels of the pallet rack provide $3 \times 540 = 1,620$ pallet reserve positions. This number exceeds the 1,040 required; this is a consequence of the hybrid configuration where one floor-level position means three positions in the upper levels.

TABLE 8.7 Piece Pick Operations, Fast Movers, Carton Flow Rack, Daily Summary

Order Size	Window	Number of Orders	Lines per Order	Total Lines	Lines per Hour	Labor Hours	Pick Method	Qty. per Line	Total Pieces	Number of Trips
Small	W1	20	6	120	30	4.0	SWP	2	240	2
	W2	22	6	132	30	4.4	SWP	2	264	2
Large	W1	8	9	72	30	2.4	SOP	6	432	8
	W2	10	9	90	30	3.0	SOP	6	540	10
Total		60		414		13.8			1476	22

5 aisles on left are for C items.
Portion of typical pick tour shown in these aisles.

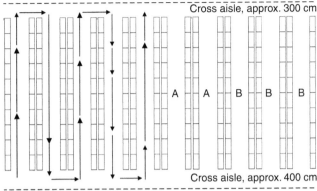

10 aisles, length 31.5 m, 20 frames per aisle.
Pick aisle is approx. 380 cm. System is 4 levels high.

FIGURE 8.12   Pallet rack area layout.

The structure of a pallet rack frame is shown in Figure 8.6. The frame width accommodates three pallets on each level, and the frames are connected back-to-back for stability. Because of the relatively low activity, the products are classified by adjusted turnover, and the assignment of classes is shown in Figure 8.12, which shows a typical layout of the pallet rack area. Each aisle contains 20 frames and contains $20 \times 3 = 60$ pick locations. Within an activity class, the assignment is by product number. The number of frames needed is $540/3 = 180$, corresponding to 9 aisles. The actual layout has 10 aisles. Specifying such a large area inevitably means that some adjustment and fitting must be done so that the aisles do not contain building columns, there is sufficient space for main circulation aisles, and so forth.

The retrieval process for small orders is similar to that in the carton flow area: SWP using a pallet jack to select items for five orders at a time, resulting in four trips during the A.M. window and 5 in the P.M. window. For large orders, the SOP method is used, resulting in 16 trips (two per order) in the A.M. and 20 (two per order) in the P.M. window, as shown in Table 8.8. The assignment of products into classes (A, B, and C) by adjusted turnover has some benefit here, but not as much as would be expected. When batch picking is used for small orders, the number of stops on a pick list increases, and this reduces the benefit from activity-based storage assignment. (It is a coincidence that the daily sum of 1,476 cartons is the same as the daily sum of 1,476 pieces in the carton flow rack.)

Replenishment activity occurs when a low-level pick location is empty or will become empty during the next time window. The warehouse management system (WMS) triggers a replenishment move from an upper level to the low level and the removal of the empty pallet base. Table 8.9 reflects this activity.

TABLE 8.8   Carton Pick Operations, Lower Level of Pallet Rack, Daily Summary

Order Size	Window	Number of Orders	Lines per Order	Total Lines	Lines per Hour	Labor Hours	Pick Method	Qty. per Line	Total Cartons	Number of Trips
Small	W1	20	2	40	15	2.7	SWP	1.5	60	4
	W2	22	2	44	15	2.9	SWP	1.5	66	5
Large	W1	8	30	240	15	16.0	SOP	3	600	16
	W2	10	30	300	15	20.0	SOP	2.5	750	20
Total		60		624		41.6			1476	45

TABLE 8.9   Pallet Handling, Internal, Daily Summary

Operation Type	Equivalent Pick Units	Units Handled	Pallets per Hour	Labor Hours
Pick customer orders	22 pallets	22	15	1.5
Replenish carton pick slots	1476 cartons	50.4	20	2.5
Total		72.4		4.0

8.3.4   Pallet Floor Storage

The pallet floor storage area is for products that move in pallet quantities and that can be stacked in pallets; these products do not need to be stored in pallet racks, although that is always an option. There is a requirement for storing a maximum of 140 products and 1,560 total pallets, with an average stacking height of 2.5. Using lanes that are 3 pallets deep (see Fig. 8.13), each lane holds 7.5 pallets. Thus, 208 lanes are needed. The actual area assigned has considerably more space, to allow for future activity increase. As is the situation for a pallet rack, the large area means that adjustment and fitting must be made to avoid structural columns, allow for main circulation aisles, and so forth. The storage assignment in this area is similar to that in the carton pick area, that is, class based by adjusted turnover. Within a class, storage assignment is based on the first available location of an empty lane: when a new lot is received, it is stored in the first available location in its activity class. The retrieval activity in this area is straightforward: the lane containing the oldest product is identified, and the first accessible pallet is removed and taken to the outbound staging area. The activity in this area is included in Table 8.9.

8.4   Sorting, Packing, Consolidation, and Staging Descriptions

The description given is for the A.M. time window; the P.M. window is similar but has slightly higher volume. To gain efficiency in the retrieval process, the SWP method is used extensively. This means that the items placed into the SWP carts must then be sorted, consolidated, and packed by order. The sorting of pieces is really more like consolidation: the items in the carts are not mixed since each compartment holds items for only one order or part of an order. There are only 4 carts and 4 pallets that undergo this process (see Table 8.10):

> 1 cart from bin shelving for small orders
> 2 carts from carton flow for small orders
> 4 pallets from carton pick for small orders
> 1 cart from bin shelving for large orders

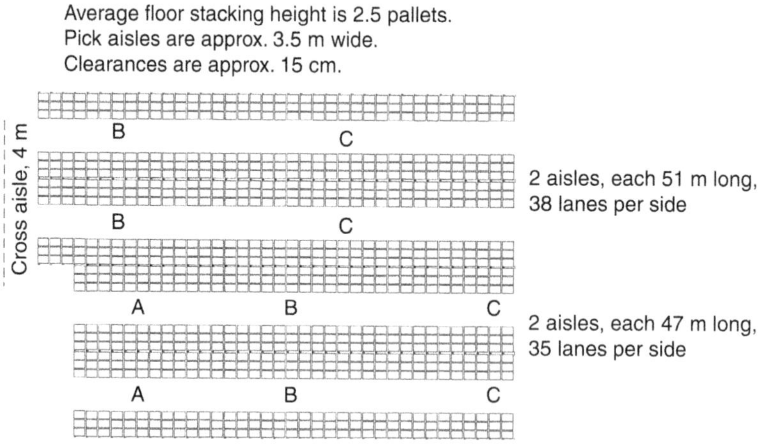

FIGURE 8.13   Pallet floor stacking area layout.

TABLE 8.10 Sort, Pack, Consolidate, Stage, Daily Summary

Incoming Vehicles	Order Size	Window	Pick Area, Method	Optn. Type	Number of Orders	Lines per Order	Total Lines	Qty. per Line	Total Qty.	Rate	Apply to	Labor Hours
1 cart	Small	W1	Bin shelving, SWP	Sort, pack	20	2	40	2	80	30	Lines	1.3
2 carts	Small	W1	Carton flow, SWP	Sort, pack	20	6	120	2	240	30	Lines	4.0
4 pallets	Small	W1	Carton pick, SWP	Sort, cons.	20	2	40	1.5	60	40	Cartons	1.5
1 cart	Large	W1	Bin shelving, SWP	Sort, pack	8	3	24	6	144	30	Lines	0.8
8 carts	Large	W1	Carton flow, SOP	Cons. pack	8	9	72	6	432	30	Lines	2.4
16 pallets	Large	W1	Carton pick, SOP	Cons.	8	30	240	2.5	600	40	Cartons	15.0
AM Window subtotal					28		536		1556			25.0
1 cart	Small	W2	Bin shelving, SWP	Sort, pack	22	2	44	2	88	30	Lines	1.5
2 carts	Small	W2	Carton flow, SWP	Sort, pack	22	6	132	2	264	30	Lines	4.4
5 pallets	Small	W2	Carton pick, SWP	Sort, cons.	22	2	44	1.5	66	40	Cartons	1.7
1 cart	Large	W2	Bin shelving, SWP	Sort, pack	10	3	30	6	180	30	Lines	1.0
10 carts	Large	W2	Carton flow, SOP	Cons. pack	10	9	90	6	540	30	Lines	3.0
20 pallets	Large	W2	Carton pick, SOP	Cons.	10	30	300	2.5	750	40	Cartons	18.8
PM Window subtotal					32		640		1888			30.3

The three carts for small orders are staged before the pack stations, and the items for the different orders are removed and packed into overpack cartons. These overpack cartons are then consolidated with regular (single product) cartons from the four pallets. Since most small orders are shipped by package courier, the cartons (overpack and full product) for those orders then move to the staging area for package courier. In Figure 8.14 this flow is to the right.

The one cart for the large orders is staged before a pack station, and the items are removed and packed into overpack cartons. These overpack cartons then move to the left. Also to the left are:

> 8 carts for individual, large orders (SOP)
> 16 pallets for individual, large orders (SOP, 2 per order)

Again, the sorting of pieces in the carts is more like consolidation, since the items for different orders are not mixed in the same vehicle. The items in the carts are packed into overpack cartons, and then all three flows are consolidated onto pallets and staged for the LTL carriers:

> Overpack cartons for items from bin shelving, from the 1 cart
> Overpack cartons from carton flow, from the 8 carts
> Full cartons from carton pick, from the 16 pallets

It should be mentioned here that the nature of the work in this area depends on the way that items are picked. If the SOP and SWP methods are used, then the work is mainly packing and consolidation. On the other hand, if batch picking is used, where an order picker selects items for more than one order into a container or onto a conveyor, then items must be sorted, either manually or mechanically. The choice of which method(s) to use is not always obvious. In many situations, there is more than one cost-effective solution, whereas in others a detailed comparison of alternatives is needed.

8.5 Warehouse Management

The operation of the warehouse requires careful and constant management. The scanning of received products is just one example of the functions performed by the WMS. It is beyond the scope of this chapter to present details of a typical WMS. However, some main features should be mentioned here. The tracking of flows throughout the warehouse is one of the basic functions of a WMS. This can be done manually, but most facilities today use barcode scanners, and many use barcode scanners integrated with radio-frequency transmitters (RFID) to allow for real-time updates of the underlying database. A typical WMS enables the functions listed below. These requirements are not inclusive, but only indicate the types of functions desired. Further details are in Sharp (2001).

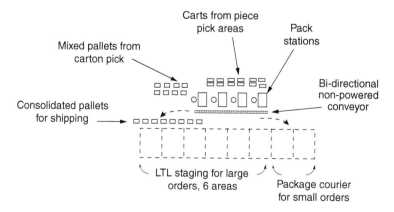

FIGURE 8.14 Sorting, packing, consolidation, and staging area layout.

The WMS should enable scheduling of personnel, including regular full-time employees and temporary and part-time employees. Tracking of employee productivity is useful for training and workload balancing. Workload scheduling should be linked to forecast information, and the conversion of product volumes should be automatically translated to labor hours by function and employee productivity.

In the receiving function, the WMS should have online verification of expected receipts; it should flag out-of-stock conditions, process partial receipts, and quarantine products requiring inspection. It should generate labels for pallets and cartons with data on SKU (unique product type), description, date received, lot or purchase order number, expiration code(s), and location code(s). It should assign storage location recognizing physical characteristics of product, physical characteristics of location, environmental restrictions, and stock rotation. It should also have the ability to send products directly to outbound vehicles (cross-docking). The ability to schedule trucks and assign them to docks is also useful.

Control of storage and inventory, one of the most important functions of a WMS, includes confirmation of stow (storage) action, updating of inventory upon stow, stock reservation capability, and provision for cycle counting. The WMS should support more than one location per SKU and more than one SKU per location. Report generation should include stock activity reports (fast, medium, slow, dead), empty location reports, and anticipated replenishment of forward pick areas.

An important function of the WMS is order processing. The WMS should support online verification of item availability, online verification of customer credit status, and inventory reservation at time of order entry. It should validate quantity restrictions, suggest the next quantity multiple, support quantity price breaks, and allow for flexibility in pricing by customer and order type. It should record priority and shipping methods, generate invoices, have flexibility for partial and split shipments, and have flexibility for shipping charges (customer pays or warehouse pays).

Order picking usually involves the largest labor component in a warehouse and offers the greatest opportunity for savings. Because of the potential complexity of order picking, this area is one of the most crucial aspects of a WMS. At a minimum, the WMS should support SOP, SWP, and batch picking, with flexibility for changing from one mode to another. Batch picking may require grouping of orders based on criteria like shipping deadline, truck route, and storage locations. Orders might be picked in waves corresponding to time windows. Consolidated pick documents need to be generated, considering route optimization, container capacities, and workload balancing among pickers. Often, labels need to be generated and packing instructions issued. Last, truck loading instructions need to be generated.

Hardware requirements and compatibility present further questions, such as processor type (PC, main frame), operating system (Windows, Unix, Linux), network compatibility, support for RF terminals, support for pick-to-light displays, support for voice prompt and voice recognition systems, and support for RF tags. Summarizing, the selection and implementation of a WMS is a major decision that requires time, money, and expert advice.

8.6   Facility Layout and Flows

8.6.1   Translation of Abstract Flow Diagram to Layout

Warehouse layout planning differs from traditional factory layout planning in several respects. First, one or two large storage departments usually account for more than half the total space. Second, the locations of the receiving and shipping docks are often dictated by the surrounding roads and site topography. These first two factors mean that often there are only a few ways the layout can be arranged. Third, except for pallets, the actual cost of moving product from one department to another is relatively small compared to the cost of processing within departments. This means that it is not so important where these departments, especially those for piece picking, are located. Fourth, unlike some manufacturing equipment, many storage media can be configured in a variety of ways without greatly affecting

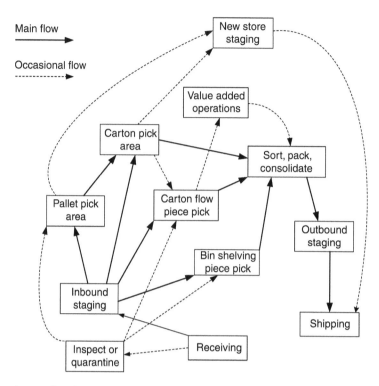

FIGURE 8.15 Abstract flow diagram for facility layout.

the equipment cost or operating efficiency. These last two factors give the designer more flexibility in determining the final layout without having to be too concerned about efficiency of flow between departments.

Using the descriptions of product flows given earlier, an abstract flow diagram is constructed, as shown in Figure 8.15. Solid lines indicate regular, daily flow, while dashed lines show occasional flow. It is possible to construct this abstract diagram so that no product flows cross. This suggests that a layout can be constructed with the same characteristic, that is, with no product flows crossing. The resulting area layout is shown in Figure 8.16. By using major circulation aisles, it is possible to keep product flows from crossing. In addition to department boundaries, some outlines of equipment units are shown, as well as major aisles. More detail is shown in the earlier illustrations for the individual departments. Inevitably, some departments were enlarged so that boundaries would follow column lines or major aisles.

8.6.2 General Building Description

The overall building is a rectangle of dimensions 80×100 m, with a column grid on 20×20 m spacing. The receiving and shipping docks are combined, with a total of 8 dock doors. Receiving is on the left side, LTL shipping in the center and right, and package courier shipping on the extreme right. The lower left section is devoid of storage media: most of the area is for pallet floor storage, with some small sections for inbound staging, inspection and/or quarantine, and new store staging. The pallet rack area (carton pick on lower level, reserve pallet storage on upper levels) occupies most of the upper part of the layout. Along the right side is an area for office, toilets, and break room; these areas are not shown in detail. However, it is preferable for the toilets and the break room to be near the dock so visiting truck drivers can easily access them. The forklift battery charging room and maintenance area are in the upper right, mainly because this reduces the length of expensive electric conductor from the nearest utility

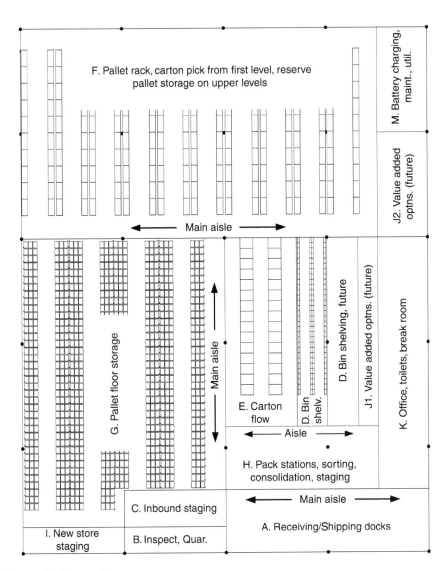

FIGURE 8.16 Facility area layout.

pole. It also keeps the room out of the way of product flows. The gap between the office and the battery charging room is designated for future value-added activities. It could also be used for expansion of the office.

The area to the left of the office is where most of the action is in this warehouse. The piece pick areas are vertically aligned so that the output from those areas flows down to the pack stations, and then to the shipping dock. The flow layout of the pack stations is described earlier and shown in Figure 8.14. There is space allocated for a doubling of the bin-shelving area. The narrow strip to the left of the office is only 4 m wide. It could possibly be used for value-added operations in the future, but at least part of it needs to be a personnel aisle. Table 8.11 presents a summary of department areas. The major circulation aisles are not separately specified but included in the large departments they serve. Table 8.12 is a summary of storage capacity by department.

TABLE 8.11 Department Area Summary

Dept. ID	Department	Dimensions, m	Area, sq m
A	Receiving/shipping	40 × 12	480
B	Inspect, quarantine	20 × 5	100
C	Inbound staging	20 × 7	140
D1	Bin shelving, piece pick, slow movers	36 × 6	216
D2	Bin shelving, future	36 × 6	216
E	Carton flow rack, fast movers	36 × 14	504
F	Pallet rack	70 × 40	2800
G	Pallet floor storage (rectangle includes B, C, I)	40 × 60	2060
H	Pack stations, sorting, consolidation, staging	30 × 12	360
I	New store staging	20 × 5	100
J1	Value-added operations, future	36 × 4	144
J2	Value-added operations, future	20 × 10	200
K	Offices, toilets, break room	48 × 10	480
M	Battery charging, maintenance, utilities	20 × 10	200
		Total	8000

Department areas include circulation space.

8.6.3 Flows and Circulation

The general flow of product within the building is clockwise, starting at the receiving dock. There are no product flow crossings except within the pack stations, where it is unavoidable. At the back of the dock is a horizontal circulation aisle (shown by a two-headed arrow). There is a vertical circulation aisle from the dock to the pallet rack area and a horizontal aisle along the lower edge of the pallet rack. Between the piece pick areas and the pack stations there is a horizontal aisle.

8.7 Performance and Cost Analyses

Evaluating warehouse operations is done from three perspectives: inventory turnover, productivity, and cost. The first perspective reveals opportunities for improving the purchasing function in the organization. Warehouses that have high turnover usually have higher productivity and lower unit costs. Productivity is usually based on labor hours required to process orders and order lines. Cost follows from capital assets and labor productivity.

These types of analyses have many potential pitfalls. For example, the inventory turnover at a warehouse may depend on purchasing decisions made at the corporate level, and this is often beyond the control of the warehouse manager. In a multi-level distribution system, the lowest level usually stocks

TABLE 8.12 Storage Capacity Summary

Department	Storage Media	Width, cm	Depth, cm	Height, cm	Number of Media Units	Unit Stored	Capacity, Units Stored	Unit Picked	Capacity, Units Picked
Bin shelving, slow piece pick	Shelving unit	100	40	180	120	Carton	1217	Piece	59,258
Carton flow, fast piece pick	Carton flow frame	200	220	180	30	Carton	3000	Piece	146,100
Carton pick, including reserve storage above	Pallet rack frame	315	140	465	200	Pallet	2400	Carton	70,320
Pallet floor storage	Floor storage lane	135	85	420	292	Pallet	2190	Pallet	2190

Note: Pallet floor storage has an average height of 2.5 pallets.

only the faster-moving products while the regional and national levels stock slower-moving items. Data for two large facilities that had 10% or more of the products not sold during a 12-month period was verified. In most situations, those products would be considered "dead" and candidates for removal. However, in both situations the mission of the warehouse was to stock spare parts for expensive industrial equipment that had a useful life of 20 years. From that perspective, the slow overall turnover was unavoidable. In another facility, a global warehouse for a large manufacturer of construction and earth-moving equipment with sales and support services around the world, 10% of the orders were rush orders. These were for products that were not stocked at local, regional, or national distributors. Clearly, the high fraction of rush orders leads to higher overall costs per order. From the perspective of global logistics, however, the overall approach seems sensible.

Several benchmarking studies have been made of warehouse operations (Schefczyk, 1993; Hackman et al. 2001; Chen, 2004; Frazelle, 2006). Some of these studies include extremely wide ranges of parameters. For example, in one study, the lines shipped per product per year ranged from 1,000 to 900,000; the number of products ranged from 250 to 225,000; and the inventory turns per year ranged from 2 to 60. This diversity poses challenges in interpreting any comparisons.

8.7.1 Turnover Analysis

We will perform the turnover analysis by estimating the average inventory for each product set to be half the design capacity. This is an approximate method to get some quick results. A more detailed method would require actual data on inventory. The operation is scheduled to move from an old facility into the one being described. Any inventory data from the old facility reflects constraints on purchasing decisions and thus is not directly usable. Similarly, capacity that exceeds the design requirements can lead to purchasing decisions that take advantage of special discounts; such action can "fill" the available capacity. Another way to estimate average inventory would be to establish the safety stock or reorder point for each product and use that information with a mathematical inventory model (Nahmias, 2005).

The bin-shelving area for slow piece picking requires 100 shelving units, each of which holds on average 10.14 cartons. The resulting 1,014 cartons correspond to a maximum inventory of $1,014 \times 48.7 = 49,382$ pieces an average 24,691. With 250 operating days per year, the daily sales of 492 pieces results in an average time in storage of $24,691/492 = 50$ days. This corresponds to 5.0 inventory turns per year. The carton flow area for fast piece picking requires 27 frames. Since each frame holds 20 lanes $\times$ 5 cartons, the maximum inventory is 2,700 cartons, and the average is 1,350 cartons, or $1,350 \times 48.7 = 65,745$ pieces. Daily sales of 1,476 pieces results in an average storage time of 44.5 days, and 5.6 turns per year. The pallet rack area for carton pick needs 1,620 locations. The average inventory in cartons is $1,620 \times 29.3 \times 0.5 = 23,733$. Daily sales of 1,476 cartons results in an average storage time of 16 days and 16 turns per year. The pallet rack area needs a maximum storage capacity of 1,560 pallets, or average of 780, so the daily sales of 22 results in an average storage time of 35 days and 7 turns per year. These results are summarized in Table 8.13.

TABLE 8.13   Turnover Analysis

Department	Units	Average Inventory	Daily Sales	Avg. Time in Storage, Days	Turns per Year
Bin shelving, slow piece pick	Piece	24,692	492	50.2	5.0
Carton flow, fast piece pick	Piece	65,745	1476	44.5	5.6
Carton pick, lower level of pallet rack	Carton	23,733	1476	16	16
Pallet floor storage	Pallet	780	22	35	7

8.7.2 Productivity Analysis

The performance analysis is done at the department level for pick operations and at the warehouse and facility levels for the entire operation. Detailed performance analysis could be done for each individual operation, such as unloading trucks, stowing products, packing orders, and so forth. Since the operation will move into the facility in a few months, only data for the planned operation are available. Thus, detailed performance is reflected in the productivity rates used in the tables. These include pallet trips at 15–20 per hour, carton stow at 30 per hour, piece line retrieval at 30 per hour, carton line retrieval at 15 per hour, line pack at 30 per hour, and carton sort and consolidate at 40 per hour.

At the department level, we obtain productivity per order, per line, and per piece, carton, or pallet. These results are shown in Table 8.14. For example, in the bin-shelving area, each labor hour corresponds to 13.0 orders, 30.0 lines, and 107 pieces. At the warehouse level, we reflect all direct labor, including that used for unloading and loading trucks (11.6 h), stowing (included in 11.6 h), replenishing (4.0 h), and sorting and packing (55.3 h). Another factor that must be considered is that in planning the operation, the labor hours represent effective hours. A warehouse employee typically works 6.5 effective hours on an 8-h shift. The rest of the time is spent on preparing to receive instructions, meetings, breaks, and idle time due to the irregular schedule of activities. Further, employees are paid for holidays. Thus, the value of 80 direct labor hours for bin shelving, carton flow, carton pick, and pallet floor stack reflects this ratio of 8 paid hours for 6.5 effective hours applied to the 64.0 h, rounded to an integer number of 10 people. Including the warehouse indirect labor increases this number to 176 h, reflecting an additional 12 people. Productivity at the warehouse (total labor) level is 0.3 orders per hour and 6.8 lines per hour.

At the facility level, we also reflect management labor of 13 people, which consists of supervisory, maintenance, and sales staff (see Table 8.15). It is not unusual for the administrative labor to be more than the direct labor for a small operation like this one. These values can then be used for benchmarking the operation with other facilities.

8.7.3 Cost Analysis

The natural extension of productivity analysis is to cost analysis. Table 8.15 shows the investment costs for building and equipment, their annual maintenance costs, and the translation into annual costs, with and without the time value of money (TMV) of 15% per year. These costs reflect only the storage requirements for the immediate future in the pallet rack, carton flow, and bin-shelving areas, based on the design requirements. Labor costs for the facility are as follows:

Order pickers	10 @ $45,000	$450,000
Other WH labor	12 @ $38,000	$456,000
Administrative	13 @ $61,000	$793,000
	Total	$1,699,000

TABLE 8.14 Performance Analysis, Daily Average

Department	Hours	Orders	Orders per Hour	Lines	Lines per Hour	Unit Type	Units	Units per Hour
Bin shelving	4.6	60	13.0	138	30.0	Piece	492	107.0
Carton flow	13.8	60	4.3	414	30.0	Piece	1476	107.0
Carton pick	41.6	60	1.4	624	15.0	Carton	1476	35.5
Pallet floor stack	4.0	22	5.5	22	5.5	Pallet	22	5.5
Warehouse, direct	80	60	0.8	1198	15.0			
Warehouse, total	176	60	0.3	1198	6.8			
Facility, total	280	60	0.2	1198	4.3			

TABLE 8.15   Cost Data

Item	Qty.	Unit Price	Initial Investment	Life-Time	Annual Cost, no TMV	Maint., Annual	Total Annual Cost, no TMV	Total Annual Cost, w. TMV 15%
Building, sq.m.	8000	350	2,800,000	40	70,000	56,000	126,000	477,574
Pallet rack	1620	60	97,200	20	4860	972	5832	16,501
Carton flow rack	30	5000	150,000	20	7500	1500	9000	25,464
Bin shelving	100	500	50,000	20	2500	500	3000	8488
Forklift truck	3	30,000	90,000	5	18,000	18,000	36,000	44,848
Pallet jack	3	2500	7500	5	1500	1125	2625	3362
Pick cart	10	1000	10,000	5	2000	1000	3000	3983
Pallet base, extra	1000	20	20,000	5	4000	3000	7000	8966
Pack stations	4	2000	8000	10	800	1200	2000	2794
Other, misc.	1	20,000	20000	5	4000	3000	7000	8966
Totals			3,252,700		115,160	86,297	201,457	600,948

In addition, there is $200,000 in annual costs for utilities and other administrative expenses. Considering the fixed investment costs, the annual costs of labor and equipment maintenance, utilities, administration, and the time value of money, the total cost per order line is $8.35, and $167 per order. These costs are on the high side compared to other facilities, but they reflect the relatively low volume of operations, with anticipated growth, and the nature of the high-quality product line. Further, they include all costs of the facility operation, whereas many benchmark figures report only direct labor in the warehouse.

8.8   Summary

Warehouse operations are much more complex than they appear at first glance. Profiling (partitioning) of products and orders leads to a potential multitude of warehouses inside the warehouse. The ingenuity of manufacturers to develop new technology, along with rapid advances in data processing (WMS) and mobile communications (RFID) present an ever-changing set of alternatives for storing products and retrieving items for customer orders (Kulwiec, 1982). This chapter is an attempt to present an introduction to warehousing using a case example with sufficient detail to illustrate the main concepts.

The variety of storage and retrieval technologies makes the equipment selection process difficult for the designer. At the same time, the variety of storage assignment and retrieval methods presents a challenge to both the facility designer and operator. In most circumstances, it is not possible within the limits of time and budget to investigate all possible alternatives. Instead, a guided selection process for functional departments and retrieval processes is recommended (McGinnis et al., 2005).

Acknowledgments

This research was supported by the National Science Foundation under grant EEC-9872701, the W. M. Keck Foundation, the Ford Motor Company, and The Logistics Institute at the Georgia Institute of Technology. Discussions with my colleagues Leon McGinnis and Marc Goetschalckx have provided valuable advice in the approach to the warehouse design problem.

References

Bartholdi, J.B. and S. Hackman, (2006), *Warehousing and Distribution Science*, text in progress, Atlanta, Georgia.
Chen, W.C., (2004), Available at: http://www2.isye.gatech.edu/ideas/

Francis, R.L., L.F. McGinnis, Jr., and J.A. White, (1992), *Facility Layout and Location: An Analytical Approach,* 2nd Ed., Prentice Hall, Englewood Cliffs, NJ.

Frazelle, E.H., (2006), Available at: http://www.logisticsvillage.com/MediaCenter/Documents/Presenta tions/2005WarehouseBenchmarkingRpt.pdf

Goetschalckx, M. and H.D. Ratliff, (1990), Shared Storage Policies Based on the Duration Stay of Unit Loads, *Management Science,* Vol. 36–9: 53–62.

Hackman, S.T., E.H. Frazelle, P.M. Griffin, S.O. Griffin, and D.A. Vlatsa, (2001), Benchmarking Warehousing and Distribution Operations: An Input-Output Approach, *Journal of Productivity Analysis,* Vol. 16: 79–100.

Kulwiec, R.A., (1982), *Material Handling Handbook,* John Wiley & Sons, New York.

McGinnis, L.F., D. Bodner, M. Goetschalckx, T. Govindaraj, and G. Sharp, (2005), Toward a Comprehensive Descriptive Model for Warehouses, in Meller, R., et al. (Eds.), *Progress in Material Handling Research: 2004,* MHI, Charlotte, pp. 265–284.

Schefczyk, M., (1993), Operational Performance of Airlines: An Extension of Traditional Measurement Paradigms, *Strategic Management Journal,* Vol.14 (4): 301.

Sharp, G.P., (2001), Warehouse Management, in Salvendy, G., (Ed.), *Handbook of Industrial Engineering,* 3rd Ed., John Wiley & Sons, New York, pp. 2083–2109.

Steven, N., (2005), *Production & Operations Analysis,* 5th Ed., McGraw-Hill/Irwin, New York.

Tompkins, J.A., J.A. White, Y.A. Bozer, E.H. Frazelle, J.M.A. Tanchoco, and J. Trevino, (1996), *Facilities Planning,* John Wiley & Sons, New York.

9

Distribution System Design

9.1	Introduction	**9**-1
9.2	Engineering Design Principles for Distribution System Design	**9**-3
	Heterogeneous Data for Distribution System Design • Engineering Design Principles	
9.3	Data Analysis and Synthesis	**9**-5
	Logistics Data Components	
9.4	Distribution System Design Models	**9**-8
	K-Median Model • Location-Allocation Model • Warehouse Location Problem • Geoffrion and Graves Distribution System Design Model • Geoffrion and Graves Formulation	
9.5	Sensitivity and Risk Analysis	**9**-15
9.6	Distribution Design Case	**9**-16
9.7	Conclusions	**9**-18
	References	**9**-18

Marc Goetschalckx
Georgia Institute of Technology

Distribution system design is the strategic design of the logistics infrastructure and logistics strategies to deliver products from one or more sources to the customers. Because of the long-term impact of the distribution system, the interrelated design decisions, and the different objectives of the various stakeholders, designing a distribution system is a highly complex and data-intensive engineering design effort. A large variety of mathematical programming models has been developed to provide decision support to the design engineer. The results of the models and tools have to be very carefully validated. The uncertainty of the forecasted data has to be explicitly incorporated through sensitivity and risk analysis. The final configuration is often based on the balance between many different factors, and many alternative configurations may exist. However, modeling-based design is the only available method to generate high-quality distribution system configurations with quantifiable performance measures.

9.1 Introduction

In today's rapidly changing world, corporations face the continuing challenge to constantly evaluate and configure their production and distribution systems and strategies to provide the desired customer service at the lowest possible cost. Distribution system design focuses on the strategic design of the logistics infrastructure and logistics strategies to deliver the products from one or more sources to its customers at the required customer service level. Typically, it is assumed that the products, the sources of the products (manufacturing plants, vendors, and import ports), the destinations of the products (customers), and the required service levels are not part of the design decisions but constitute constraints or parameters for the system. Distribution system design focuses on the following five interrelated decisions:

1. Determining the appropriate number of distribution centers
2. Determining the location of each distribution center
3. Determining the customer allocation to each distribution center
4. Determining the product allocation to each distribution center
5. Determining the throughput and storage capacity of each distribution center

A schematic illustration of the questions in distribution system design is shown in Figure 9.1. Decisions on delivery by direct shipping and transportation mode selection are part of the overall distribution system design.

The objective of the distribution system design is to minimize the time-discounted total system cost over the planning horizon subject to service-level requirements. The total system cost includes facility costs, inventory costs, and transportation costs. It should be noted that the detailed inventory and transportation planning decisions are made at the tactical or even operational level, but that aggregate values for the corresponding costs and capacity parameters are used in the strategic design. The facility costs include labor, facility leasing or ownership, material handling and storage equipment, and taxes.

It is clear from the description that designing a distribution system involves making numerous trade-offs. Let us assume that transportation from the manufacturing facilities to the distribution center occurs in relatively larger quantities at a relatively lower cost and that delivery from the distribution center to the customer occurs in smaller quantities at a higher cost rate. Increasing the number of distribution centers typically has the following consequences:

- Customer service levels improve because the average transportation time to the customers is smaller.
- Outbound transportation costs decrease because the local delivery area for each distribution center is smaller.
- Inbound transportation costs increase because the economies of scale of the transportation to the distribution centers are reduced.

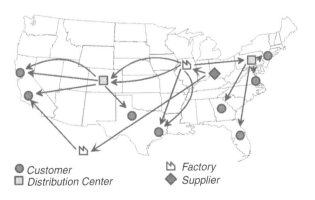

● *Customer*   ◪ *Factory*
◻ *Distribution Center*   ◆ *Supplier*

FIGURE 9.1 Distribution system schematic.

- Inventory costs increase because there are more inventory stocking locations and there is less opportunity for risk pooling so that the required safety stocks increase.
- Facility costs increase because of the overhead associated with each facility and increased handling costs because the economies of scale of handling inside the distribution centers are reduced.

9.2 Engineering Design Principles for Distribution System Design

9.2.1 Heterogeneous Data for Distribution System Design

Engineering design of any system is based on data and models for the particular system area, and the design of distribution systems is no different. However, because of the large number and variety of participants in the system, the long planning horizon, and the large variety of possible distribution systems and strategies, the data for distribution system design are highly diverse and highly uncertain. This is in contrast with the more focused data and models in other engineering disciplines, such as for the design of a bridge in civil engineering, a pump in mechanical engineering, or an integrated circuit in electrical engineering.

To make the proper trade-offs, a large amount of data from a variety of sources is required. This includes:

1. Data on the customer demand for products for all the time periods in the planning horizon.
2. Product characteristics such as monetary value and physical dimensions.
3. Geographical location data for all the product sources such as manufacturing facilities and import ports, for the distribution center candidate locations, and for the customers.
4. Transportation cost rates by transportation mode and by origin and destination point.
5. Fixed facility operating costs associated with the distribution centers. Different costs can be caused by different land and construction costs in function of location or by different equipment costs in function of technology and size of the center.
6. Variable facility operating costs associated with labor and material handling costs inside the distribution centers.
7. Order processing and information technology costs associated with each distribution center.
8. Capacity constraints on the throughput and storage of various possible sizes of the distribution centers.
9. Required service levels by customer and product combination. This may include maximum delivery time to the customer from the distribution center or minimum acceptable fill rate in the distribution center.

These data have to be extracted from a variety of sources. A basic list of data sources is given next in order of decreasing data specificity and accuracy. The most relevant and accurate data are based on the in-house databases of historical transactions. Prime examples of such databases are customer sales orders, customer data, facility data, and freight bills. While relevance, in-house availability, and accuracy are the main advantages of this type of data, data volume, historical time frame, and availability for the current system only are the main disadvantages. The detailed information in these databases can be overwhelming and it has to be aggregated in order for it to be used in a strategic design model. Fundamentally, the data provide highly detailed information on what the corporation did in the past. This type of data is most suited for the restructuring of an existing distribution system. A second source of data is contained in corporate documents such as the annual report and the corporate strategic plan. These documents contain aggregate data such as the cost of capital to the corporation or the corporate service level goals.

The previous two types of data sources are specific to the logistics organization or corporation itself. The next sources of data are reports and databases on the general business area. They include logistics performance ratios for "best-of-class" corporations, databases of the aggregate industry, and detailed forecast reports for the industry. Corporations typically have access to such data through membership in trade organizations. Their membership may oblige them in turn to report their activities to the association. Another source of this type of data is provided by specialized consulting or trade organizations that produce either reports for sale or publish annual rankings and reviews [see, e.g., Trunick (2006)]. This type of data allows the corporation to compare or benchmark their own logistics operations and costs against their competitors and provides them with aggregate data on opportunities for business expansion. These data are very useful in a distribution design project for a new product, new customer group, or new geographical area. However, these data may be very expensive to acquire.

The last class of data sources is provided by governmental organizations and provides data on the overall status and characteristics of the economy and the population. In the United States, the Department of Commerce and the Census collect large amounts of data and provide statistical summaries free of charge or for a modest price.

Validation and reconciliation of data may expose significant incompatibilities and inconsistencies between various stakeholder organizations in the corporation. The process of assembling a single data set on which the design will be based and on which all stakeholder groups agree is time-consuming and expensive. It is not unusual that 60% to 80% of the design project cost and duration is spent on collecting, validating, and aggregating the data. A single point of authority and funding is required to bring the data collection phase to a successful completion.

It is crucial for the success of the designed distribution system to realize during the design that the distribution system will be constructed and implemented in the near future, while it is intended to operate and serve for an extended period into the far future. Virtually all the data used in the design project are based on forecasts of economic, commercial, industrial, and population parameters. The error ratios of these forecasts may easily be thousands of percentages. Corporations and design engineers are under enormous pressure to design a system that will operate with minimum cost. However, the resulting designs often are lacking in flexibility and robustness. Selecting the trade-off between efficiency and robustness is typically done by the senior management of the corporation. However, providing the decision-makers with performance metrics for the various designs is the task of the engineering design group. Failing to incorporate the inherent uncertainty of the data in those evaluations may expose the design engineers to liability.

9.2.2 Engineering Design Principles

Three well-established principles are essential for the successful completion of a design project for a distribution system: (*i*) data synthesis and validation, (*ii*) successive model refinement, and (*iii*) sensitivity and risk analysis reporting. The essence of the first principle is captured by the popular acronym GIGO, which stands for "garbage-in, garbage-out." The distribution system design based on faulty data will not satisfy the design requirements regardless of the sophistication and validity of the design model. The essence of the second principle is captured by the popular acronym KISS, which stands for "keep it simple stupid." The distribution system design generated by an integrated and comprehensive model is nearly impossible to validate, unless simpler models can be used. For example, in prior work, a model with more than 1.2 million variables was executed repeatedly to select one preferred configuration of the distribution system [see Santoso et al. (2005)]. A completely separate program was created to check the feasibility and cost of the generated designs. To the author's knowledge, a similar catchy acronym does not exist for the third principle of analysis that explicitly incorporates data uncertainty.

These three principles are further explored in detail in the following sections. Distribution network design also is discussed in several books on logistics and supply chains such as Simchi-Levy et al. (2003), Shapiro (2001), Ballou (2004), Wood et al. (1999), Stadtler and Kilger (2003), and Robeson and Copacino (1994).

9.3 Data Analysis and Synthesis

The data analysis and synthesis phase of the design project will be highly dependent on the individual project. The overall goal of this phase is to create a data set that contains valid and agreed-upon data for all the major components in the design project. For many objects in the data set there may be only a single data value, for example, the longitude and latitude coordinates of a city. For other objects, the data can only be described by statistical distributions and their characteristics in the function of possible scenarios. For example, the demand of a particular customer area for a particular product may be stored as its statistical distribution type, mean, and standard deviation for the worst-case, best-guess, and best-case scenarios.

9.3.1 Logistics Data Components

All the data for a distribution system design project are typically stored in a single database. Using a database allows the use of database validation tools and consistency checks. The data are organized in the function of objects and their characteristics. Similar objects are collected in classes. The most important objects in a distribution system design project and some of their characteristics are described in the following.

9.3.1.1 Time Periods

Planning and design of logistics systems occurs at the strategic, tactical, and operational levels. The different planning levels are distinguished by their duration. The various time period(s) are fundamental components in the logistics planning. If only a single time period exists, the planning or model is said to be static. If there exist multiple periods, the model is said to be dynamic. For a strategic planning project such as the design of a distribution system, often there are five periods of one year, corresponding to a five-year strategic plan. For a tactical planning model the periods are often months, quarters, or semesters. If the system is highly seasonal, the strategic design may be based on five cycles of seasons.

9.3.1.2 Geographical Locations

Logistics components exist at a particular location in a geographical or spatial area. Typically, the geographical areas become larger in correspondence to longer planning periods. For a strategic model, the areas may be countries or states in the United States. If there exists only a single country, the system is said to be domestic; if more than one country exists; the system is said to be global.

The combination of a country and a yearly period is used very often to capture the financial performance of a logistics system. The combination typically has financial characteristics such as budget limitations, taxation, depreciation, total system cost, and net cash flow.

9.3.1.3 Products

The material being managed, stored, transformed, or transported is called a *product*. An equivalent term is *commodity*. It should be noted that the term *material* is here applied very loosely and applies to discrete, fluid, and gaseous materials; livestock; and even extends to people. If only a single material is defined, the system is said to be single commodity. If multiple materials are defined, the model is said to be multi-commodity.

It is very important to determine the type of material being considered. A first-level classification is into people, livestock, and products. The products are then further classified as commodity, standard, or specialty. Different types of products will have different service level requirements, which in turn dictate the overall structure of the distribution system. A product is said to be a commodity if there are no distinguishable characteristics between quantities of the same product manufactured by different producers. Examples of commodities are lowfat milk, gasoline, office paper, and polyethylene. Consumers acquire products solely on the basis of price and logistics factors such as availability and convenience.

A product is said to be a standard product if there exist comparable and competing products from different manufacturers. However, the products of different producers may have differences in functionality and quality. Examples are cars, personal computers, and forklift trucks. Consumers make acquisitions based on trade-offs between functionality, value, price, and logistics factors. A product is said to be a specialty or custom product if it is produced to the exact and unique specifications of the customer. Examples are machines, printing presses, and conveyor networks. The product is described by a technical specification and the supplier is selected by reputation, price, and logistics factors.

If one or more products are transformed into or extracted from another product, the products are said to have a bill of materials. The presence of a bill of materials makes the design problem significantly more complicated. While value-adding operations such as labeling are common in distribution centers, bills of materials are more common in supply chain design projects that also include the configuration of the manufacturing system.

9.3.1.4 Facilities

The locations in the logistics network where material can enter, leave, or be transformed are called facilities and are typically represented by the nodes of the logistics network. Suppliers are the source of materials and customers are the sink for materials. The internal operation of suppliers or customers is not considered to be relevant to the planning problem. The other facilities are called transformation facilities.

9.3.1.5 Customers

The customer facilities in the network have the fundamental characteristic that they are the final sink for materials. What happens to the material after it reaches the customer is not considered relevant to the planning problem. The customer facilities can be different from the end customers that use the product, such as the single distribution center for the product in a country, the dealer, or the retailer.

For every combination of products, periods, and customers there may exist a customer demand. A demand has a pattern, be it constant, with linear trend, or seasonal. Service level requirements are a complicating characteristic of customers in logistics planning. Prominent service level requirements are single sourcing, minimum fill rate, maximum lead time, and maximum distance to the serving distribution center. The single sourcing service requirement requires that all goods of a single product group or manufacturer are delivered in a single shipment to the customer. Single sourcing makes it easier to check the accuracy of the delivery versus the customer order and it reduces the number of carriers at the customer facility where loading and unloading space often is at a premium. As a consequence, single sourcing is a very common requirement. A customer may have a required fill rate, which is the minimum acceptable fraction of goods in the customer order that are delivered from on-hand inventory at the immediate distribution center for this customer. Finally, there may be a limit on the lead time between order and delivery based on competitive pressures. This limits, in turn, the maximum distance between a customer and the distribution center that services this customer. But this maximum distance depends on the selected transportation mode.

9.3.1.6 Suppliers

The supplier facilities in the network have the fundamental characteristic that they are the original source of the materials. What happens to the material before it reaches the supplier and inside the supplier facility is not considered relevant to the planning problem. The supplier facilities can be different from the raw material suppliers that produce the product, such as the single distribution center for the product in a country. For every combination of supplier facility, product, and time period there may exist an available supply.

Quantity discounts are one of the complicating characteristics of suppliers in logistics planning. A supplier may sell a product at a lower price if this product is purchased in larger quantities during the corresponding period. This leads to concave (nonlinear) cost curves in the function of the quantity purchased.

9.3.1.7 Transformation Facilities

The transformation facilities in the network have the fundamental characteristic that they have incoming and outgoing material flow and that there exists conservation of flow over space (transportation) and time (inventory) in the facility. Major examples of transformation facilities are manufacturing and distribution facilities, where the latter are also denoted as warehouses.

For every combination of transformation facility, time period, and product there may exist incoming flow, outgoing flow, inventory, consumption of component flow, and creation of assembly flow. All of these are collectively known as the production and inventory flows. A facility may have individual limits on each of these flows.

Transformation facilities have two types of subcomponents: machines and resources. Machines represent major transformation equipment such as bottling lines, assembly lines, and process lines, and are more common in supply chain design projects. A resource is a multi-product capacity limitation. Typical examples of resources are machine hours, labor hours, and material handling hours.

9.3.1.8 Transportation Channels

Transportation channels, or *channels* for short, are transportation resources that connect the various facilities in the logistics system. Examples are over-the-road trucks operating in either full truck load (TL) or less-than-truck-load (LTL) mode, ocean-going and inland ships, and railroad trains.

For every combination of transportation channel, time period, and product there may exist a transported flow. A channel may have individual limits on each of these flows. A major characteristic of a channel is its conservation of flow; that is, the amount of flow by period and by product entering the channel at the origin facility equals the amount of flow exiting the channel at the destination facility. A second conservation of flow relates channel flows to facility throughput flow and storage. The sum of all incoming flow plus the inventory from the previous period equals the sum of all outgoing flow plus the inventory to the next period. The channels represent material flow in space, while the inventory arcs represent material flow in time. Note that such period-to-period inventory is extremely rare in strategic logistics systems and should only be included for highly seasonal systems that use seasons as strategic time periods.

A channel has two types of subcomponents: carriers and resources. A carrier is an individual moving container in the channel. The move from origin to destination facility has a fixed cost, regardless of the capacity utilization of the carrier; that is, the cost is by carriers and not by the quantities of material moved on the carrier. Examples are a truck, intermodal container, or ship. A carrier may have individual capacities for each individual product or multi-product weight or volume capacities. A resource is a multi-product capacity limitation. Examples of resources are cubic feet (meters) for volume, tons for weight, or pallets. Truck transportation may be modeled as a carrier if a small number of trucks are moved and cost is per truck movement, or it may be modeled as a resource if the cost is per product quantity and a large or fractional number of trucks are allowed.

There exist several complicating characteristics for modeling transportation channels. The first one is the presence of economies of scale for transportation costs. The second one is the requirement that an integer number of carriers have to be used. Since typically a very large number of channels exist, this creates a large number of integer variables. Less common is the third complicating factor, which requires a minimum number of carriers or a minimum amount of flow if the channel is to be used.

All of the logistics components described so far have characteristics. For example, most of the facilities, channels, machines and their combinations with products and periods have a cost characteristic. Sales have a revenue characteristic. The financial quantities achieved in a particular country and during a particular period are another example of characteristics.

9.3.1.9 Scenarios

So far, all the logistics components described were physical entities in the logistics system. A scenario is a component used in the characterization of uncertainty.

Many of the parameters used in the planning of logistics systems are not known with certainty but, rather, have a probability distribution and are said to be stochastic. For example, demand for a particular product during a period by a particular customer may be approximated by a normal distribution with a certain mean and standard deviation. In a typical logistics planning problem there may be thousands of stochastic parameters. The combination of a single realization or sample of each stochastic parameter with all the deterministic parameters is called a scenario. Each scenario has a major characteristic, which is its probability of occurring. However, this probability may not be known or even be computable.

A very large amount of data for the logistics objects defined earlier and their attributes has to be generated, collected, and validated. To reduce the data acquisition and management burden, to reduce the forecast errors, and to provide better insight, the logistics objects have to be aggregated. Customers are aggregated by customer class and then by geographical proximity. Products are aggregated by physical characteristics and demand patterns. A very small number of transportation modes are considered, such as TL and LTL. Often sufficient accuracy can be achieved with a few hundred customers and a few tens of products.

9.4 Distribution System Design Models

Once the data have been collected, validated, and aggregated, the next task is to determine high-quality configurations for the distribution system. Because of the large size of the data set and the heterogeneous nature of the requirements and objectives, an objective engineering design has to use design models.

A meta-model is an explicit model of the components and rules required to build specific models within a domain of interest. A logistics planning meta-model can be considered as a model template for the domain of activity planning for logistics systems. The planning models for the design of distribution centers belong to the logistics planning meta-models. They have the following general structure:

Decide on	1) Transportation activities, resources, and infrastructure
	2) Inventory levels, resources, and infrastructure
	3) Transformation activities, resources, and infrastructure
	4) Information technology systems
Objective	1) Minimize the risk-adjusted total system cost over the planning horizon
Subject to	1) Capacity constraints such as demand, infrastructure, budget, implementation time
	2) Service level constraints such as fraction of demand satisfied, fill rate, cycle times, and response times
	3) Conservation of flow constraints in space, over time, and including bill of materials
	4) Additional extraneous constraints, which are often mandated by corporate policy
	5) Equations for the calculation of intermediate variables such as the safety inventory, achieved fill rate, and other performance measures

The two major types of design decisions are (*i*) the status of a particular facility and relationships or allocations during a specific planning period and (*ii*) the product flows and storage quantities (inventory) in the distribution system during a planning period. For example, if the binary variable y_{klt} equals one, this may indicate that a facility of type l is established and functioning at location or site k during time period t. Similarly, the binary variable w_{pklts} indicates if product p is assigned or allocated to a distribution center at site k and of type l during period t in scenario s or not. The continuous variable x_{ijmpts} indicates the product flow of product p from facility i to facility j using transportation channel m during time period t in scenario s.

Distribution systems are typically designed to minimize the time-discounted total system cost over the planning horizon, denoted by *NPVTSC*. Often, the system is designed with an expected value objective and then evaluated with respect to more complicated objectives through simulation. Let cdf_t denote the capital discount factor for a period t and $\mathbf{E}[\cdot]$ denote the expectation operator. Then the objective of the strategic distribution system design is $\min\{E[NPVTSC]\}$.

If the capital discount factor remains constant over the planning horizon, the expression for the *NPVTSC* simplifies to

$$NPVTSC = \sum_{t=1}^{T} TSC_t \cdot (1 + cdf)^{-t} = \sum_t \left(\sum_{c \in C} \frac{TSC_{ct}}{er_{ct}} \right) \cdot (1 + cdf)^{-t} \tag{9.1}$$

TSC_{ct} is the total system cost for a country in the currency of the country during a particular time period (year), and it is based on all the facilities in operation or being established in that country during that time period. er_{ct} is the exchange rate for the currency of country *c* expressed in the currency of the home country. If only one country is involved in the distribution system, the first expression can be used to calculate the *NPVTSC*.

The strategic design of distribution and supply chains is based on the application of a sequence of models with increasing realism and complexity. The results of using the previous model in the sequence are used to validate the current model. The first two models are the K-median and Location-Allocation (LA) models. The next model is called the Warehouse Location Problem (WLP). A more comprehensive variant of the WLP was first published by Geoffrion and Graves (1974). Finally, a number of comprehensive models for single country and global logistics have been developed. See, for example, Dogan and Goetschalckx (1999), Vidal and Goetschalckx (2001), and Santoso et al. (2005).

9.4.1 K-Median Model

The K-median model is used to determine the number and location of distribution centers and the customer allocations with respect to a set of customers in order to minimize the total system cost.

9.4.1.1 K-Median Formulation

$$Min \sum_{i=1}^{N} \sum_{j=1}^{N} c_{ij} x_{ij} \tag{9.2}$$

$$s.t. \sum_{j=1}^{N} x_{ij} = 1 \quad i = 1 \dots N \tag{9.3}$$

$$\sum_{j=1}^{N} y_j \leq K \quad j = 1 \dots N \tag{9.4}$$

$$x_{ij} \leq y_j \quad i, j = 1 \dots N \tag{9.5}$$

$$x_{ij} \geq 0, y_j \in \{0,1\} \quad i, j = 1 \dots N \tag{9.6}$$

where
- y_i 1 if a distribution center is established at customer location *j*, zero otherwise.
- x_{ij} 1 if customer *i* is serviced from the distribution center at location *j*.
- c_{ij} Cost to service customer *i* completely from the center at location *j*.
- K Maximum number of distribution centers to be established.

The objective of Constraint 9.2 is to minimize the sum of the costs to service the customers (median problem). There are two types of decisions: the first one selects which distribution centers will be established, and the second one assigns customers to the centers. Constraint 9.3 ensures that each customer has to be served from a center. Constraint 9.4 ensures that the number of distribution centers is no larger than the maximum number (K). Constraint 9.5 allows customer i only to be served from location j if the center at location j is established. The distribution centers are assumed to have no capacity restrictions. It is also assumed that the set of customers covers all of the distribution area. The status of a distribution center is a binary variable since a center cannot be fractionally open and, thus, this problem has to be solved with a mixed-integer programming solver. It should be noted that the customer assignment variable x is modeled as a nonzero continuous variable without upper bound, but the optimal solution of this uncapacitated problem will yield automatically zero and one values for the assignments barring alternative optimal solutions. If a fractional optimal solution is generated, the assignment variables can also be declared as binary variables with the lowest branching priority in the mixed integer programming solver. The K-median problem has been studied extensively [see, e.g., Francis et al. (1992)], and can be solved reasonably efficiently for realistic problem sizes. Observe that an upper bound on the number of established centers is required; otherwise, the optimal solution would be to open a center at every customer location. Often determining this upper bound is part of the design project. This can be achieved by running the model for a series of acceptable upper bounds and comparing the resulting configurations. The formulation has the advantage that the assignment costs c are completely under the designer's control. They can be proportional to the customer demand size, transportation distance between center and customer, the product of the two, or any problem-specific   value.

The formulation has the disadvantage that no site-specific costs can be incorporated. The model is highly aggregate and usually only a single time period is used.

To yield reasonable configurations, the formulation assumes that a customer exists in every section of the design area, so that a center could be established there. If no such coverage of the design area exists, then the following LA formulation is more appropriate.

9.4.2   Location-Allocation Model

The LA model considers manufacturing facilities (plants), customers, and distribution centers (depots). It determines the location of the distribution centers and the allocation of customers to distribution centers based on transportation costs only. The distribution centers can be capacitated and flows between the distribution centers are allowed.

The algorithm starts with an initial solution in which the initial location of the distribution centers is specified. This initial location can be random, specified by the user, or the result of another algorithm. Based on this initial location, the network flow algorithm computes the transportation distances d and then assigns each customer to a distribution center with sufficient capacity by solving the following network flow problem.

9.4.2.1   LA Formulation (Allocation Phase)

$$Min \sum_{i=1}^{M}\sum_{j=1}^{N} c_{ij}d_{ij}w_{ij} + \sum_{j=1}^{N}\sum_{k=1}^{L} c_{jk}d_{jk}v_{jk} \tag{9.7}$$

$$s.t. \sum_{j=1}^{N} v_{jk} = dem_k \qquad k=1\dots L \tag{9.8}$$

$$\sum_{j=1}^{N} w_{ij} \le cap_i \qquad i=1\dots M \tag{9.9}$$

$$\sum_{k=1}^{L} v_{jk} \leq cap_j \quad j = 1 \dots N \tag{9.10}$$

$$\sum_{i=1}^{M} w_{ij} - \sum_{k=1}^{N} v_{jk} = 0 \quad j = 1 \dots N \tag{9.11}$$

$$w_{ij} \geq 0, v_{jk} \geq 0 \tag{9.12}$$

where

w_{ij}, v_{jk} The product flows from plant i to distribution center j and from distribution center j to customer k, respectively.

c_{ij}, c_{jk} The transportation costs per unit flow and per unit distance from plant i to distribution center j and from distribution center j to customer k, respectively.

d_{ij}, d_{jk} The inter-facility transportation distances from plant i to distribution center j and from distribution center j to customer k, respectively.

cap_i, cap_j Throughput capacity of plant i and distribution center j, respectively.

dem_k Demand of customer k.

Constraint 9.8 ensures that each customer receives its full demand. Constraints 9.9 and 9.10 ensure that the capacity of the plants and distribution centers is observed. Constraint 9.11 ensures that the total inflow into a distribution center is equal to the total outflow, that is, that conservation of flow is maintained. This network flow formulation can be very efficiently solved by a linear programming solver for all realistic problem sizes. The result of the allocation phase is the assignment of customers to distribution centers as given by the flow variables.

After all the customers have been allocated to a distribution center with available capacity, a second sub-algorithm locates the distribution centers so that the sum of the weighted distances between each source and sink facility is minimized for the given flows. This problem is formulated as a continuous, multiple-facility weighted Euclidean minisum location problem.

9.4.2.2 LA Formulation (Location Phase)

$$Min \ f(x, y) = \sum_{i=1}^{M} \sum_{j=1}^{N} c_{ij} w_{ij} \sqrt{(x_j - a_i)^2 + (y_j - b_i)^2}$$
$$+ \sum_{j=1}^{N} \sum_{k=1}^{L} c_{jk} v_{jk} \sqrt{(x_j - a_k)^2 + (y_j - b_k)^2} \tag{9.13}$$

where

$(a_i, b_i), (a_k, b_k)$ The (known) Cartesian location coordinates of customers i and plants k.

(x_j, y_j) The location coordinate variables of distribution center j.

$d_{ij} = \sqrt{(x_j - a_i)^2 + (y_j - b_i)^2}$ Euclidean distance norm.

The solution to this unconstrained continuous optimization problem can be found by setting the partial derivatives equal to zero and solving the resulting equations iteratively [see Francis et al. (1992) for further details]. However, an approximate solution can be obtained by computing the center of gravity solution.

$$x_j = \frac{\sum_{i=1}^{M} c_{ij} w_{ij} a_i + \sum_{k=1}^{L} c_{jk} v_{jk} a_k}{\sum_{i=1}^{M} c_{ij} w_{ij} + \sum_{k=1}^{L} c_{jk} v_{jk}} \tag{9.14}$$

$$y_j = \frac{\sum_{i=1}^{M} c_{ij} w_{ij} b_i + \sum_{k=1}^{L} c_{jk} v_{jk} b_k}{\sum_{i=1}^{M} c_{ij} w_{ij} + \sum_{k=1}^{L} c_{jk} v_{jk}} \tag{9.15}$$

The solution provided by Equations 9.14 and 9.15 is optimal with respect to the squared Euclidean distance norm, and it provides sufficient accuracy at this level of a strategic design project for the Euclidean distance norm. It should be noted that the iterative solution algorithm based on the partial differential equations is usually started with this center of gravity solution as the starting point. The location phase provides new locations for the distribution centers. Based on these new locations, the distances between the various facilities can be updated.

The algorithm iteratively cycles through its allocation and location phase until the network flows remain the same between subsequent iterations. The obtained solution is dependent on the initial starting locations for the distribution centers, so several different starting configurations should be used and the best final solution retained.

This model is again highly aggregate and usually only a single time period is modeled. The model has the advantage that it can locate distribution centers in locations where no customers are present. Capacities of the plants and distribution centers can be incorporated. The model assumes that distribution centers can be located anywhere within the boundaries of the feasible domain, which may not be feasible because of geographically infeasible regions such as oceans, lakes, and mountain ranges. The model has the disadvantage that no site-dependent costs can be incorporated. The solutions are only approximate and indicate a general area for the location of the distribution centers. This model is called a site-generating model since it creates the solution locations.

9.4.3 Warehouse Location Problem

In the WLP model the distribution centers can only be established in a finite number of given locations. The model is called a site-selection model since it selects center locations from a list of candidate locations. Because the candidate locations are known in advance, site-dependent costs can now be included in the model. The number of warehouses to establish is based on the cost trade-off between fixed facility costs and variable transportation costs. Establishing an additional distribution center yields higher fixed facility costs and lower variable transportation costs.

9.4.3.1 Warehouse Location Problem Formulation

$$Min\ z = \sum_{j=1}^{N} \left(f_j y_j + \sum_{i=1}^{M} c_{ij} x_{ij} \right) \tag{9.16}$$

$$s.t.\ \sum_{j=1}^{N} x_{ij} = 1 \quad i = 1 \ldots M \tag{9.17}$$

$$\sum_{i=1}^{M} x_{ij} - My_j \leq 0 \qquad i = 1 \dots M, \ j = 1 \dots N$$

(9.18)

$$y_j \in \{0,1\}, x_{ij} \geq 0 \qquad j = 1 \dots N, \ i = 1 \dots M$$

(9.19)

where, in addition to the definitions for the K-median problem, the following parameter is defined:

f_j fixed cost for establishing a distribution center at candidate location j.

The objective of Constraint 9.16 is to minimize the sum of the costs of the facilities and the costs to service the customers. There are two types of decisions: the first one selects which distribution centers will be established, and the second one assigns customers to the centers. Constraint 9.17 ensures that each customer has to be served from a center. Constraint 9.18 allows customer i only to be served from location j if the center at location j is established. The distribution centers are assumed to have no capacity restrictions.

An alternative formulation for the WLP replaces Constraint 9.18 with a larger number of the following constraints, where each constraint has fewer variables:

$$x_{ij} - y_j \leq 0 \qquad j = 1 \dots N, i = 1 \dots M$$

(9.20)

Historically, this has yielded faster solution times, but contemporary mixed-integer programming solvers recognize the structure of constraints of type 9.18 and have optimized their solution algorithms so that the differences in solution times have become negligible.

This formulation has the advantage that site-dependent costs can be incorporated. But the formulation only makes the trade-off between facility costs and the transportation costs. The throughput capacities are not incorporated.

Based on a currently existing configuration or a baseline design configuration, it is possible to evaluate the relative savings of establishing a new distribution center based on its site-relative cost $\rho_j (\mathbf{U})$.

$$\rho_j(\mathbf{U}) = f_j + \sum_{i=1}^{M} \min\left\{0, c_{ij} - u_i\right\}$$

(9.21)

where

u_i Current cost for servicing customer i.

$\rho_j (\mathbf{U})$ Site-relative cost for opening warehouse j based on the current customer service cost u_i.

Note that both u_i and c_{ij} are the cost for servicing the total demand of a customer. Candidate sites with a large negative cost, which is equivalent to large positive savings, are highly desirable sites for establishing a distribution center. Candidate sites with a large positive cost are undesirable for a new distribution center. The current cost for servicing customer i is the sum of its transportation cost and its allocated share of the fixed cost of the center that currently services it. A common cost allocation is to make the cost shares proportional to the annual demand of the customers serviced by the center. The site-relative cost provides an efficient mechanism to rank potential candidate locations, without having to resolve the base WLP. Further information can be found in Francis et al. (1992).

9.4.4   Geoffrion and Graves Distribution System Design Model

The K-median and the WLP models ignore the capacity restrictions of distribution centers. All of the previous models considered only a single product and this ignores the single-sourcing customer service constraints. Geoffrion and Graves (1974) developed a model that incorporated both capacity

Arc-based   Path-based

FIGURE 9.2   Illustration of arc- and path-based transportation flows.

and single-sourcing constraints. One of its fundamental characteristics was that the flow was modeled along a complete path from the supplier, through the distribution center, and to the customer by a single flow variable. Formulations of that type are called path-based. If a flow variable exists for each transportation move, then the formulations are said to be arc-based. The difference between path-based and arc-based formulations is illustrated in Figure 9.2. Path-based formulations have many more variables than arc-based formulation for the equivalent system. On the other hand, arc-based formulations have to include the conservation of flow equations for each commodity and each intermediate node of the logistics network.

9.4.5 Geoffrion and Graves Formulation

$$Min \sum_{ijkp} c_{ijkp} x_{ijkp} + \sum_j \left(f_j z_j + h_j \sum_{kp} dem_{kp} y_{jk} \right) \tag{9.22}$$

$$s.t. \sum_{jk} x_{ijkp} \leq cap_{ip} \quad \forall ip \tag{9.23}$$

$$\sum_i x_{ijkp} = dem_{kp} y_{jk} \quad \forall jkp \tag{9.24}$$

$$\sum_j y_{jk} = 1 \quad \forall k \tag{9.25}$$

$$TL_j z_j \leq \sum_{pk} dem_{kp} y_{jk} \leq TU_j z_j \quad \forall j \tag{9.26}$$

$$x_{ijkp} \geq 0, y_{jk} \in \{0,1\}, z_j \in \{0,1\} \tag{9.27}$$

where the following notation is used:

c_{ijkp} Unit transportation cost of servicing customer k from supplier i through depot j for product p.

f_j Fixed cost for establishing a distribution center at candidate location j.

h_j Unit handling cost for distribution center at candidate location j.

cap_{ip} Supply availability (capacity) of product p at supplier i.

dem_{kp} Demand for product p by customer k.

TL_j, TU_j Lower and upper bounds on the flow throughput of distribution center at candidate location j.

z_j Status variable for distribution center at candidate location j, equal to 1 if it is established, zero otherwise.

y_{jk} Assignment variable of customer k to distribution center at candidate location j, equal to 1 if the customer is single-sourced from the center, zero otherwise.

x_{ijkp} Amount of flow shipped by supplier i through distribution center j to customer k of product p.

The objective of Constraint 9.22 minimizes the sum of the transportation cost, fixed facility costs, and distribution center handling costs. Constraint 9.23 ensures sufficient product availability at the suppliers. Constraint 9.24 ensures that the customer demand is met for each product and ensures conservation of flow for each product at the distribution centers. Constraint 9.25 forces every customer to be assigned to a distribution center. Constraint 9.26 ensures that the flow through the distribution centers does not exceed the throughput capacity and that, if a distribution center is established, it handles a minimum amount of flow.

The above formulation captured many of the real-world constraints and objectives of distribution system design. The formulation can be solved with an efficient but complex solution algorithm based on Benders' decomposition that requires significant experience in mathematical programming and computer programming. It allows the solution of real-world problem instances with limited computational resources. At the current time, very sophisticated commercially available mixed-integer programming solvers and powerful computer processors have made use of the Benders' decomposition algorithm unnecessary except for all of the largest problem instances. Using a path-based or arc-based formulation for distribution systems design has become largely a matter of designer preference.

The Benders' decomposition solution algorithm is still used when the designer wants to incorporate data uncertainty explicitly in the model through the use of scenarios. Instead of having a single demand value per customer and product, a number of demand scenarios are included in the model. Common choices for scenarios are best-guess (the most likely scenario), best-case, worst-case, and so on. In the formulation stated earlier, a scenario is represented by an additional subscript s for all parameters and variables except the facility status variables z. The objective for the scenario-based model becomes

$$Min \ \sum_s p_s \left[\sum_{ijkp} c_{ijkps} x_{ijkps} + \sum_j \left(f_{js} z_j + h_{js} \sum_{kp} dem_{kps} y_{jks} \right) \right] \qquad (9.28)$$

with

p_s Probability of scenario s.

It is often very difficult to determine the scenario probabilities accurately. The values may be based on imprecise managerial judgment. From the modeling point of view, the scenario probabilities have to satisfy the following constraint:

$$\sum_s p_s = 1 \qquad (9.29)$$

9.5   Sensitivity and Risk Analysis

In addition to the scenarios discussed earlier, the distribution system configuration should be further evaluated to measure its response to small variations in the parameter values. The data sets for this evaluation are created by random sampling from the probability distribution for each of the data parameters. The material flows are then determined for the given distribution system configuration and a particular sampled data set by a minimum cost network flow optimization. The formulation for the network flow problem is identical to the Geoffrion and Graves model, but with the status and assignment variables fixed by the given configuration.

Based on the sensitivity analysis discussed, a particular configuration of the distribution system has a certain expected value and standard deviation of the *NPVTSC*. A classical risk analysis graph can be plotted where each candidate configuration is placed according to two dimensions: one axis representing the expected value and the other axis the variability or risk measure. Often the corporation does not

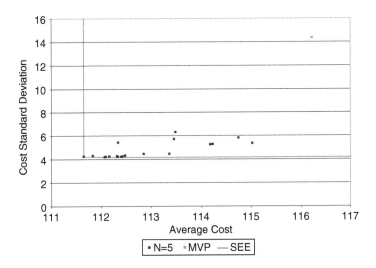

FIGURE 9.3   Risk analysis graph for distribution system design.

know explicitly its risk preferences and is interested in identifying several alternative high-quality distribution system configurations with various risk performances. The efficiency frontier is the collection of distribution system configurations that are not Pareto-dominated by any other configuration. For any efficient or non-Pareto-dominated configuration, no configuration exists that has simultaneously a smaller expected value and a smaller variability value. For a given set or sample of distribution system configurations that are located in the risk analysis graph, the sample efficiency envelope (SEE) of those configurations can be determined by connecting efficient configurations. This SEE is an approximation of the efficiency frontier. The risk analysis graph for an industrial case with the standard deviation chosen as risk measure and including the SEE is shown in Figure 9.3 [see Santoso et al. (2005)]. Note that the best distribution system configuration for the most likely value of the parameters (MVP scenario) is indicated by the square. The performance of the MVP in this example illustrates the often-observed fact that the best (optimal) distribution system configuration for the best-guess value of the parameters may have a performance far away from the efficiency frontier. The risk analysis graph is a very powerful communications tool with corporate executives since it displays in a concise manner the expected yield and risk of several possible candidates. It is the function of the design engineer to perform all the calculations, optimizations, and simulations that are then synthesized into this graph. The preferred distribution system configuration is then selected by senior management from the configurations close to the SEE.

9.6   Distribution Design Case

The following distribution design case is based on a real-world design project; however, the company name and some of the details have been obscured or changed to protect confidentiality. MedSup, a subsidiary of a larger corporation, delivers medical supplies to primary care providers in the continental United States, which include general and specialized physician offices, small surgery centers, and specialty clinics. Hospitals and large surgery centers as well as home care and long-term care facilities are not part of the customer base since they are served by other subsidiaries. MedSup has a current distribution system with 13 distribution centers, four of which are exclusively used for the primary care customers, while the others are shared with the other subsidiaries that deliver to the other customer classes. Competitive pressures have established next-day delivery as the required customer service standard. The system has to handle a large number of relatively small customer orders with a very short turnaround time. MedSup anticipates an increased demand from primary care providers for their products based on the aging of the general population with its corresponding increase in health care requirements. MedSup also expects that the current demographic relocations will become even more pronounced in the future. The objective was to design a distribution system that maintains the customer service standard at the lowest possible cost for the current and future customer base. Specific questions to be answered by the design project are the number and locations of new distribution centers and the identification of any current distribution centers to be closed.

The design utilized two primary types of data. The first type is the current and future population distribution in the continental United States. This data was obtained from Microsoft MapPoint, which contains U.S. Census population data from 1998, 2000, and 2002. The second data set is the geographical distribution of primary care practices in the United States, where the practices are categorized by medical specialty. In addition, the current configuration of the distribution system is provided.

MedSup decided to focus in the first phase on the general configuration of the distribution system, since no site-specific data for distribution center establishment and operation were immediately available. To support this high-level view, it was decided to aggregate customers by 3-digit zip codes (ZIP3). There are 878 ZIP3 zones in the continental United States. The K-median model was used as objective to the sum of the weighted distances. The K-median model was used rather than the LA model because the ZIP3 zones sufficiently covered the continental United States and because of the reduced programming requirements for the solution algorithm.

The distances were computed with the great circle distance norm between central locations in each ZIP3 zone. The great circle distance norm was used because the location data were available as longitude and latitude coordinates. The great circle distance norm computes the distance along a great circle on the surface of the earth between two points with latitude and longitude coordinates (lat_i, lon_i) and (lat_j, lon_j) with the following formula, where R denotes the world radius. The earth radius is approximately 6366.2 km or 3955.8 miles.

$$d_{ij}^{GC} = R \cdot arccos(cos(lat_i)cos(lat_j)cos(lon_i - lon_j) + sin(lat_i)sin(lat_j)) \tag{9.30}$$

The exact computation method for the weights for the K-median formulation is case specific and different formulas should be used during the sensitivity analysis. The exact weight formula for this case is proprietary. The weight is proportional to the population, the number of primary care practices, plus an additional weight for specific types of practices in the ZIP3. Parallel to the modeling effort, the marketing and operations organizations in MedSup were interviewed to identify possible locations for new centers. The K-median model was first solved without and then later with the current distribution configuration as constraints. The model was solved with a commercial mixed-integer programming solver and required about 30 min of computation time per run. The model contained 878 binary variables, 770,884 continuous variables and terms in the objective function, and 771,763 constraints. The maximum number of distribution centers (K) varied from 12 through 16. When K was systematically increased, the majority of distribution centers remained in the same ZIP3 zones and the splits of customer zones appeared logically to MedSup. The system configurations were compared with the system configurations determined by the marketing and operations departments. The configurations were nearly identical, if center locations that were in different ZIP3 zones in the two configurations were considered identical if they were located in the same metropolitan area. Finally, candidate locations were ranked by how many times they appeared in the model solution, by preference of management, by population growth, and by practice count. Sensitivity analysis was performed on the relative weights of those factors. Three metropolitan areas (Houston, TX; Chicago, IL; and Oakland, CA) ranked consistently first through third, but no single location was preferred for all values of the weight factors. The objective function value decreased from 8.3% to 4.5% if a distribution center in those three locations was established in addition to the 13 currently existing centers. The next phase of the design project will require the collection of detailed site-specific cost and capacity data for those three locations and a more comprehensive model such as Geoffrion and Graves.

9.7   Conclusions

This design project illustrated again the following observations about strategic distribution system design. First, without modeling-based decision support, the configuration of a distribution system is essentially reduced to intuition or guesswork. Second, the concept of a single "optimal" distribution system configuration generated by deterministic optimization is an illusion. Third, through careful modeling-based sensitivity analysis, a limited number of high-quality candidate configurations can be identified and submitted for final selection.

Several major factors such as cycle and safety inventory and taxation have not been discussed so far. More comprehensive models that incorporate these factors have been developed, but such models must be used with extreme care and typically have a steep learning curve. Their use can be only recommended if the models will be used repeatedly.

Three phases are essential for the successful completion of a distribution system design project. During the first phase, data from a variety source are collected, validated, aggregated, and synthesized. This activity is time- and resource-consuming, but it provides the foundation on which the rest of the design project is based. In the second phase, a series of design models is formulated and solved. The models become increasingly comprehensive, require more sophisticated and computationally expensive algorithms, become more difficult to validate, and the results become more difficult to interpret. Validation and interpretation of the current model must be completed before the next-level model can be used. In the third phase, sensitivity analysis is used extensively. The models are solved with a large variety of data values and the results are statistically analyzed. In the end, a limited number of high-quality configurations are identified and presented to the upper management for final selection.

Clearly, a strategic distribution design project is a time- and resource-intensive activity. But a properly executed project can reduce the distribution costs by 5% to 10%.

References

Ballou, R. H., (2004). *Business Logistics Management*, 5th Edition, Pearson Education, Upper Saddle River, New Jersey.

Dogan, K. and M. Goetschalckx, (1999). A Primal Decomposition Method for the Integrated Design of Multi-Period Production-Distribution Systems, *IIE Transactions*, Vol. 31, No. 11, pp. 1027–1036.

Francis, R. L., L. F. McGinnis, and J. A. White, (1992). *Facility Layout and Location: An Analytical Approach*, 2nd Edition, Prentice-Hall, Englewood Cliffs, New Jersey.

Geoffrion, A. M. and G. W. Graves, (1974). Multicommodity Distribution System Design by Benders Decomposition, *Management Science*, Vol. 20, No. 5, pp. 822–844.

Robeson, J. and W. Copacino, (Eds.), (1994). *The Logistics Handbook*. Free Press, New York, New York.

Santoso, T., S. Ahmed, M. Goetschalckx, and A. Shapiro, (2005). A Stochastic Programming Approach to Designing Strategic Supply Chains Under Uncertainty, *European Journal of Operational Research*, Vol. 167, No. 1, pp. 96–115.

Shapiro, J. F., (2001). *Modeling the Supply Chain*, Duxbury Press, Pacific Grove, California.

Simchi-Levi, D., P. Kaminsky, and E. Simchi-Levi, (2003). *Designing and Managing the Supply Chain: Concepts, Strategies, and Case Studies*, 2nd Edition, McGraw-Hill, New York, New York.

Stadtler, H. and C. Kilger, (2003). *Supply Chain Management and Advanced Planning*, 3rd Edition, Springer, Heidelberg, Germany.

Trunick, P. A., (2006). Don't Oversimplify Site Selection, *Logistics Today*, March 2006, pp. 28–30.

Vidal C. and M. Goetschalckx, (2001). A Global Supply Chain Model with Transfer Pricing and Transportation Cost Allocation, *European Journal of Operational Research*, Vol. 129, No. 1, pp. 134–158.

Wood, D. F., D. L. Wardlow, P. R. Murphy, J. Johnson, and J. C. Johnson, (1999). *Contemporary Logistics*, 7th Edition, Prentice-Hall, Englewood Cliffs, New Jersey.

10

Transportation Systems Overview

10.1 Introduction and Motivation **10**-1
10.2 Moving People versus Moving Goods **10**-2
  Differences and Similarities in Systems • Differences and
  Similarities in Performance Measures • Shared
  Systems • System Design Challenges
10.3 Transportation Modes **10**-5
  Mode Characteristics • Mode Selection • Multi-Mode
  Transportation
10.4 Importance of Transportation Infrastructure **10**-8
  • Scope of Transportation Solutions Provided
  • Consolidation versus Operational Frequency
  • Domestic and International Infrastructure
  • Global Infrastructure Example: FedEx
10.5 Difficulties in Forecasting Freight Demand **10**-11
10.6 Case Study: Dutch Railway Infrastructure
  Decisions **10**-12
10.7 Concluding Remarks **10**-13
References **10**-14

Joseph Geunes
University of Florida

Kevin Taaffe
Clemson University

Transportation systems form a vital backbone of economic activity, enabling the movement of people and goods required for providing goods and services. Effective creation and management of transportation systems can provide a substantial competitive advantage for a firm in the private sector, and can drastically influence a nation's productivity and global competitiveness from a public-sector perspective. This chapter provides a foundation for understanding critical factors in efficient transportation system development, as well as the complexities that lead to challenging decision problems in transportation service delivery.

10.1   Introduction and Motivation

Transportation systems, broadly defined, encompass the collective infrastructure, equipment, and processes utilized in the movement of people and goods among different geographic locations. The relative economic importance of transportation systems is evidenced by the fact that between 1990 and 2001, the cost of transportation equipment, service, and infrastructure ranged between 10.2% and 10.9% of the United States Gross Domestic Product (GDP), with transportation's contribution to the GDP totaling more than $1 trillion per year since 1999 (in 2005 dollar value).* Passenger transportation expenditures in 1999 exceeded $936 billion, while freight expenditures topped $560 billion.† Transportation, therefore, accounts for a significant portion of the U.S. economy, and the same holds for the majority of industrially developed nations. This investment in transportation is a substantial factor in enabling the United States to lead the world in real GDP per capita.‡ In addition to the impacts of transportation systems on productivity, these systems also contribute to the quality of life of consumers in the form of leisure travel (tourism-related goods and services recently topped $1 trillion annually in the United States§).

The focus of logistics engineering in this domain is on identifying the most efficient methods for establishing and utilizing transportation infrastructure and equipment. The chapters in the following section of this handbook discuss methods for a variety of transportation planning decision contexts and problems. The intent of this chapter is to provide an overarching foundation for the scope of relevant issues in the study of transportation systems and to characterize the range of decision types in this field.

Within the transportation context, it is important to distinguish between the roles and functions of carriers and shippers. A carrier performs the transportation function and must therefore concern itself with issues such as managing and operating a transportation fleet and associated support equipment and facilities. A shipper, on the other hand, has a need to move a good from place to place, but does not perform the transportation function (except in cases where the shipper and carrier are the same; that is, a shipper maintains and manages an internal fleet of vehicles for goods transport). The shipper is therefore concerned with the cost, quality, responsiveness, and reliability of the transportation service (which is provided by a carrier or a set of carriers). This distinction will play an important role in characterizing the relevant issues an organization faces with respect to transportation systems.

The organization of this chapter is as follows: Section 10.2 begins by characterizing the important differences in transportation systems that cater to transporting people versus those that focus on moving freight. There we identify the factors that differentiate the challenges faced in designing and operating these distinct types of transportation systems. Transporters face the challenge of determining the most effective mode for moving a good, which we discuss in Section 10.3. Section 10.4 then considers the importance of transportation infrastructure in enabling productivity and competitiveness in a global economy. For transporters of both people and goods, forecasts of transportation demands drive transportation investment, as well as the ultimate utilization of the resulting transportation equipment and infrastructure. These factors in turn directly impact the return on transportation investment as well as the efficiency (and congestion, or associated loss of efficiency) of the transportation system. We consider the complexities involved in forecasting transportation demands in Section 10.5. Section 10.6 presents a case example highlighting the importance of transportation systems planning in practice, and Section 10.7 provides concluding remarks.

10.2   Moving People versus Moving Goods

When considering the movement of people (as opposed to goods), the distinction between carriers and shippers does not play an important role. In this context, we focus on transportation carriers, who

---

* *Source*: U.S. Department of Transportation, Bureau of Transportation Statistics: www.bts.gov.

† *Source*: Eno Transportation Foundation, Inc., Transportation in America, 2001 (Washington, DC, 2000).

‡ Data as of 2004. *Source*: U.S. Department of Labor, Bureau of Labor Statistics, Office of Productivity and Technology: www.bls.gov/fls/

§ *Source*: U.S. Bureau of Economic Analysis: www.bea.gov.

typically offer one mode of transport (e.g., an airline, rail company, bus company), and their concerns lie in providing an efficient (and profitable) means of moving these individuals within their system network. An exception to this would be regional mass transit systems, which can offer several modes for people to travel within an urban area. Hensher and Button (2000) and Hall (2002) characterize key issues in modeling transportation systems. While these references also offer an introduction into freight transportation [a subject which is covered in greater detail by, e.g., Friesz (2000) and Crainic (2002)], they provide a much more thorough treatment of passenger transportation (or movement of people).

In contrast, when considering the movement of goods or freight, carriers and shippers have differing and unique roles. Carriers and shippers engage in cooperative partnerships (to varying degrees, depending on the context), much like retailers and suppliers in a supply chain setting. Each must remain competitive within its own line of business, yet they often depend on each other to achieve their desired levels of performance. An exception to this would be those companies who own their transport fleet for moving their goods. Most freight transportation modeling in the operations literature adopts the viewpoint of the carrier, where the focus is on determining an appropriately designed system that can provide transport for a wide range of consumers or shippers.

Transportation problems have been studied for many decades. Applications in the airline industry, for example, have led to the introduction of a number of operations research techniques for solving various types of transportation problems, including schedule generation and fleet assignment [e.g., Lohatepanont and Barnhart (2004)], crew scheduling [e.g., Hoffman and Padberg (1993)], and yield management problems [e.g., McGill and van Ryzin (1999)]. The majority of this work focuses on systems that transport people or passengers. Barnhart and Talluri (1997) detail an excellent introduction to this field. Barnhart et al. (2003) also provide a recent survey of operations research applications in the airline industry. Until recently, less attention has been paid to the cargo and freight side of the airline industry. As one would imagine, many similar issues exist, especially since the same fleet is used for transporting both passengers and cargo. Still, modeling air cargo decisions introduces new and different objectives and constraining factors. For freight operations that operate independently of any systems that move people, there are even clearer distinctions in the associated transport systems. We address these broader issues and design challenges in this section.

10.2.1 Differences and Similarities in Systems

The clearest distinction between transportation systems that move people versus freight would most likely be evident in regional mass transit systems. Here, a transportation system offers regularly scheduled operations with many intersecting routes, allowing people to easily connect to other routes in reaching their final destinations. The system provides a daily capacity (based on scheduled routes using assigned vehicles), and individuals typically do not purchase advance tickets that reserve them any specific portion of this capacity. However, they are free to ride on any part of the system at any time that they choose, provided that capacity is available. The closest analog in the movement of goods would likely be the transportation networks of parcel and package delivery firms. These firms have the flexibility to determine the routing of the items in their networks (except for the origin and destination points), while the items themselves (people) determine the routes they take in regional mass transit systems.

Regional mass transit systems often include a mix of modes, such as rail, light rail, elevated/underground trains, and buses, and the design of these systems must take into consideration the needs of passengers. Capacity is typically measured in terms of the number of passengers that can be accommodated (e.g., the number of passengers that can be moved between an origin–destination pair per unit time). In contrast, systems that move goods may have different temperature and space usage requirements, and can utilize space more efficiently in the movement of inanimate and durable objects. Moreover, capacity in this context is often measured in terms of the volume (or weight) of freight that can be moved between locations per unit time.

When traveling outside urban areas without private vehicles, various scheduled transportation systems typically carry passengers, nearly all of which require reservations of space on the mode of travel. For long-distance travel, available mode choices include roadway (bus or car), rail, sea, or air; however, the predominant means of travel across long distances is either rail or air. The choice of mode is not only influenced by cost and convenience, but also by the transportation infrastructure within the particular country or region, as well as the prevailing culture within that region.

Freight transportation modes are slightly more diverse, involving roadway (truck), rail, inland waterways, pipelines, sea, or air. Road and rail transportation represents a significant percentage of total freight moved, and the use of road transportation has steadily increased in recent years [see UNECE (2001) and Eurostat (2002)]. In the European Union (EU) countries, for example, 77% of freight was moved by roadway and 15% was moved by rail in 1999.

While ocean transport for moving people typically only applies to leisure travel, it is extremely important for shipping materials from heavy industries (where ocean transport is the only viable alternative) and for shipping low-cost items over long distances. Barge transportation on inland waterways provides a similar service as an alternative to road or rail for cross-country transports. For each mode of freight travel, the carrier typically has a volume capacity, and depending on the size and weight of the products to be shipped, each system's capacity can vary. Thus, in addition to its origin and destination, a particular good may also dictate the mode of travel based on its size, weight, and value.

10.2.2   Differences and Similarities in Performance Measures

Clearly, moving people and goods involve differing performance measures and objectives. Passengers typically would like to spend as little time as possible in a transportation system, although they recognize the trade-offs between cost and convenience. For example, the cost of a cross-country bus ticket in the United States is much lower than that of a flight, although the former may require days while the latter can be completed in less than half a day. The individual must therefore consider the overall utility gained from a bus trip versus a flight when determining how to go across the country. The transportation carrier's performance when transporting people is often a function of individuals' perceptions of the overall value of a form of transportation. A number of elements determine this overall value, including safety, monetary cost, time, and value-added services.

When it comes to freight, on the other hand, the items being moved do not experience the trip, and the key performance measures involve cost, trip duration, and reliability. Unlike people, goods in transport accrue "pipeline" inventory holding costs that are typically proportional to the duration of time in the transportation system (which might be roughly analogous to the value of time for a person in transit; in either case, an investment opportunity cost is incurred). The shipper must therefore consider the trade-off between transportation and inventory costs when making transportation mode decisions. While transportation modes with long lead times are often less costly, they lead to higher pipeline inventory costs. Moreover, for mass merchandise with uncertain consumer demand, longer transportation lead times imply greater inventory safety stock costs to buffer against uncertain lead-time demand. An additional complicating factor affecting inventory cost, which is discussed in greater detail in Section 10.3, is the reliability of a given transportation mode. Less reliable modes (where reliability might be measured, e.g., by on-time performance or by the standard deviation from the average delivery lead-time value) naturally lead to increased buffer stocks to provide insurance in cases where deliveries are late.

A somewhat unique performance factor within freight shipping is the notion of "empty-balancing." Due to trade imbalances between countries and geographic regions within a particular country, vehicles sometimes need to travel empty in order to rebalance the system. This need exists at a much smaller level in passenger transport systems, in the form of "dead-heading" crew or vehicles to realign the system. For example, while there may be imbalances in mass transit travel between the morning (into the city) and evening (out of the city) rush hours, people generally return to their points of origin at some point during their journey. The same is true for passengers using air or rail transportation for work or leisure. Freight transporters often seek out shippers who can utilize excess capacity in return trips, while passenger transporters may utilize pricing to increasing utilization on under-utilized trips.

10.2.3 Shared Systems

The most common form of a shared transportation system for people and goods is commercial aviation. While the system network is designed to provide passengers a means of reaching their destinations in an acceptable travel time, the airlines can also provide cargo capacity on these same flights for those products that have a time-sensitive component. As previously mentioned, the airline industry has been developing mathematical programming solutions for passenger travel and cargo for the past few decades. However, cargo research has gained more interest in recent years as the airlines attempt to identify new revenue streams.

To a lesser degree, there is some shared travel by rail and sea. In particular, cruise ships can provide some point-to-point freight capacity as these ships travel between their ports of call.

10.2.4 System Design Challenges

Given the differences and similarities in how people and goods prefer to travel, several challenges arise when designing a transportation system. Adopting the designs for a passenger transport system will not apply in many cases for freight transportation. For logistics companies deciding on what type of system to provide, the choice will often depend on whether it wants to offer high weight capacity, express deliveries, custom routing, or door-to-door services. Each of these may drive a different set of customers, so the logistics provider must have a comprehensive understanding of the needs and preferences of the customer base that it wishes to serve.

For example, overnight shipping providers (or carriers) have similar objectives to those in passenger travel. While the intermediate destinations are not necessarily important, the freight must reach its destination by specific time-sensitive deadlines. Other carriers may focus on providing shipping without time-sensitive freight, and such carriers are primarily concerned with meeting promised delivery dates. These characteristics of the customer needs and expectations can make a substantial difference in the requirements of the fleet capacity.

From fleet capacity and route structure to empty-balancing and multi-mode solutions, there are many issues that face any potential freight carrier, and many of these solutions will be unique to the freight industry. The models that are designed to provide such tactical design solutions will also be reliant on quality freight forecast data. We will address the issues of modes, infrastructure, and demand forecasting in the remaining sections.

10.3 Transportation Modes

As discussed in Section 10.2.1, there are many modes of transport available for freight: road, rail, maritime, air, and pipelines. We briefly discuss the characteristics of each mode, as well as the situations in which one of these modes would be considered the preferred method for transporting freight. Then, we motivate the need for multi-mode infrastructure solutions in successful logistics engineering.

10.3.1 Mode Characteristics

Let us briefly examine the modes of travel available for people and freight, and discuss the characteristics of each of these modes. As stated earlier, available options for transporting people include roadway, rail, sea, or air. Within each mode, varying levels of flexibility exist. Roadway and air offer the highest level of flexibility in terms of schedule options. Private automobiles can provide virtually any door-to-door service they desire. Buses still offer flexibility based on the number of stops included in their route structure. Since highway networks are remarkably well connected and developed, many destinations can be reached. However, the most rapidly growing mode has been air travel. Travelers can reach a growing number of destinations by air, which dramatically reduces trip times in many instances. When using a travel mode such as air, however, it is likely that the passenger will require a multi-mode solution to reach his or her final destination. This could include light rail, bus, mass transit, rental cars, or taxis.

In Section 10.2.1, we also noted the available modes of travel when shipping freight, which include roadway, rail, inland waterways, pipelines, sea, or air. Road and rail transportation represent significant percentages of total freight moved, and the use of road transportation has steadily increased in recent years [see UNECE (2001) and Eurostat (2002)]. As we noted previously, as of 1999, for example, 77% of freight was moved by roadway and 15% was moved by rail in the EU countries.

Road and rail transportation require capital-intensive projects to expand existing networks. Road networks offer high flexibility, and are primarily used for light industries, which require frequent, timely deliveries. Rail networks are not quite as flexible, yet the ability to containerize goods has allowed this industry to connect to sea or maritime transportation. Maritime and rail transportation are typically associated with heavy industries, and due to the volume of goods shipped by sea, this is another reason why connecting these modes is advantageous. As an example, excluding Mexico and Canada, over 95% of U.S. foreign trade tonnage is shipped by sea, and 14% of U.S. inter-city freight is transported by water [U.S. House Subcommittee (2001)]. Compared to other modes of transportation, shipment by waterways is generally less expensive, safer, and less polluting. Still, there can be substantial costs associated with port terminal operations, mostly due to port charges for shipping/receiving and inventory costs.

Air transportation can offer a method for transporting freight with either a time-sensitive nature or high value associated with it. Due to the high cost of this mode and the relatively limited capacity per vehicle (when compared to rail or water transport options), it is still used in low volumes compared to other shipping options, although it has the highest reliability among transportation mode choices.

10.3.2 Mode Selection

Transportation mode decisions for personal transport are a function of individual preferences and resources, that is, what the individual is able to afford, how the individual values his or her time, and the degree of utility the individual derives from the travel itself. We therefore focus our discussion on transportation mode decisions for goods in this section.

A highly stylized and simplified analysis of the mode decision for point-to-point delivery of a single good would proceed as follows. Suppose we manage a stock of goods that requires periodic replenishment from a supplier, and we must meet demand that occurs at a constant rate of l units per year. We have M possible modes of transportation from which to choose, and we pay for items at the time they are shipped, in addition to the shipping cost (here a mode might imply any multi-mode transportation solution). Selecting mode m implies that Q_m units will be delivered at equally spaced time intervals of Q_m/l (equivalently, on average we receive l/Q_m deliveries per year). The delivery lead time of mode m is L_m time units, which we assume for simplicity is a constant. The per shipment cost of mode m is f_m (independent of the quantity delivered), and we are also charged c_m per unit in transportation cost. Thus, our average annual transportation cost for mode m is given by

$$\left(\frac{f_m}{Q_m} + c_m\right)\lambda. \tag{10.1}$$

Because we receive a shipment of size Q_m every Q_m/l time units, and because it is optimal to receive these shipments precisely when our inventory on-hand hits zero, on average we will carry $Q_m/2$ units of cycle stock in inventory. If H denotes the cost to hold a unit of inventory on-hand for one year, then our average annual cycle stock cost for holding inventory locally when using mode m is given by

$$\frac{HQ_m}{2}. \tag{10.2}$$

Every unit of demand we meet in a year requires transportation from our supplier and spends L_m time units in the pipeline. If H_{pl} denotes the holding cost per unit of item in transit (or in the pipeline) per year, then the average annual pipeline inventory cost for mode m is given by

$$H_{pl}L_m\lambda. \tag{10.3}$$

If, for example, Q_m denotes the *economic order quantity** (EOQ) associated with mode m, that is, $Q_m = EOQ_m = \sqrt{2f_m l/H}$, then the average annual cost per unit associated with mode m (which equals total cost divided by annual demand, l) can be written as

$$c_m + H_{pl}L_m + \sqrt{\frac{2f_m H}{\lambda}}. \tag{10.4}$$

Equation 10.4 illustrates a basic trade-off in mode choices, as we would like to select the mode m from among the M choices that minimizes (10.4). In particular, those modes with short lead times (L_m) typically have high shipping costs, as reflected in the fixed (f_m) and/or variable (c_m) shipping costs. Longer lead-time mode choices, on the other hand, increase the pipeline holding cost term, while reducing the shipping cost terms.

To introduce the effects of uncertain demand without obscuring the analysis too greatly, we suppose that a positive safety stock level is required at the stocking point, and that the safety stock level is proportional to the standard deviation (uncertainty) of demand during the replenishment lead time {this is not an uncommon approach to setting safety stock levels in practice where, for example, we set some minimum level on the probability of not stocking out in any replenishment cycle; the associated probability is sometimes referred to as a cycle service level [see, e.g., Chopra and Meindl (2004)]}. In this setting, l denotes the average annual demand; safety stock is set equal to $k\sigma_L$, where k is a prescribed safety factor corresponding to the desired cycle service level; and σ_L is the standard deviation of demand during the replenishment lead time. If σ is the standard deviation of annual demand (and this annual demand is composed of a contiguous and statistically independent demand interval), then we can write the standard deviation of demand during lead time as $\sqrt{L_m}\sigma$, and then the average annual safety stock cost equals $Hk\sqrt{L_m}\sigma$. Defining $cv \equiv \sigma/l$ as the coefficient of variation, our average cost per unit becomes

$$c_m + H_{pl}L_m + \sqrt{\frac{2f_m H}{\lambda}} + Hk\sqrt{L_m}cv. \tag{10.5}$$

Equation 10.5 illustrates that long lead-time values not only increase pipeline holding costs, but also increase the required safety stock holding cost for meeting a prescribed service level at the stocking point. This effect is compounded by products with high coefficient of variation values (or equivalently, products with a high degree of demand uncertainty).

The stylized model we used to illustrate important trade-offs in selecting a transportation mode employs a number of simplifying assumptions, including that of a constant lead time. With less reliable

---

* This analysis assumes that the EOQ is feasible, or less than any capacity limit associated with the mode. Similar insights apply under capacity limits, although our intent here is to highlight the trade-offs associated with costs and lead times, and how these drive mode choices.

transportation modes, where the lead time itself is unpredictable, we tend to see an increased uncertainty of lead-time demand, which increases the impact on safety stock cost incurred in meeting a desired service level.

Our analysis also considered transportation of a single product that can be shipped in batches equal to the economic order quantity, whereas practical contexts often call for multiple products sharing shipping costs and capacity limits. Additional practical factors include risk of damage, trade tariffs and duties in international transport, and nonstationary product demands. While in principle the analysis can be generalized to account for such assumptions, the basic trade-offs between transportation and inventory costs when making mode decisions remain essentially the same and include transportation costs, inventory costs due to economies of scale in shipping (cycle stock), and inventory costs due to uncertainty in demand and less-than-perfect reliability (safety stock).

10.3.3 Multi-Mode Transportation

In seeking an end-to-end transportation solution, the most cost-effective option often involves a mix of different transportation modes. Economies of scale in transportation often lead to highly utilized transportation equipment and links between major metropolitan areas, although the metropolitan areas themselves are often not the origin and/or destination points in the end-to-end solution sought. Transportation to and from regions surrounding major metropolitan areas is then accomplished by regional transporters who focus on the economics of regional transportation. Therefore, different organizations focus on efficiency within a different piece of the multi-mode puzzle, which permits finding cost-effective door-to-door solutions.

When considering the transport systems available for people, we focus on two areas: mass transit solutions for commuting, and air travel for business or leisure. Mass transit systems typically will include one or more of the following modes of travel: bus, subway, light rail, and commuter rail. In larger cities, these systems are designed in such a way that the commuter has the ability to easily connect between one system and another (e.g., a commuter from a suburb can take a commuter rail to the city or business district and transfer to either a bus or subway to reach a particular destination). While a commuter rail can only serve a small set of station locations in a region, bus and subway systems still provide commuters access to the majority of locations in a region. For air travel, passengers commute to an airport via personal vehicles, bus/rail, or other ground transportation options.

Freight transportation also involves logical multi-mode options. As previously mentioned in Section 10.3.1, freight can be moved between rail and barges/ships through port terminals that can handle containerized goods. Again, this allows heavy goods that travel by sea to reach various land-based locations by rail. Similarly, there is a logical connection between air cargo and trucking. For heavy freight, these connections often occur as handoffs between different firms who specialize in managing and operating a single transportation mode. Coordination among these different carriers is often achieved by logistics service providers such as TNT Logistics, who often typically do not own transportation equipment, but serve to ensure that producers and distributors can achieve economical door-to-door deliveries. For small packages, this multi-mode service is most often seen with express overnight carriers such as FedEx and UPS, each of which owns a fleet of ground and air vehicles and provides door-to-door transportation solutions.

10.4   Importance of Transportation Infrastructure

An effective regional transportation infrastructure allows businesses in that region to compete in a global economy and allows consumers of the region to access goods from the rest of the world. Given the existence of free trade zones and additional markets being opened for the first time, it is extremely important for a region to determine the degree of transportation infrastructure to provide in order to connect to various parts of the world. Without adequate infrastructure, carriers cannot provide the type of service customers demand in a global economy, which can leave regional suppliers at a severe competitive disadvantage. How different locations are connected can vary greatly, depending on the regional demographics and industry, as well as the economic goals of the carrier(s) providing service in the region. Because governments, consumers, and regional firms have a stake in the overall public transportation infrastructure (e.g., public roadways), it is natural to have conflicting economic and service objectives. While this public infrastructure enables commerce, it also has environmental as well as quality of life impacts for a region. Thus, the collective interests of the stakeholders in a region as well as the economic and social trade-offs must be weighed when making public infrastructure investment decisions. For these reasons, it is extremely important to accurately assess both public and private needs across the system. Numerous political and social factors affect public transportation infrastructure decisions, which partially determine the transportation capabilities of private transportation firms. Because of this, this section focuses on the infrastructure decisions over which a private firm has control, examining important considerations when designing private infrastructure in a transportation network. The case study in Section 10.6 provides an interesting illustration of the potential for conflicting objectives and diverse interests involved in public infrastructure investment decisions.

10.4.1   Scope of Transportation Solutions Provided

The nature of the transportation solutions offered by a firm affects its need for infrastructure investment. When a firm has a product to ship from destination A to destination B, it may be presented with several options for choosing the method of shipment. Regardless of the method, the firm needs to complete the entire transaction. Some carriers may actually provide the entire freight shipping service, depending on the two locations of A and B and the type of business that the carrier intends to provide. Point-to-point shipping is defined as moving goods between any two "major" locations. Often, these locations will be warehousing or cross-docking facilities that serve many local destinations. Certain providers will focus on providing transportation between these point-to-point trips, and their system infrastructure will reflect this. That is, their infrastructure investments will focus on equipment and facilities that provide high economies of scale in shipping and terminal operations.

Other carriers focus on providing door-to-door shipping service. Such carriers are responsible for picking up the product at the shipping location of a customer (which does not need to be a centrally located warehouse with consolidated goods), and delivering it to that customer's destination of choice. Providing door-to-door service can certainly drive a different business model for freight carriers. Multi-mode express freight carriers such as FedEx and UPS tend to build extremely large system networks, with separate fleets (trucks and aircraft) to expedite the delivery of product and meet the service requirements of customers. The infrastructure provided by each of these companies allows their customers to ship products door-to-door around the world in one to two days. Other door-to-door carriers may not have the same massive infrastructure in place, so they may need to wait for enough demand to materialize before consolidating on a vehicle for transport. Such carriers have a different business model, usually moving less time-sensitive materials. However, they still must meet customer service goals, which will include predetermined delivery due dates. Note that this door-to-door service need not necessarily be provided by only one carrier. In fact, several carriers may be involved in the shipment of a good from destination A to destination B, although the shipper interfaces with a single carrier. Such carriers manage the coordination among a number of firms involved in the actual transport. The infrastructure investment for such firms might involve regional transportation networks and equipment, interfacing with the networks of long-distance carriers who manage networks that interconnect major metropolitan areas.

10.4.2   Consolidation versus Operational Frequency

As mentioned in the previous section, carriers must consider how frequently to provide service between any two points. This delivery frequency has important implications for investment in equipment and facilities. One approach is to consolidate demands from several points until this accumulated demand reaches the capacity of the transporting vehicle. This "on-demand" approach is desirable for carriers because it ensures low unit transportation costs and high capacity utilization. The investment in vehicle capacity is therefore lower than would be required when shipping partially loaded vehicles at prescheduled times. This approach may, however, be quite undesirable for customers with time-sensitive delivery requirements or with high-value goods that have high associated inventory holding costs.

In this "on-demand" context, the carrier may have the flexibility of delaying individual customer delivery requests until the carrier can generate sufficient revenue to warrant the entire shipment. The unscheduled nature of such shipments can also cause problems depending on the mode of travel. For example, the freight and passenger rail industry often share the same service network (i.e., the same track). As the number of scheduled operations that use a common infrastructure increases, it becomes increasingly critical to know when to expect these "on-demand" shipments. System capacity simply may not be able to accommodate them at the point in time when the vehicle capacity is reached and the "on-demand" shipment is ready for transport.

Another option is for the carrier to provide scheduled operations that match its customer shipping requirements. In other words, the carrier can schedule a particular delivery once a day, once a week, and so on. Uncertainty in shipping requirements will often result in under-utilized capacity under this approach, which will require a higher capacity investment (as compared to "on-demand" shipping). At the same time, the carrier will have a predictable schedule of operations, and hence anticipated shipping arrival dates will be more accurate. In either of these two cases ("on-demand" or scheduled frequency), their ability to achieve efficiency in transportation is driven by the accuracy of freight demand forecasts and how these forecasts are used in making capacity investments. The issue of forecasting is addressed in Section 10.5.

10.4.3   Domestic and International Infrastructure

Domestic infrastructure in developed regions of the world is typically quite effective in enabling transportation to virtually anywhere in the region. Customers have the ability to move goods between almost any two points that they desire, and numerous carriers offer services to do this. The level of customer service provided by these carriers must meet the standards expected for the particular country, based on the cultural and political nature of that country. When connecting domestic infrastructure with international infrastructure, a host of new issues arises.

From a carrier point-of-view, not all carriers are equipped to expand their businesses internationally. If their primary mode of transport is roadway and rail, they may be limited to providing additional long-haul services to land-based destinations (i.e., it may be difficult to venture into maritime or air travel). A carrier's success may be driven by the unique environment of its domestic operation. Serving new markets may require a change in the carrier's economic and customer service objectives.

Carriers who are willing to make such changes must also now deal with the additional logistical issues with moving goods across borders. In general, it can be much harder to provide accurate delivery dates when the goods must be cleared through customs. Each country has its own rules for how this process works, and likewise, the modeling requirements for determining appropriate transport system requirements can be case-specific. Nonetheless, there is an enormous market for companies that provide international logistics solutions; in particular, freight forwarders, who specialize in moving goods between countries, can serve as valuable partners for firms seeking global expansion.

10.4.4   Global Infrastructure Example: FedEx

The FedEx Corporation provides an excellent example of a global logistics network.* FedEx coordinates deliveries throughout a global network broken down into five regions:

- Asia-Pacific
- Canada

---

* *Source:* FedEx Corporate Web Site: www.fedex.com.

- Europe, Middle East, and Africa
- Latin America-Caribbean
- United States

The FedEx operation began by providing service to 25 cities in the United States via a single hub in Memphis using a fleet of 14 planes in 1973. Packages were flown to and from Memphis from each of the connecting cities on a daily basis, and couriers transported packages on the ground within a 25-mile radius of each connecting airport. In 1977, FedEx was successful in lobbying the U.S. Congress to allow private cargo airlines to purchase larger planes, which led to the purchase of seven Boeing 727 aircraft that year, and paved the way for the unprecedented growth in air cargo that followed.

In 1981, FedEx began international service to Canada in cooperation with Cansica, a Canadian licensee. By the end of the 1980s the company had purchased three Canadian air cargo firms and was providing full service to Canada. Its 1984 purchase of Gelco expanded operations into Europe and Asia. In 1987, FedEx acquired Island Airlines, which provided air cargo service to the Caribbean. FedEx subsequently acquired Tiger International (owner of the air cargo firm Flying Tigers) in 1989, becoming the world's largest cargo airline. Flying Tigers' existing business throughout the world led to a substantial global expansion for FedEx, who was then the largest air cargo carrier in South America and was providing service to Europe and Asia.

Today, Fedex serves over 220 countries and employs roughly 260,000 employees and contractors throughout the world. Their express package delivery network (FedEx Express) reaches 375 airports using 10 air express hubs and 677 aircraft, with approximately 43,000 motor vehicles. FedEx Ground uses ground transportation for package delivery in the United States, Canada, and Puerto Rico. The ground fleet contains 18,000 motor vehicles connecting 29 ground hubs and 500 pickup/delivery terminals. For larger packages, FedEx Freight provides trucking services using more than 10,000 tractors throughout the United States, Canada, Mexico, South America, Europe, and Asia, with 321 service centers. Its collective global network and companies handle approximately 6 million deliveries per day throughout the world.

Clearly, FedEx provides an example of an organization that has successfully confronted the complexities discussed throughout this chapter. As a door-to-door delivery service provider who serves countries all over the globe, the company must achieve economies of scale between metropolitan areas using a combination of air and truck travel modes, while effectively scheduling time-sensitive local deliveries in any region using smaller motor vehicles. Its massive global infrastructure permits reaching almost anywhere in the developed world in a very short time and serves as a vital component of the supply chains of firms in many countries.

10.5   Difficulties in Forecasting Freight Demand

Transportation infrastructure planning and development are primarily driven by forecasted demand for transportation. Governments have faced this challenge for many years in developing public transportation infrastructure, and this has driven a continuous stream of transportation pattern studies as well as research on methods for predicting transportation demands. A greater amount of public information is available concerning methods for predicting passenger travel than freight demand, as passenger travel studies are often sponsored by governments, while competitive firms do not necessarily make their freight demand studies available to the public. As a result, and because well-developed large-scale freight transportation networks do not have an extremely long history, research on freight demand modeling remains at a relatively early stage as a discipline.

Because transportation of freight forms the backbone of a large percentage of economic activity, forecasting freight demand can be as complex as forecasting the performance of an economy. For example, if we could accurately forecast freight flows between two countries, we could then likely provide an accurate estimate of the trade balance between the two countries. It is no surprise, therefore, that advanced methods for predicting freight demand between countries utilize well-developed methods from international economics [see Haralambides and Veenstra (1998)] and time-series forecasting. Zlatoper and Austrian (1989) provide an excellent characterization of econometric models for transportation forecasting.

Winston (1983) characterizes predictive freight flow models as either aggregate or disaggregate models. The aggregate models consider a geographic region and attempt to predict the percentage of total flow that utilizes a given mode in the region. Such aggregate models found in the literature typically employ log-linear regression, with a mode's relative market share dependent on relative price and other independent quantitative and qualitative factors, including population demographics and economic indices [see Regan and Garrido (2001) for a more detailed discussion of these methods, in addition to an excellent survey of freight demand modeling research]. Disaggregate models focus on the individual decision-maker's choice of mode and incorporate individual decision factors in forecasting methods. Clearly, the disaggregate models require substantially more data and understanding of individual utility factors. These models characterize a probability distribution of the utility value an individual derives from a given mode, assuming some deterministic utility factor (e.g., that depends on the corresponding good and industry), and a stochastic error term that accounts for variations among different consumer preferences. The probability distributions of mode utilities are then used to characterize the probability one mode will be preferred to another by a typical shipper (or passenger).

As the foregoing discussion indicates, freight transport demand modeling requires an understanding of economic, behavioral, and demographic factors, as well as advanced statistical forecasting methods. The number of factors affecting transportation demand and the complexity of the interrelationships among these factors can make accurate freight modeling as difficult as predicting the stock market. This section highlighted the complexities inherent in freight demand modeling and provided some basic characterizations of effective approaches. [For more details on freight demand modeling, see Regan and Garrido (2001).]

10.6　Case Study: Dutch Railway Infrastructure Decisions

The Dutch railway system in the mid-1990s provides an excellent example of the conflicting objectives and trade-offs inherent in transportation systems decisions.* Highly congested roadways in the 1990s, which had negative implications for the economy and the environment, forced the Dutch government to explore ways to improve its transportation system. Increasing the use of rail for both passengers and freight was an attractive alternative in terms of reducing the environmental impact of transportation, but the rail system capacity was hardly able to handle the impact. Moreover, long delays (due to insufficient capacity) and relatively high prices for rail travel discouraged both passengers and freight shippers from choosing this option.

Control of railway infrastructure development and funding rested with the government, and prior to 1997 the monopolist firm Nederlandse Spoorwegen (NS) operated the railway with a goal of maximizing profit (subject to certain restrictions on infrastructure and service areas). Therefore, the government's infrastructure investment decisions (and government requirements to keep certain lines open whether or not they were profitable) constrained NS's profitability. In 1997, the EU also required privatizing rail operations and allowing competition in this industry. This created added complexity for the government, which now had to assign infrastructure to multiple competing firms.

To address its transportation infrastructure shortcomings, the Dutch government allocated about $9 billion to improve rail infrastructure between 1985 and 2010. In addition to reducing roadway congestion, the government had several additional priorities, including:

- Stimulating regional economies by providing rail connections to large metropolitan areas.
- Increasing the amount of freight carried via rail.
- Reducing the need for short flights to other countries in Europe.

---

* A detailed discussion of this case context can be found in Hooghiemstra et al. (1999).

Private rail operating firms, on the other hand, are interested in profit maximization, and therefore would like to see faster travel times and reduced delays on their networks, which could be achieved to a significant degree through equipment and infrastructure (both rail-line and energy) improvements. The Dutch government relied on an independent organization called Railned to provide recommendations on how best to invest the funds allocated to rail infrastructure improvement. A team of government, Railned, and private operating firm representatives assembled to tackle the problem of maximizing the return on infrastructure investment. The team developed three sets of project portfolio options (that they called cocktails), each emphasizing a different investment focus. These three focus areas were (*i*) metropolitan area, (*ii*) main port, and (*iii*) regional (non-metropolitan) development. To evaluate the investment options, they drew on a sequence of decision support models.

Given an investment option (which consists of a set of potential projects), the first step required estimating the demand for rail service if a given set of projects were implemented. For this they used the Dutch National Mobility Model (DNMM), which uses econometric regression to determine the demand for travel using a mode, given the service level provided by the mode, population sizes, and various economic factors associated with regions served by the mode. After estimating travel demands, these demands provide input to an optimization model that determines trip frequencies and equipment required to meet demands on the rail network links at minimum operating cost. Inherent in this optimization model is a utility value for each transportation alternative that is used to estimate the percentage of total passengers who will choose a given route. Although our discussion here greatly simplifies the description of the forecasting and optimization models employed (which include a large-scale integer-linear programming model), it is a quite complex system and requires a heuristic solution in order to obtain a good feasible solution in reasonable computing time.

Because of the number of qualitative factors affecting the attractiveness of a solution (from a government and social perspective), further evaluation (beyond profitability) was required subsequent to implementing the forecasting and optimization models. The team of analysts performed a cost-benefit analysis for each investment alternative by assigning a monetary value to each of the important qualitative factors. This analysis allowed the team to quantify the value and utility of each of the possible investment alternatives, and to recommend a course of action to the government. In the end, the government selected the second-ranked alternative (involving metropolitan area development), because of the apparent perception on the part of the government that the profitability of private firms received too high a weight in the cost-benefit analysis, while the qualitative impacts of metropolitan area congestion received too low a weight. The operations research–based analysis served an extremely valuable purpose in this context, providing the team with a methodologically based tool for evaluating a set of extremely complex decision alternatives. Moreover, the system continues to pay dividends through repeated analysis of additional transportation infrastructure investment options.

10.7 Concluding Remarks

Our goal in this chapter was to provide a general framework for understanding the issues and trade-offs inherent in transportation systems decisions. We have taken a necessarily broad overview in this discussion, highlighting many of the qualitative factors that lead to conflicting objectives, and make transportation systems decisions a complex field of study. Because transportation systems affect the economic performance of a region, as well as the daily lives of nearly all people, our discussion focused on the systems and infrastructure and how they affect the movement of goods and passengers.

The FedEx global network discussed in Section 10.4 provides a nice snapshot of the progress that has been made in developing transportation systems in the last 30 years. The ability to ship a package anywhere in the United States within 24 h, a luxury many now take for granted, is quite remarkable, particularly in light of the fact that this reach extends far beyond the U.S. borders. This progress could not have been accomplished without the collective infrastructure investment made by developed and developing countries during this time frame. The willingness of the people and governments of countries around the world to invest in both transportation and information infrastructure has certainly led to a tighter economic integration among countries, and has opened new markets. The continued development of new markets in Asia, Eastern Europe, South America, and Africa will likely lead to changes in transportation systems over the next 30 years that are as interesting as those of the past three decades.

References

Barnhart, C., P. Belobaba, and A.R. Odoni. Applications of operations research in the air transport industry. *Transportation Science*, 37(4):368–391, 2003.

Barnhart, C. and K. Talluri. Airline operations research. In C. ReVelle and A.E. McGarity, editors, *Design and Operation of Civil and Environmental Engineering Systems*, pages 435–469. Wiley, 1997.

Chopra, S. and P. Meindl. *Supply Chain Management: Strategy, Planning, and Operations.* Prentice-Hall, Upper Saddle River, New Jersey, 2nd edition, 2004.

Crainic, T.G. Long-haul freight transportation. In R.W. Hall, editor, *Handbook of Transportation Science*. Kluwer Academic Publishers, Boston, Massachusetts, 2nd edition, 2002.

Eurostat. *Transport and environment: statistics for the transport and environment reporting mechanism (TERM) for the European Union*, January 2002.

Friesz, T.L. Strategic freight network planning models. In D.A. Hensher and K.J. Button, editors, *Handbook of Transport Modelling*, chapter 32, pages 181–195. Pergamon, 2000.

Hall, R.W., editor. *Handbook of Transportation Science*. Kluwer Academic Publishers, Boston, Massachusetts, 2nd edition, 2002.

Haralambides, H. and A. Veenstra. Multivariate autoregressive models in commodity trades. In *8th World Conference on Transportation Research*, Antwerp, Belgium, 1998.

Hensher, D.A. and K.J. Button, editors. *Handbook of Transport Modelling*. Pergamon, New York, New York, 1st edition, 2000.

Hoffman, K.L. and M. Padberg. Solving airline crew scheduling problems by branch-and-cut. *Management Science*, 39(6):657–682, 1993.

Hooghiemstra, J.S., L.G. Kroon, M.A. Odijk, M. Salomon, and P.J. Zwaneveld. Decision support systems support the search for win–win solutions in railway network design. *Interfaces*, 29(2):15–32, March–April 1999.

Lohatepanont, M. and C. Barnhart. Airline schedule planning: Integrated models and algorithms for schedule design and fleet assignment. *Transportation Science*, 38(1):19–32, 2004.

McGill, J.I. and G.J. van Ryzin. Revenue management: Research overview and prospects. *Transportation Science*, 33:233–256, 1999.

Regan, A.C. and R.A. Garrido. Modeling freight demand and shipper behaviour: State of the art, future directions. In D. Hensher, editor, *Travel Behaviour Research: The Leading Edge*, pages 185–216. Pergamon Press, Oxford, 2001.

United Nations Economic Commission for Europe (UNECE). *Annual bulletin of transport statistics for Europe and North America*, July 2001.

U.S. House Subcommittee on Coast Guard and Maritime Transportation. *Joint Hearing on Port and Maritime Transportation Congestion*, May 2001.

Winston, C. The demand for freight transportation: Models and applications. *Transportation Research Part A*, 17(6):419–427, 1983.

Zlatoper, T. and Z. Austrian. Freight transportation demand: A survey of recent econometric studies. *Transportation*, 16:27–46, 1989.

11

Logistics in Service Industries

11.1 Introduction 11-1
11.2 Within-the-Facility Logistics 11-3
Case Study: Mobile Robot Delivery Systems in Hospitals
11.3 Between-the-Facility Logistics 11-10
11.4 Summary 11-14
Acknowledgments 11-15
References 11-15

Manuel D. Rossetti
University of Arkansas

11.1 Introduction

According to the U.S. Department of Commerce, the service sector has been the fastest-growing section of the U.S. economy during the last 50 years. In fact, as of 1999, the service sector accounted for up to 80% of the U.S. economy. This remarkable statistic is related to structural shifts first from agriculture, to manufacturing, and now to services within the United States over the last century. To understand the application of logistics to the service sector, we must first examine the types of firms that constitute the service sector and the types of services they provide. Cook et al. (1999) provide a comprehensive review of the ways in which the service industry has been classified over the last 50 years. The U.S. Census Bureau conducts a survey of the service sector to understand revenues, growth, and the effect of the service sector on the U.S. economy. The primary classification used for the service sector is based on the National American Industry Classification System (NAICS). The primary categories for the service sector include:

- Transportation and warehousing
- Information
- Finance and insurance
- Real estate and rental leasing
- Professional, scientific, and technical services
- Administrative and support, and waste management and remediation services
- Health care and social assistance
- Arts, entertainment, and recreation
- Other services (except for public administration)

From this classification, it is easy to see why the service sector constitutes such a large part of the U.S. Gross Domestic Product. This handbook concentrates on many aspects of the transportation and warehousing category. This chapter will discuss issues related to logistics applied to other service sector areas.

Murdick et al. (1990) offer one of the more useful definitions of a service: "Services can be defined as economic activities that produce time, place, form, or psychological utilities." This definition allows us to conceptualize services as nontangible deliverables; however, services are often inseparable from actual physical products. For example, when a doctor performs an operation to replace a hip in a patient, the doctor is performing a service; however, the service cannot be provided without the artificial hip. This connection to the physical delivery of a product has allowed many firms that were previously purely manufacturing oriented to move into the delivery of services—services that add value or utility to their customer base. A classic example of this is International Business Machines. While still a leading manufacturer, IBM is now arguably one of the most competitive providers of service (maintenance, repair, software, training, consulting, etc.).

While all business activity is customer focused, the service industry's primary focus is on the delivery of service directly to the customer. That is, within a service transaction, the customer is often involved directly in the experience. For example, in the entertainment industry (e.g., theme parks), it is the customer's direct interaction that provides the "entertainment" service. Service sector firms have a special need to address the following questions:

1. Who are my customers? How do they demand service?
2. What are the elements of the service provided to the customer? In other words, define the service content for the customer. In addition, how will we measure customer satisfaction with the services?
3. What operating strategies are important to providing service to the customer?
4. How should the service be delivered to the customer? What should constitute the delivery system and what are the capacity characteristics of the system?

The first question must be answered so that the firm can begin to address the latter questions. In answering question 1, the firm must define the characteristics of the customer, for example, what are their attributes, demographics and requirements. In addition, the firm must understand how customer demand will be realized. The demand for services may be highly variable and stochastic in nature. Characterizing the behavior of customers and their resulting demand over time through forecasting is important in any industry, but may be especially difficult in service industries since the delivered product is often intangible. The second question forces firms to try to make the intangible, tangible. That is, the more the service content can be described and measured, the easier it will be to decide on how to deliver the service, which is the key to questions 3 and 4. In question 4, the firm needs to organize its operating strategies around the customer and the service content, and in question 5, the firm begins to answer how the service will be delivered. It is this latter question, especially the issue of capacity planning, which is critically important to logistics planners.

All these questions imply that it is critically important for service industries to know their customers and to design and implement their logistics delivery structure based on customer requirements that may change over time. Within service industries is not always possible to build up inventory, and it may take a long time to create increased capacity through new facilities. For example, the airline industry has widely varying demand patterns, but can react to demand changes only by adding flights, crews, planes, etc., all of which are discrete capacity changes. These increments are only possible in a timely manner if excess capacity already exists. The hotel and car rental industries also experience similar demand and capacity requirements. In addition, the time needed to react is longer, especially if it involves the movement or location of a facility to meet customer demand. For example, in the financial services industry it takes time to build a banking infrastructure in order to serve a growing population area, and there is the risk that the customer demand may not materialize.

As a simple example for the four questions, we might consider the theme park industry. The first question requires an understanding of the customer. The customers of a theme park are primarily families or groups of people in the younger age demographic who are willing to spend dollars on a leisure activity that they can do together. An operative concept is doing the activity together. Families or small groups will want to move together, ride together, eat together, etc. The service delivery must be designed to facilitate the delivery of service to these groups. Customer demand is time varying and stochastic within this industry both at a seasonal level but also on an hourly basis. To address question 2, we need to understand the total service content. While in the park, the customers not only want to be entertained, but they also have many other needs that must be met (e.g. food, water, transportation, rest areas, and medical response). The last two questions begin to involve questions concerning logistics. For example, the decision of where to locate rides, restrooms, food services, etc. are all predicated on understanding the demand for these items and the customer's trade-off in walking to and competing for such facilities. In addition, if the park decides to have a light rail system for moving customers within the park, the operating characteristics of the transport system must be designed. Moving customers from point to point within the park has direct customer contact; however, there will also be all the other logistical support activities to get the food, supplies, costumes, etc. into the park so that the customer can have a "fun" experience. Mielke et al. (1998) discusses the application of simulation to theme park management as well as some of the issues mentioned. In summary, within the service industry we start with customer needs and then develop logistics delivery mechanisms to meet those needs.

While service industries are unique in many respects, they also have similar characteristics to standard manufacturing and distribution systems, especially in the scope of the delivery processes. In the following sections, we discuss how the design of service logistics can be considered at two levels: within-the-facility logistics and between-the-facility logistics. We provide examples of applications at each of these two levels.

11.2   Within-the-Facility Logistics

Within-the-facility logistics involves the design and operation of the physical plant with respect to logistical goals and objectives. The techniques and issues of within-the-facility logistics are discussed elsewhere in this volume. In Chapters 11 and 12 we present an overview of some of these issues as they relate to service industries. In general, the issues involved in within-the-facility logistics may include:

1. Sizing and planning the capacity of the logistical functions
2. Designing the layout and flow of the logistics
3. Material handling system selection and operation
4. Efficiently operating the logistics system

While the flow of items (food, medicine, inventory, etc.) is a critical aspect of service logistics, systems involving service have additional requirements involving customers. When considering the customer within intra-facility logistics, there are two main issues to consider: (1) how the logistics system may indirectly affect the delivery of service to the customer, and (2) how the logistics system directly affects the customer. Keeping with our theme park example, we know the rides must be maintained and repaired. If the service parts logistics is not designed properly, then the guests may not get their rides. The guest's service is affected indirectly. Even though his or her service is affected, the guest is not directly interacting with the logistics system, that is, the service parts supply chain. In the second issue, often the major purpose of the logistics system is to move the customers. For example, within an amusement park, the service is entertainment; however, a critical logistical issue is how to get the "guests" to and from the attractions. This involves the design and operation of the guest handling systems, such as elevators, people movers, shuttle busses, and monorails. In this case, the logistics system is directly interacting with the customers and directly affecting their service.

The rental car industry is a service industry that requires careful layout of the facilities with which the customers directly and indirectly interact. For example, in Johnson (1999) the layout of the check-out areas, parking and washing areas, and check-in areas were examined for the impact on customer waiting and service efficiency. Simulation was used to examine multiple layout scenarios and to determine the impact on customer service. In other situations, such as public bus transport and air travel, the service is the transport of people. In these instances, the waiting time of the customers using the logistics system becomes even more important. In Takakuwa and Oyama (2003), airport terminal design is examined via simulation to understand the waiting time of passengers using the service. Such models require the detailed analysis of internal flows as customers utilize the service. This often necessitates an analysis of the capacity of the system and the scheduling of the availability of the staff or handling systems. Thus, the design and layout questions go together with tactical planning issues such as staffing and scheduling of service. For example, Rossetti and Turitto (1998) examined the use of dynamic hold points within a transit system to prevent the phenomenon of "bunching" within bus schedules, that is, buses catching up with each other on a route and then traveling together. This is not only highly detrimental to operating efficiency, but also causes poor service because the scheduled arrival times are not met.

While many service systems directly involve the movement of people, many others such as retail stores, hospitals, and banks rely on people, but their service is not the movement of people *per se*. In these situations, the logistics engineer must consider how the logistics system may indirectly affect the customer. For example, within a retail system, the main purpose of the logistics system is to move the goods to the store for sale to the customer. While there are many logistical decisions to get the items to the store, we must also consider the effect on the customer in the store. The "back room" logistics in retail stores becomes an important consideration, especially how and when to move the items to the shelf so as to minimize the disruption of customer shopping. This consideration of the disruption to customer shopping causes many (if not most) retailers to have goods delivered in the late evening, with shelf stocking occurring in the overnight hours. This customer service decision drives the replenishment processes, the truck delivery schedule, and ultimately distribution center operations. To further illustrate how the customer affects logistics system design and operation, we will examine the implications of using mobile robots to perform delivery functions within a hospital with a case study. As we will see, in considering service systems it is important to consider the entire system, especially the customer.

11.2.1 Case Study: Mobile Robot Delivery Systems in Hospitals

In previous research, the author was asked to analyze the delivery functions within a hospital, and in particular, examine whether or not autonomous mobile robotic carriers could be utilized in such an environment. Portions of this discussion are based on Rossetti et al. (2000)[*] and Rossetti and Seldanari (2001). The hospital selected for analysis was the University of Virginia Medical Center (UVA-MC) located in Charlottesville, Virginia. At the time of the case study, the UVA-MC was a 591-bed, eight-floor complex and represents a medium-to-large-size hospital facility that, at the time, handled about a 454 daily bed census with close to $420 million in annual operating expenses. The hospital has two elevator banks. One elevator bank is located on the west side of the hospital, while the other is located on the east side of the hospital. Each bank of elevators consists of two rows of three elevators each. For each elevator bank, one row of three is reserved for visitors and the other row is reserved for hospital personnel. Figure 11.1 illustrates the basic layout of the floors of the hospital.

As illustrated in Figure 11.2, the hospital has many delivery components. This case study focuses on the use of mobile robots for pharmacy and clinical laboratory deliveries. The mobile robot examined (see Figure 11.3), in this case study is manufactured and sold by Cardinal Health Inc. (www.cardinal.com/pyxis). The Pyxis HelpMate robotic courier is a fully autonomous robot capable of carrying out delivery missions between hospital departments and nursing stations. The Pyxis HelpMate robotic system uses a specific world model for both mission planning and local navigation. The world is represented as a network of links (hallways) and an elemental move for the robot is navigating in a single

---

[*] With kind permission of Springer Science and Business Media.

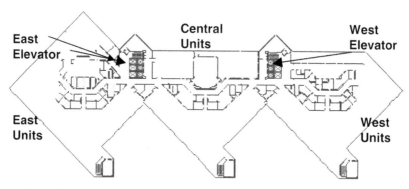

FIGURE 11.1   Generic hospital floor layout.

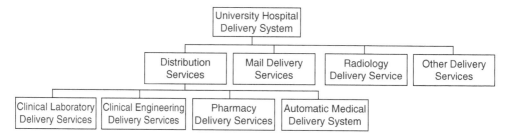

FIGURE 11.2   Components of a hospital delivery system.

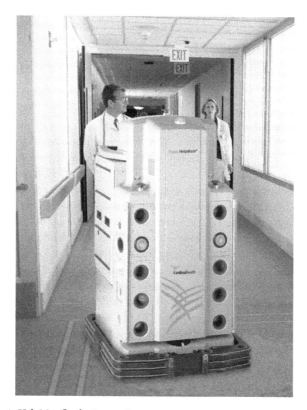

FIGURE 11.3   The Pyxis HelpMate® robotic courier.

hallway, avoiding people and other obstacles. In situations where more than one robot is present, a computerized supervisor properly spaces the robots along the hallways, since the robots compete for space and for the elevators.

The pharmacy and clinical laboratory delivery processes utilize human workers to complete the deliveries. Each hospital floor consists of a number of hospital units. Each hospital unit collects specimens during the course of its operation. The specimens require transport to the clinical laboratory located on the second floor of the hospital, where they are tested. The results of the tests are reported back to the hospital units via the hospital's laboratory information network.

The clinical laboratory process collects specimens that are placed on floors 3 to 8 from the 29 medical units of the hospital. For routine pickups and deliveries, the human courier follows a predefined route. Each courier is assigned two floors: one person for the third and fourth floors, a second person for the fifth and the sixth floors, and a third person for the seventh and eighth floors. Couriers wait in the personnel lounge until it is time to start the shift. At the beginning of the shift, couriers make their way to the top floor of their route and visit each unit assigned to their route on their way to the clinical laboratory. If they have picked up items during the route, they deliver the items to the clinical laboratory; otherwise, they repeat their route. During the shift, there are three breaks that are scheduled for couriers: two breaks of 15 min each and 1 break of 30 min. When a specimen requires STAT delivery, the courier picks up the specimen and then takes the best direct route to the clinical laboratory for delivery. Any items that have already been picked up along the route are also dropped off at the laboratory. The courier then travels back to the unit that was next on the route before he or she responded to the STAT delivery. The determination of whether or not a specimen is STAT is dependent on the nurses or the doctors and their determination of the patient's medical needs.

Courier delivery for pharmaceuticals is broken into two distinct delivery processes. These are the delivery of routine pharmacy medicines and the delivery of STAT pharmacy medicines. Three couriers are assigned to deliver medicines to the appropriate units. Couriers performing routine deliveries are each assigned three floors: (3, 4, 5) for one courier and (6, 7, 8) for another courier. One courier performs STAT delivery to all the floors. The delivery process for routine pharmacy is similar to the clinical laboratory delivery process. The courier picks up the medicines at the central pharmacy located in the basement of the hospital and destined for units along his or her route. The courier uses the elevator to travel to the top floor of the route and then visits each unit on the route. At the nursing station within the unit, a box is kept for pickups and deliveries. The courier drops off the medicines at his or her destinations and picks up any unused medicine for return to the pharmacy. After completing all the floors on the route, the courier returns to the pharmacy to drop off unused medicines and to pick up a new batch of medicines for delivery. Figure 11.4 illustrates that the demand for pharmacy delivery services varies significantly by hospital unit and by time of day. The clinical laboratory demand characteristics are similar, but with less variability. These sorts of demand processes are often characteristic of service systems and complicate the scheduling of the staffing requirements. In addition, for a hospital, we must take into account the fact that the facility must provide service 24 hours per day, 7 days a week, 365 days per year.

As indicated, the human courier process is labor intensive and requires low-skilled labor. This sort of process is a prime candidate for automation. The four key issues of within-the-facility logistics must be addressed. The solution approach for this case study involved material handling design and selection. Because of the service nature of this system, special care was taken to ensure that capacity and performance can meet customer requirements. In order to understand whether or not mobile robots would be beneficial for this situation, we used simulation modeling and cost analysis to compare alternatives involving robots to the current operating situation. The simulation models were built using the Arena simulation environment. Both the robotic couriers and the human couriers were modeled using the guided transporter modeling constructs available within Arena. To analyze the delivery processes, four models were developed. The first model described the current system with human couriers. The second and third models described the operation of the system with mobile robots serving as the primary delivery mechanism with independent operation of clinical laboratory and pharmaceutical deliveries.

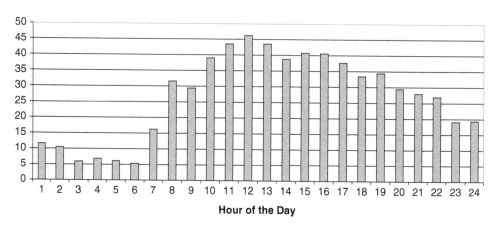

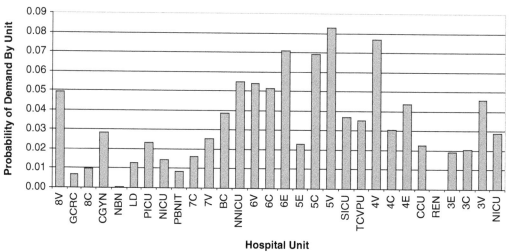

FIGURE 11.4   Pharmacy time demand process. (From Rossetti, M.D., Kumar, A., and Felder, R. *Health Care Management Service*, Vol. 3, pp. 201–213, 2000; Springer Sciences and Business Media. With permission.)

The mobile robot models acted essentially the same as the human courier model except for minor changes to accommodate the speed of the robots, the amount of time they wait at the hospital units to be loaded by nurses, and elevator interactions. The fourth model combined the processes associated with clinical laboratory and pharmaceutical deliveries and utilized mobile robots as the delivery mechanism.

Within this simulation modeling paradigm, each floor consists of a network of links and nodes. Links and nodes have a capacity limiting the maximum number of robots allowed to occupy them at any time. A careful choice of link and node capacity makes it possible to avoid deadlocks in floors where more than one robot can travel simultaneously; moreover, it enables the modeling of a "space cushion" between two consecutive robots in the same hallway. The length of the space cushion can be set by choosing the appropriate minimum length for each link. Only three relevant differences distinguish human couriers from robots: human couriers do not compete for space while they are walking along the hospital hallways. During an 8-h shift each courier is allowed to take three breaks: two breaks are short (15 min), one break is long (30 min), and robots must use the elevators to move from one floor to another. One elevator in each bank is retrofitted for use by the robots. If a robot needs to use the elevator, the elevator's calling mechanism ensures that the elevator does not respond to any other floor calls.

In addition, the elevator has a weight sensor to indicate if all riders have disembarked. Essentially, these rules ensure that no people are in the elevator when the robot is using it. If more than one robot is requesting the elevator, the robots are served according to FIFO logic, regardless of their current floor. Human couriers use the elevators as any other human passenger since they do not have a special priority. All the other human courier behaviors are modeled the same as the robot's behaviors except for traveling speed.

Robot speed modeling deserves particular mention: it was modeled as a triangular random variable to account for possible interference with humans in the hospital hallways (if a robot is blocked, it automatically stops and computes a path around the obstruction) and to account for velocity changes because of curves and long, straight hallways. A new value of the velocity for a link is assigned every time the robot traverses the link. The parameters for the triangular distribution (min, mode, max) were (0.274 m/s, 0.508 m/s, 0.63 m/s) based on vendor recommendations. Human courier velocity was modeled in the same way to account for courier unscheduled breaks or fatigue. The parameters for the triangular distribution (min, mode, max) for humans were (0.381 m/s, 0.762 m/s, 0.875 m/s) based on standardized data and observation of the human couriers. The model data and the model were verified and validated using standard statistical simulation techniques; see, for example, Chapter 11 of Banks et al. (1996). Further details of the simulation models are given in Rossetti et al. (2000) and Rossetti and Seldanari (2001).

In comparing the system with human carriers to the systems involving robotic carriers, a number of key issues needed to be examined. The first issue was to determine whether or not the robotic carriers were competitive in terms of cost, both in operating cost and in terms of any capital investment. The second issue was that the robotic carriers be competitive in terms of performance. The key performance measures included (but were not limited to) turnaround time, delivery time variability, cycle time, and utilization. Turnaround time refers to the time from when the delivery is requested to when it is completed. Delivery time variability is the standard deviation of the turnaround time. Cycle time is the time it takes a courier to complete one cycle of its assigned route, and utilization refers to the percentage of total time spent carrying items for delivery.

The simulation models were run and the performance measures collected. As indicated in Tables 11.1 and 11.2, robotic couriers are extremely cost-effective. The cost of the robots included the cost of the equipment (robots, batteries, carrying compartments, robot communication system, door actuators, etc.), cost of installation, and cost of operation and maintenance (service contract, monitoring personnel, energy, etc.). A net present value calculation was performed over a five-year planning horizon. The cost of the courier system was based on a loaded hourly rate of $10.26/h for 24 h/day and 365 days/year. In order to obtain full yearly coverage over sick days, vacations, etc., one person is considered equivalent to 1.4 FTE.

The two-robot alternative has lower cost, but it has difficulty matching the performance of the three-courier model. A one-for-one replacement of the couriers with robots reduces the cost by roughly 74%, with only an approximate 20% increase in turnaround time. The six-robot alternative dominates the other alternatives by maintaining low cost and significantly improving the turnaround time and the delivery variability. For the combined model in Table 11.3, robots perform both pharmacy and clinical laboratory delivery. The combined delivery had a 75% decrease in cost, a 34% decrease in turnaround

TABLE 11.1 Clinical Laboratory Summary of Performance Measures

	Two Robots	Three Robots	Six Robots	Courier
COST	$81,110	$107,605	$178,027	$407,614
TAT (min)	47.28 (1.97)	33.54 (1.07)	18.9 (0.44)	28.08 (2.16)
DV (min)	24.77 (1.87)	16.67 (0.82)	8.63 (0.04)	20.72 (2.83)
CT (min)	67.03 (2.01)	42.25 (0.87)	20.72 (0.33)	26.3 (1.57)
UTIL	92.50% (0.44)	91.90% (0.63)	81.70% (1.52)	88.33% (0.68)

Source: Rossetti, M.D., Kumar, A., and Felder, R. *Health Care Management Science*, Vol. 3, pp. 201–213, 2000; Springer Sciences and Business Media. With permission.

TABLE 11.2 Pharmacy Model Summary Results

Performance Index	Alternatives		
	Two Robots	Three Robots	Courier
Cost	$86,141.00	$104,579	$281,742
Turnaround time	102.25 (15.06)	71.16 min (13.25)	55.87 (9.21)
Delivery variability	86.88 (22.97)	57.87 min (19.724)	49.22 (13.86)
Average cycle time	57.37 (2.11)	42.35 min (1.255)	30.86 (1.11)
Utilization	13.28% (3.22)	56.87% (7.323)	11.69% (1.97)

Source: Rossetti, M.D., Kumar, A., and Felder, R. *Health Care Management Science,* Vol. 3, pp. 201–213, 2000; Springer Sciences and Business Media. With permission.

time, and a 38% decrease in delivery variability while virtually matching the cycle time and utilization performance of the courier-based system.

From these results, we can see that mobile robots are a highly competitive alternative to human couriers. To further explore the indirect effects of such a system, an extensive sensitivity analysis involving the trade-offs between both quantitative and qualitative factors using the analytical hierarchy process (AHP) was performed. The AHP [see Saaty (1977, 1994, 1997)] is a technique that can be applied to address problems that have multiple conflicting performance measures. Several examples of the application of AHP to transportation system planning and automation introduction in manufacturing can be found in the literature [see e.g., Albayrakoglu (1986); Khasnabis (1994); and Mouette and Fernandes (1997)]. AHP is based on the analysis of a hierarchy structure. Decision analysis techniques like AHP are especially relevant within service industries because these techniques attempt to incorporate nonquantitative measure of performance into the decision process.

Figure 11.5 presents the complete AHP tree used in the analysis. As can be seen in the figure, both quantitative performance measures are captured as well as qualitative performance measures. In particular, there are other important considerations such as safety, noise, and technical innovation that are important to both the users of the delivery system (doctors/nurses/administrators) and to the customers of the service system (patients). For example, this analysis incorporates the effect of additional elevator delay on the patients and their families caused by the use of the elevator by the robots. In addition, the robots make noise as they actuate the hospital unit doors. These issues are important from a patient's point of view.

The stability of the system response after modifications in the decision-maker preference structure affecting the AHP Global Priorities was checked through a sensitivity analysis. The analysis investigated whether or not preference structure modifications could benefit the human-based solution. The analysis showed that the robotic delivery system is preferable with respect to the human-based system with an overall confidence level of 99.986% based on the preference structure of the decision-makers. In addition, while some changes to the priorities can allow the human-based system to be preferable, the priority

TABLE 11.3 Summary for Combined Delivery

Performance Index	Results in Absolute Terms	
	Six Robots	Courier
Cost	$178,076	$689,356
Turnaround time	28.14 (1.461)	42.69 (5.055)
Delivery variability	12.30 (1.404)	20.01 (2.963)
Average cycle time	28.97 (0.722)	28.70 (0.937)
Utilization	89.69% (0.508)	86.72% (1.031)

Source: Rossetti, M.D., Kumar, A., and Felder, R. *Health Care Management Science*, Vol. 3, pp. 201–213, 2000; Springer Sciences and Business Media. With permission.

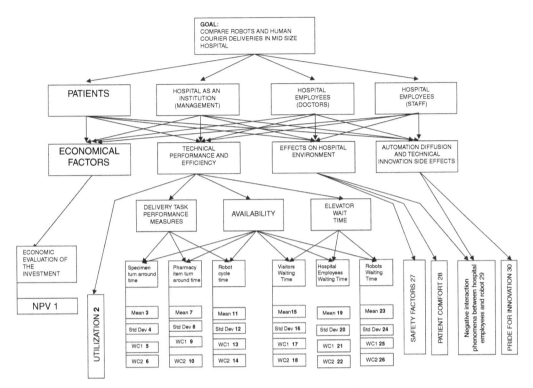

FIGURE 11.5   AHP hierarchy structure. (From Rossetti, M.D. and Seldanari, F., *Computers and Industrial Engineering*, 41, 309–333, 2001. With permission from Elsevier.)

weights were not realistic for practical situations. A complete discussion of the sensitivity analysis is found in Rossetti and Seldanari (2001).

In this case study, we illustrated that when analyzing service systems, there is a critical need to consider all relevant factors that may directly or indirectly affect the customer. In this case study, we saw that through a systems perspective, considering the current system, alternatives, and a complete understanding of the socio-technical factors, we can make strong decisions involving logistics systems that support service systems. Examining the effect of logistics alternatives in this manner is especially important in within-the-facility logistics because of the proximity of the customer to the logistics solution. In the following section, we briefly discuss a case study involving between-the-facility logistics or supply chain management. In these types of situations, the end customer often drives the overall requirements for the logistics solution, but the actual solution involves rather typical supply chain decision-making concepts.

11.3   Between-the-Facility Logistics

Between-the-facility logistics refers to the structure and processes required to move materials between facilities in order to meet customer requirements. The structure and processes constitute the logistical network. Logistical network design involves determining the number, location, type, and capacity of the facilities. In addition, the network connectivity (who supplies what to whom) and the inventory requirements become important issues in creating time and place utility for customers. Supply chain management constitutes the key paradigm by which logistics network designs will impact customer service. A large amount of research has been conducted on supply chain modeling, inventory policy determination, and operations research techniques applied to logistics. This volume highlights many of these supply chain issues.

Within the area of logistical network design, the specialized area of facility location (discussed elsewhere in this volume) has been able to make significant contributions to how service facilities are located. As an example, extensive research has been performed on questions of network design for which service coverage is the main issue. The p-center, p-median, and other covering problems all are specific examples of models that are motivated from service coverage. For example, the p-center problem attempts to locate p facilities on a network in such a way as to minimize the maximum travel time from a user to the closest facility. This type of model is important in locating emergency response teams, hospital facilities, schools, etc. A review of network design problems can be found in Daskin (1995). In addition, a comprehensive review of the application of operations research techniques to emergency services planning can be found in Goldberg (2004). Min and Melachrinoudis (2001) examine the use of location-allocation models within the banking industry. In location-allocation problems, we jointly solve the location of the facilities as well as the allocation of the services that the facilities provide. An interesting aspect of many service systems is the hierarchical nature of the systems. For example, banking customers can receive different services from full-service banks, satellite banks, automatic teller machines, and drive-through facilities. Min and Melachrinoudis (2001) developed an optimization model for deciding which services to offer at which levels of the hierarchy while maximizing the profitability and customer response to services and while minimizing the risks associated with offering the services. These types of problems often require the dynamic and stochastic modeling of customer behavior [see, e.g., Wang et al. (2002), who incorporate queuing analysis into the location of automatic teller machines within a service network].

Location analysis provides the network structure within which logistic processes must be allocated and then operated. It is the operation of these processes within the supply chain that motivates service industries to consider two main alternatives: in-house logistics versus outsourced logistics. For many service industries like banking and health care, logistics is simply a support activity. Because of this, two key questions that firms in the service industry face are (*i*) whether or not logistics strategy is a key to their success with their customers, and (*ii*) whether or not they should execute the logistic processes internally or whether they should rely on an external provider of logistics.

We will illustrate these concepts with a discussion of logistics within the healthcare industry. According to Burns and Wharton School Colleagues (2002), the health care value chain consists of the producers of medical products, such as pharmaceuticals and medical devices; the purchasers of these products, such as wholesalers and group purchasing organizations; the providers including hospitals, physicians, and pharmacies; the fiscal intermediaries (insurers, HMOs, etc.); and, finally, the payers (patients, employers, government, etc.). Within the healthcare value chain (see Figure 11.6), physical products (drugs, devices, supplies, etc.) are transported, stored, and eventually transformed into health care services for the patient. Burns and Wharton School Colleagues (2002) discuss a number of configurations of health care value chains that are used to manage the inventory supply process from producer to payer. Each configuration will result in its own performance in terms of cost and reliability of service in delivering the medical services.

Hospital executives must make decisions regarding whether to maintain inventory and distribution functions in-house, to outsource them to external firms, or to engage in collaborative ventures with such external firms. Figure 11.6 illustrates the health care value chain with the key decision of in-house logistics and outsourced logistics contrasted. This decision is typically a major issue in corporate strategy for many service industries. For example, in recent years, hospitals have formed large "integrated delivery networks" (IDNs) that combine multiple hospitals and often, large physician groups. A chief intent of these strategies has been to achieve economies of scale to reduce rising health care costs, although such economies have proved elusive (Burns and Pauly, 2002). Initially, IDNs sought these economies by consolidating finance and planning functions, yielding little cost savings. IDNs have only recently begun to pursue these economies through integration of their supply chain activities. Because such activities (products, services, and handling) constitute up to 30% of a hospital's cost structure, the potential for cost savings through consolidation and scale economies seems more promising.

Based on these recent trends, it is clear that hospital service providers have recognized that efficient logistics is a key to their success in holding down costs. Even if a firm decides that logistics is a key

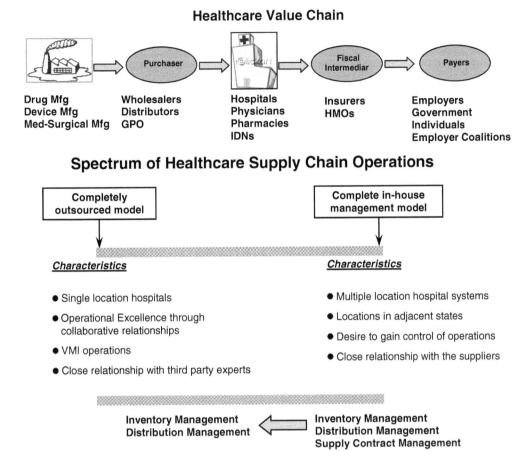

FIGURE 11.6 The health care value chain.

component of its service delivery strategy, it does not mean that the firm should perform the logistics functions internally. It may instead consider its service its key competency and perform that function exceptionally well, while delegating logistics to a firm that excels in logistics. Thus, the question of whether or not these functions should be internalized or externalized remains open and depends on many factors.

To illustrate some of these factors, we will discuss two contrasting health care providers: Mercy Health Systems and The Nebraska Medical Center. Sisters of Mercy Health Systems is a hospital system based in St. Louis, Missouri, that operates facilities and services in a seven-state area encompassing Arkansas, Kansas, Louisiana, Mississippi, Missouri, Oklahoma, and Texas (see Fig. 11.7). With a total of 26,000 coworkers, 815 integrated physicians, 3,100 medical staff members, and 3,600 volunteers. Sisters of Mercy Health Systems (hereforth referred to as Mercy) can be classified as a medium- to large-scale hospital system. Its members include 18 acute care hospitals providing more than 4,000 licensed beds, a heart hospital, a managed care subsidiary (Mercy Health Plans), physician practices, outpatient care facilities, home health programs, skilled nursing services, and long-term care facilities. Mercy is also the ninth largest Catholic health care system in the United States, based on net patient service revenue. Established in 1986, Mercy is operated through regional "Strategic Service Units" (SSUs) which enjoy mutual benefits of local management and system strength. Mercy programs and services are driven by the specific needs of each SSU's community, and local operating autonomy is valued. A key component of Mercy's service is its focus on five quality factors: information about programs that educate patients,

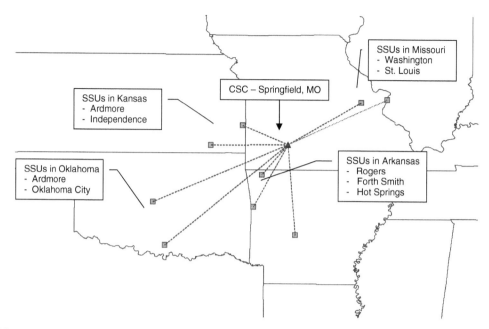

FIGURE 11.7 Mercy Health Systems strategic service units.

technology that enables and expands service, customized patient service experience, doctor quality, and nurse quality. As we can see, the key competency for Mercy is patient service, not logistics; however, Mercy recognized that logistics can be a key enabler in improving its quality factors.

As part of its quality improvement efforts, Mercy launched a program called Mercy Meds. Mercy Meds entailed a comprehensive transformation of medication administration processes that incorporated technology, supply chain management, strategic partnerships and improved work processes to enhance safety and efficiency in the delivery of medications to patients. Supply chain management activities are a key part of Mercy Meds and play an important role in the provision of high-quality service to patients. The newly designed process begins at a consolidated services center (CSC) which serves as a centralized warehouse and distribution center for the entire Mercy organization. As a part of the Mercy Meds scheme, Mercy took a unique step of becoming its own pharmaceutical distributor. Through a partnership with the nation's largest pharmaceutical wholesaler, AmerisourceBergen, the CSC purchases, stores, repackages, bar-codes, and distributes pharmaceuticals used across Mercy.

Through the Mercy Med effort, Mercy developed in-house expertise that eventually led Mercy to establish an in-house organization, Resource Optimization and Innovation (ROi), to manage its internal and external supply chain activities. ROi operates as a "for profit" internal entity which is now responsible for group purchasing functions, logistics, distribution, and other supply chain activities for the entire Mercy system. For example, Mercy now operates a private fleet in order to meet its SSU's specific delivery requirements. Mercy's success in implementing its in-house model can be attributed to its size (8 SSUs, more than 3,600 beds), location (relatively close geographically), specific logistics requirements (high fill rate requirements, specific delivery requirement), and the recognition by top management that logistics can be a key enabler.

To contrast the in-house model, we will briefly discuss how The Nebraska Medical Center (NMC) handles its inventory and logistics processes, and why it chose to outsource its logistics functions. The NMC formed in October 1997 with the merger of Clarkson Hospital, Nebraska's first hospital, and University Hospital, the teaching hospital for the University of Nebraska Medical Center (UNMC), located in Omaha. The NMC is a single-location, medium-sized hospital with 689 beds and more than 950 physicians. Prior to outsourcing its inventory functions, the NMC operated as many other similar-size hospitals. It had its own warehouse, located at a site close to the hospital, where all pharmaceutical, medical surgical, and other supplies were received and stored. Demand for various items from the hospital was fulfilled through the warehouse. The NMC purchased most of its pharmaceutical and medical surgical requirements from a large distributor of national name-brand medical/surgical supplies to hospitals.

Like many hospitals of its size, the NMC managed its own inventory and supply chain operations and placed orders with the distributor as required. Typically, hospitals do not have strong and matured inventory and supply chain management functions, as those are not core competencies of the hospitals. But these functions may have a significant impact on the overall customer service offered by the hospitals and also on the cost of operations, as almost 30% of a hospital's total operating cost are materials costs which include pharmaceutical and medical surgical supplies. The executives at NMC started thinking about improving the operations and recognized that there was a significant opportunity in inventory management and supply chain functions. The NMC wanted to focus on its core competency, health care services, and wanted to offer differentiation in health care services at a lower cost and better returns on investment. After benchmarking and studying the vendor-managed inventory program between P&G and Wal-Mart, NMC decided to look for similar opportunities in NMC operations. Following over two years of studying the intricacies of its operations, NMC identified a third-party partner (Cardinal Health) who could run its inventory and supply programs. Cardinal Health was picked because of its capabilities of offering inventory management and supply chain services by integrating its information systems with NMC's systems. Cardinal Health already had inventory hardware and software systems in place via its Pyxis inventory systems and thus already understood many of NMC's requirements.

Under the new program Cardinal would not only manage inventory of all pharmaceutical and medical surgical supplies at NMC but also own the entire inventory at NMC for a monthly fee and a share in the yearly savings. One of the main drivers for this initiative was the freeing up of capital for NMC. Typically, capital is scarce at smaller-sized hospitals, and the new initiative freed over $80 million for NMC. This capital, which was investment in inventory, could then be reinvested in other technological improvements and improvements in service offerings. Now Cardinal manages the inventory levels at NMC and is penalized for any out-of-stock situations. Inventory locations in departments are replenished four times a week, on Monday, Tuesday, Thursday, and Friday, from its distribution center in Omaha, Nebraska, which is just two miles from NMC. The replenishment lead time from this distribution center is around 9 h and orders can be placed at the piece level. Any emergency requirements are fulfilled in a shorter time if necessary. NMC also receives some supplies from Cardinal Health's distribution centers in Chicago, Illinois, and Kansas City, Kansas, which typically have a lead time of three days, but orders can be placed only in case packs. NMC's in-stock performance has improved significantly after implementation of this new program and is currently over 99.5%.

The NMC is an example of a hospital that has neither the demand levels necessary to bypass the Group Purchasing Organization (GPO) nor a matured supply chain function within the hospital. Thus, outsourcing the supply chain operations and inventory management functions to a third-party expert who could use its expertise has proven to provide significant benefits. The dynamics of supply chain operations in the health care industry are constantly changing as the roles of distributors, GPOs, and manufacturers change. Hospitals need to identify the best supply chain strategy based on size, in-house management capabilities and the presence of trustworthy and capable third-party experts in the vicinity that can reduce the operating costs and improve health care services by focusing on their core competencies. Both the examples, Mercy Health Systems and The Nebraska Medical Center, illustrate that no matter the size of the service system, logistics can become a key enabler and a strategy for success in providing improved customer service.

11.4   Summary

Service industries differ from traditional manufacturing industries in many important ways. One of the key differences is the cost structure associated with service firms. The study performed for the Council of Logistics Management [Arthur D. Little, Inc. and Penn State University (1991)] indicates that up to 75% of the total operating cost for service-oriented companies is labor and capital. That is, the majority of costs are fixed. In contrast, manufacturing firms have a different cost structure, with much more cost tied up in inventory. The cost structure of service firms presents a unique problem in that an extra customer adds little marginal cost, but may add significant revenue. In other words, the customer is truly "king." Because so much cost is tied up in labor and capital, it is critically important that service industries design their service delivery systems from a customer and cost-efficient standpoint. The optimal design of logistics functions is and will continue to be a key enabler for service companies looking for competitive advantage in the marketplace.

Acknowledgments

I would like to thank Amit Bhonsle for his excellent work on the Mercy Health Systems and The Nebraska Medical Center examples and for allowing me to utilize some of his work. In addition, I would like to thank Marty Comody at The Nebraska Medical Center and Gerald Ledlow at Mercy Health Systems for allowing us to perform our case study work. I would also like to thank Francesco Selandari for all his efforts on the original Pyxis Helpmate robotic system research that led to the case study discussed in this chapter. Finally, I would like to thank Sannathi Chaitanya for assisting in finding some of the references associated with this chapter.

References

Albayrakoglu, M.M. (1986). Justification of new manufacturing technology: a strategic approach using the analytic hierarchy process, *Production and Inventory Management Journal*, vol. 37, no.1, 71–76.

Arthur D. Little, Inc. and Penn State University (1991). *Logistics in Service Industries*, The Council for Logistics Management, Oak Brook, IL.

Banks, J., Carson, J.S. II, and Nelson, B.L. (1996). *Discrete-Event System Simulation*, Englewood Cliffs, NJ: Prentice-Hall International.

Burns, L.R. and Wharton School Colleagues (2002). *The Health Care Value Chain: Producers, Purchasers, and Providers*, San Francisco: Jossey-Bass.

Burns, L.R. and Pauly, M.V. (2002). Integrated delivery networks: a detour on the road to integrated healthcare?, *Health Affairs*, July–August, 128–143.

Cook, D.P., Goh, C.-H., and Chung, C.H. (1999). Service typologies: a state of the art survey, *Production and Operations Management*, vol. 8, no. 3, 318–338.

Daskin, M.S. (1995). *Network and Discrete Location: Models, Algorithms and Applications*, John Wiley and Sons, Inc., New York.

Goldberg, J.B. (2004). Operations research models for the deployment of emergency services vehicles, *EMS Management Journal*, vol. 1, no. 1, January–March, 20–39.

Johnson, T. (1999). Using Simulation to Choose Between Rental Car Lot Layouts, in the *Proceedings of the 1999 Winter Simulation Conference*, P.A. Farrington, H.B. Nembhard, D.T. Sturrock, and G.W. Evans, eds., IEEE, Piscataway, New Jersey.

Khasnabis, S. (1994). Prioritizing transit markets using analytic hierarchy process, *Journal of Transportation Engineering*, vol. 120, 74–93.

Mielke, R., Zahralddin, A., Padam, D., and Mastaglio, T. (1998). Simulation Applied to Theme Park Management, in the *Proceedings of the 1998 Winter Simulation Conference*, D.J. Medeiros, E.F. Watson, J.S. Carson, and M.S. Manivannan, eds., IEEE, Piscataway, New Jersey.

Min, H. and Melachrinoudis, E. (2001). The three-hierarchical location-allocation of banking facilities with risk and uncertainty, *International Transactions in Operational Research*, vol. 8, 381–401.

Mouette, D. and Fernandes, J.F.R. (1997). Evaluating goals and impacts of two metro alternatives by the AHP, *Journal of Advanced Transportation*, vol. 30, 23–35.

Murdick, R.G., Render, B., and Russell, R.S. (1990). *Service Operations Management*, Allyn and Bacon, Boston, MA.

Rossetti, M.D. and Turitto, T. (1998). Comparing static and dynamic threshold based control strategies, *Transportation Research Part A*, vol. 32, no. 8, 607–620.

Rossetti, M.D. and Seldanari, F. (2001). Multi-objective analysis of hospital delivery systems, *Computers and Industrial Engineering*, vol. 41, 309–333.

Rossetti, M.D., Kumar, A., and Felder, R. (2000). Simulation of mid-size hospital delivery processes, *Health Care Management Science*, vol. 3, 201–213.

Saaty, T.L. (1977). A scaling for priorities in hierarchy structures, *Journal of Mathematical Psychology*, vol. 15, 234–281.

Saaty, T.L. (1994). How to make a decision: the analytic hierarchy process, *Interfaces*, vol. 24, no. 6, 19–43.

Saaty, T.L. (1997). Transport planning with multiple criteria: the analytic hierarchy process applications and progress review, *Journal of Advanced Transportation*, vol. 29, no. 1, 81–126.

Takakuwa, S. and Oyama, T. (2003). Simulation Analysis of International Departure Passenger Flows in an Airport Terminal, in the *Proceedings of the 2003 Winter Simulation Conference*, S. Chick, P.J. Sanchez, D. Ferrin, and D.J. Morrice, eds., IEEE, Piscataway, New Jersey.

Wang, Q., Batta, R., and Rump, C.M. (2002). Algorithms for a facility location problem with stochastic customer demand and immobile servers, *Annals of Operations Research*, vol. 111, no. 1–4, 17–34.

12

Logistics as an Integrating System's Function

12.1 Introduction **12**-1
12.2 Logistics—Total "System's Approach" **12**-2
12.3 Logistics in the System Life Cycle **12**-4
Logistics in the System Design and Development
Phase • Logistics in the Production and/or Construction
Phase • Logistics in the System Operation and Sustaining
Support Phase • Logistics in the System Retirement and
Material Recycling/Disposal Phase
12.4 Summary and Conclusions **12**-19
12.5 Case Study—Life-Cycle Cost Analysis **12**-20
Description of the Problem • Summary

Benjamin S. Blanchard
*Virginia Polytechnic Institute and
State University*

12.1 Introduction

The objective of this chapter is to view logistics from a total system's perspective (i.e., the "total enterprise") and within the context of its entire life cycle, commencing with the initial identification of a "need" and extending through system design and development, production and/or construction, system utilization and sustaining support, and system retirement and material recycling and/or disposal.

Historically, logistics has been viewed in terms of activities associated with physical supply, materials flow, and physical distribution, primarily associated with the acquisition and processing of products through manufacturing and the follow-on distribution of such to a consumer (customer). The emphasis has been on relatively small consumable components and not on "systems" as an entity. More recently, the field of logistics has been expanding to greater proportions through the development of supply chains (SCs) and implementation of the principles and concepts of supply chain management (SCM), with logistics being a major component thereof. Even with such growth and redefinition, the emphasis has continued to be on the processing of relatively small components in relation to manufacturing and production processes and the establishment of associated supplier networks. The issues dealing with initial system and/or product design, system utilization and sustaining life-cycle support, and system retirement and material recycling and/or disposal have not been adequately addressed within the current spectrum of logistics.

An objective and challenge for the future is to address logistics in a much broader context, reflecting a total system's approach. The interfaces and interaction effects between the various elements of logistics and the many other functional elements of a system are numerous and their interrelationships could have a great impact on whether or not a given system will be able to ultimately accomplish its intended mission successfully. In this context, logistics and its supporting infrastructure, considered as a major element of a total system, can provide an effective and efficient integrating function. Further, there are logistics requirements in all phases of a typical system life cycle, and this integrating function must be life-cycle oriented, as design and management decisions made in any one phase of the life cycle can have a significant impact on the activities in any other phase. Thus, it is important to address this logistics integrating function within the context of the "whole" in order to be life-cycle complete; that is, the implementation of a system's life-cycle approach to logistics.

12.2 Logistics—Total "System's Approach"

In defining a *system*, one needs to consider all of the products, processes, and activities that are associated with the initial development, production, distribution, operation and sustaining support, and ultimate retirement and phase-out of the system and its elements. This includes not only those procurement and acquisition functions that provide the system initially, but also those subsequent maintenance and support activities that enable the system to operate successfully throughout its planned period of utilization. Thus, the makeup of a "system" should include both the prime elements directly related to the actual implementation and completion of a specific mission scenario (or series of operational scenarios) and those sustaining logistics and maintenance support functions that are necessary to ensure that the specified system operational requirements are fulfilled successfully and in response to some specified customer (consumer) need. Accordingly, the "logistics support infrastructure" should be considered (from the beginning) as a major "subsystem" and addressed as such throughout the entire system life cycle.

Referring to Figure 12.1, the various blocks reflect some of the major activities within the system life cycle. Initially, there is the identification of a specific customer/consumer need, the development of system requirements, and the accomplishment of some early marketing and planning activity (block 1). This leads to design and development, involving both the overall system developer and one or more major suppliers (blocks 2 and 3, respectively). Given an assumed design configuration, the production process commences, involving a prime manufacturer and a number of different suppliers (blocks 4 and 3, respectively). Subsequently, the system is transported and installed at the appropriate customer/user operational site(s), and different components of the system are distributed either to some warehouse or directly to the operational site (blocks 5 and 7, respectively). In essence, there is a forward (or "outward") flow of activities, that is, the flow of activities from the initial identification of a need to the point when the system first becomes operational at the user's site, which is reflected by the shaded areas in Figure 12.1.

In addition to the forward flow of activities as indicated in Figure 12.1, there is also a reverse (or "backward") flow, which covers the follow-on maintenance and support of the system after it has been initially installed and operational at the customer's (user's) site. Referring to Figure 12.1, this includes all activities associated with the accomplishment of on-site or organizational maintenance (block 7), intermediate-level maintenance (block 8), factory and/or depot-level maintenance (blocks 4 and 6), supplier maintenance (block 3), and replenishment of the necessary items to support required maintenance actions at all levels, for example, special modification kits, spare and repair parts and associated inventories, test and support equipment, personnel, facilities, data, and information. System "maintenance" in this instance refers to both the incorporation of system modifications for the purposes of improvement or enhancement (i.e., the incorporation of new "technology insertions" throughout the system life cycle), as well as the accomplishment of any scheduled (preventive) and/or unscheduled (corrective) maintenance required to ensure continued system operation. Associated with a number of the blocks as seen in Figure 12.1 are the activities pertaining to the recycling of materials for other applications and/ or disposal of such, and the supporting logistics activities as required (e.g., blocks 3, 4, 6, 7, and 8).

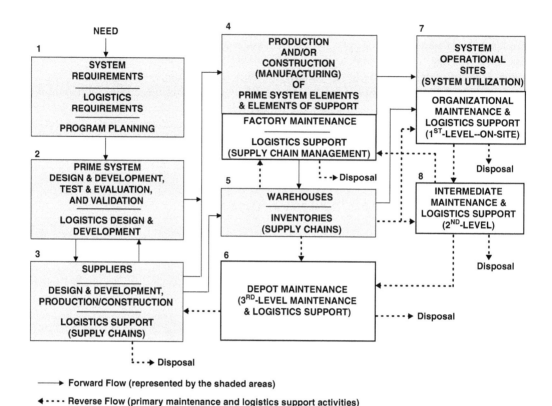

FIGURE 12.1 System operational and logistics support activities.

In the past, the various facets of logistics have been oriented primarily to the "forward" flow of activities shown in Figure 12.1 (i.e., the shaded blocks), and have not addressed the entire spectrum to include the "reverse" flow as well. This, of course, has included the different aspects of "business logistics," emphasized throughout the commercial sector and, more recently, the wide spectrum of activities pertaining to SCs and SCM. More specifically, emphasis has been on (*i*) the initial physical supply of components from the various applicable sources of supply to the manufacturer; (*ii*) the material handling, associated inventories, and flow of items throughout the production process; and (*iii*) the transportation and physical distribution of finished goods from the manufacturer to the customer's operational site(s). With the advent of SCs and SCM, the physical aspects of logistics have been expanded to include the application of modern business processes, contracting and money flow, information transfer, and related enhancements using the latest electronic commerce (EC), electronic data interchange (EDI), information technology (IT), and associated methods and models.*

In the defense sector, the field of logistics has, for the most part, included a majority of the activities identified within both the forward and reverse activity flows presented in Figure 12.1, that is, the various aspects of business logistics and sustaining system maintenance and support. The principles and concepts of integrated logistic support (ILS), introduced in the mid-1960s, emphasized a total integrated

---

* An excellent source for material dealing with the various aspects of business logistics, supply chains, and supply chain management is the Council of Supply Chain Management Professionals (CSCMP), 2805 Butterfield Road, Suite 200, Oak Brook, IL 60523 (web site: http://www.cscmp.org). Some good references include: (a) *Journal of Business Logistics (JBL)*, published by CSCMP; (b) Coyle, J.J., E.J. Bardi, and C.J. Langley, *The Management of Business Logistics*, 7th Edition, South-Western, Mason, OH, 2003; and (c) Frazelle, E.H., *Supply Chain Strategy: The Logistics of Supply Chain Management*, McGraw-Hill, New York, NY, 2002.

system-oriented life-cycle approach, with such objectives as (*i*) integrating support considerations into system and equipment design; (*ii*) developing support requirements that are related consistently to readiness objectives, to design, and to each other; (*iii*) acquiring the required support in an effective and efficient manner; and (*iv*) providing the required support during the system utilization phase at minimum overall cost. The implementation of ILS requirements greatly expanded the scope of logistics in terms of the entire system life cycle. In recent years, the advent and establishment of SCs and SCM, along with the development and application of appropriate technologies, have expanded the field even further. However, while logistics requirements, as currently being practiced in the acquisition and operation of systems, reflect some definite overall improvement, these requirements have and continue to be addressed primarily "after-the-fact," as an independent entity, and downstream in the life cycle. In other words, logistics requirements have not been treated as a major element of a given system, nor have they been adequately addressed in the design process at a time when the day-to-day technical and management decisions being made have the greatest impact on the resulting logistics and maintenance support infrastructure later on. More recently, this deficiency has been recognized and the principles and concepts of acquisition logistics have been initiated to provide additional emphasis on addressing logistics early in the system design and development process.*

At this point, there is a need to progress to the next step by integrating and implementing the best practices of each, that is, the commercial and defense sectors. More specifically, this can be facilitated by (*i*) addressing logistics from a total system's perspective, (*ii*) considering the logistics support infrastructure as a major element of that system, (*iii*) viewing logistics in the context of the entire system's life cycle, and (*iv*) by properly integrating logistics requirements into the system design process from the beginning.

In responding to the first item, it should be noted that there is both a vertical and horizontal integration process that applies here. First, one must consider a system as being included in somewhat of a "hierarchical structure." For example, there may be a need for an airplane, within the context of a higher-level airline, and as part of an overall regional air transportation capability. Logistics requirements must be properly integrated both upward and downward, as well as horizontally across the spectrum at any level. Further, and in response to the second item, the logistics requirements for any given system should be directly supportive of the mission requirements for that system and should evolve from this, and not the reverse. In this context, it is necessary to consider the logistics requirements, at any given level, as a major subsystem and in support of the system-level requirements at that level. Additionally, logistics requirements should be based on the entire life cycle of the system being addressed and, to be meaningful, should be included as an inherent part of the system design process from the beginning. These requirements should be specified from a top-down and/or bottom-up perspective and not just from an after-the-fact bottom-up approach.

12.3 Logistics in the System Life Cycle

While there may be some slight variations relative to specific wording and organization of material, it is assumed that the basic elements of logistics are as shown in Figure 12.2. The intent is to view these overall logistics requirements from both the commercial and defense sectors and to integrate such into major categories providing a "generic" approach. Referring to Figure 12.2, logistics requirements stem from higher-level system-oriented requirements, and can be properly integrated into what may be referred to as the logistics support infrastructure. Inherent within this integration process is the implementation SC and SCM concepts and principles and the application of analytical techniques and models, EC/EDI/IT methods, and so on as appropriate. Thus, the configuration reflected in Figure 12.2

---

* A broad spectrum of logistics from an "engineering" orientation is included in: Blanchard, B.S., *Logistics Engineering and Management*, 6th Edition, Pearson Prentice Hall, Upper Saddle River, NJ, 2004.

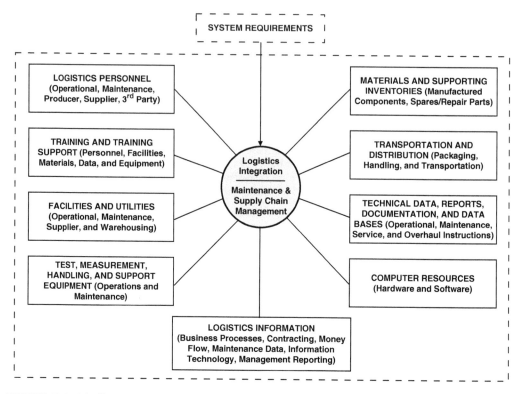

FIGURE 12.2 The "logistics support infrastructure"—a major element of a system.

includes the integration and application of products, processes, personnel, organizations, data and information, and the like, with the objective of ensuring that the system(s) in question can be effectively and efficiently supported throughout its planned life cycle.

The system "life cycle" involves different phases of activity evolving from the initial identification of a need and continuing through system design and development, production and/or construction, system operation and sustaining support, and system retirement and material recycling and/or disposal. While these phases are often considered as being strictly sequential in their relationship to each other, there is actually some concurrency required, as illustrated in Figure 12.3.

As indicated in Figure 12.3, the system life cycle goes beyond that pertaining to a specific product. It must simultaneously embrace the life cycle of the production and construction process, the life cycle of the logistics and system support capability, and the life cycle of the retirement and material recycling and disposal process. In this instance (and for the purposes of illustration), there are four concurrent life cycles progressing in parallel, and the top-down and bottom-up interfaces and interaction effects among these are numerous.

The need for the system comes into focus first. This recognition results in the initiation of a formalized design activity in response to the need, that is, conceptual design, preliminary system design, detail design and development, and so on. Then, during early system design, consideration should simultaneously be given to its production. This gives rise to a parallel life cycle for bringing a manufacturing capability into being. As shown in Figure 12.3, and of great importance, is the life cycle of the "logistics support infrastructure" needed to service the system, its production process, the associated material recycling and/or disposal process, and itself. These individual life cycles must be addressed as an integrated entity, from a top-down (and then bottom-up) perspective, and each time that a new system need is identified one should evolve through this process. This is not to infer an overall lengthy, redundant, and costly activity, but an overall process or "way of thinking." The objective is to address all of

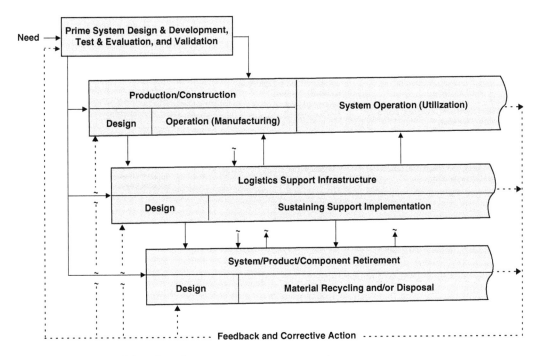

FIGURE 12.3   System life-cycle applications—a concurrent approach.

the producer, supplier, customer, and related activities (and associated resources) necessary in response to an identified need, whenever, wherever, and for as long as required. The "logistics support infrastructure" is an integral part of this requirement, and there are critical logistics activities in each phase of the life cycle.

12.3.1 Logistics in the System Design and Development Phase

Activities in this early phase of the life cycle pertain to design and development of the entire system and all of its elements, and not just limited to design of the prime mission-related components only. This phase commences with the identification of a "need" and evolves through conceptual design, preliminary system design, detail design and development, and test and evaluation, and leads to the production and/or construction phase. Inherent within these activities is the accomplishment of a feasibility analysis to determine the best "technical" approach in responding to the stated need, definition of system operational requirements and the maintenance and support concept, accomplishment of functional analysis and requirements allocation, conductance of trade-off studies and design optimization, system test and evaluation, and so on.

An important part of this early system requirements definition process is the establishment of specific quantitative and qualitative technical performance measures (TPMs), to include appropriate performance-based logistics (PBL) factors, as "design-to" requirements, that is, an input to the overall design process in the form of criteria which lead to the selection of components, equipment packaging approaches, diagnostic schemes, and so on. It is at this point when specific system design requirements are initially defined from the top down, providing design guidelines for major subsystems and lower-level elements of the system (including the logistics support infrastructure). These basic early front-end activities, which constitute an iterative process overall, are illustrated in Figure 12.4, with logistics requirements for the entire system life cycle being noted. These activities constitute an integrated composite of design functions for the individual life cycles presented in Figure 12.3.

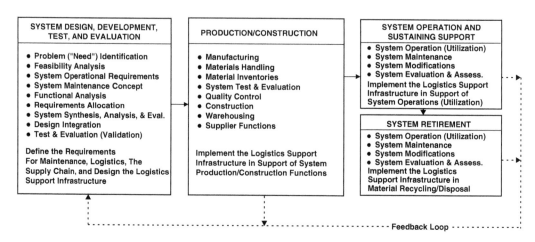

FIGURE 12.4 Logistics in the system life cycle—a system engineering emphasis.

A prime objective is to design and develop a system that not only will fulfill all of the required "operational" needs but also can be supported both effectively and efficiently throughout its planned life cycle. Inherent within the ultimate design configuration are the appropriate attributes (or characteristics) necessary to ensure that the desired functionality, reliability, maintainability, supportability (serviceability), quality, safety, producibility, disposability, and related features are incorporated, and that the system will "perform" as required. In addition to the system effectiveness side of the spectrum (i.e., the technical characteristics), one must deal with the economic factors. These, in turn, must be viewed in terms of the overall system life cycle, that is, life-cycle revenues and cost. If one is to properly assess the risks associated with the day-to-day engineering and management decision-making process throughout system design and development, the issue of "cost" must also be addressed from a total life-cycle perspective, that is, life-cycle cost (LCC). Although individual decisions may be based on some smaller aspect of cost (e.g., item procurement price), the individual(s) involved is remiss unless he or she views the consequences of those decisions in terms of total cost. Decisions made in any one phase of the life cycle will likely have an impact on the activities in each of the other phases.

In addressing the issue of "cost-effectiveness," one often finds a lack of total cost visibility, for example, the unknown factors represented by the bottom part of the traditional "iceberg." For many systems, the costs associated with design and development, construction, initial procurement and installation of capital equipment, production, and so on, are relatively well known. We deal with, and make decisions on the basis of these costs on a regular basis. However, the costs associated with the operation (utilization) and sustaining maintenance and support of a system throughout its life cycle are often hidden. This includes not only the initial acquisition and implementation of the "logistics support infrastructure" for a given system, but also the sustaining maintenance and support of that infrastructure throughout the system life cycle. The lack of total cost visibility has been particularly notable throughout the past decade or so when systems have become more complex and have been modified to include the "latest and greatest technology" without consideration of the cost impact downstream. In essence, we have been relatively successful when addressing the short-term aspects of cost but have not been very responsive to the long-term effects.

At the same time, the past is replete with instances where a large percentage of the total LCC for a given system is attributed to the downstream activities associated with system operation and sustaining support (e.g., up to 75% for some systems). When addressing "cause-and-effect" relationships, one often finds that a significant portion of this cost stems from the consequences of decisions made during the early phases of planning and design (i.e., conceptual and preliminary system design). Decisions pertaining to the selection of technologies and materials, the design of a manufacturing process, equipment packaging schemes and diagnostic routines, the performance of functions manually versus using automation, the design of maintenance and support equipment, and so forth, can have a great impact on the downstream costs and, hence, LCC. Additionally, the ultimate logistics support infrastructure selected for a system during its period of utilization can significantly affect the cost-effectiveness of that system overall. Thus, including life-cycle considerations in the decision-making process from the beginning is critical. From Figure 12.5, it can be seen that the greatest opportunities for impacting total system cost are realized during the early phases of system design. Implementing changes and system modifications later on can be quite costly.

Historically, logistics has been considered "after-the-fact," and activities associated with the "logistics support infrastructure" have not been very popular in the engineering design community, have been implemented downstream in the life cycle, and have not received the appropriate level of management attention. Although much has been done to provide an effective and efficient system support capability (with the advent of new technologies, the implementation of SC and SCM practices, the application of sophisticated analytical models and methods for analysis and evaluation purposes, etc.), accomplishing all of this after-the-fact can be an expensive approach, as system-level design boundaries have already been established without the benefits of allowing for accomplishment of the proper design-support trade-offs. Logistics requirements have been established as a consequence of design and not an integral part of the process from the beginning. Thus, and with future growth in mind, it is imperative that logistics requirements be (*i*) addressed from inception, (*ii*) established—as top-level system requirements are initially determined during conceptual design, (*iii*) developed through the establishment of design criteria ("design-to" factors) as an input to the overall system design process, (*iv*) properly integrated with the other elements of the system on an iterative basis, and (*v*) considered as an integral part of the engineering process in system design and development. This can best be facilitated through development and implementation of a logistics engineering function as an integral part of the overall system engineering process.*

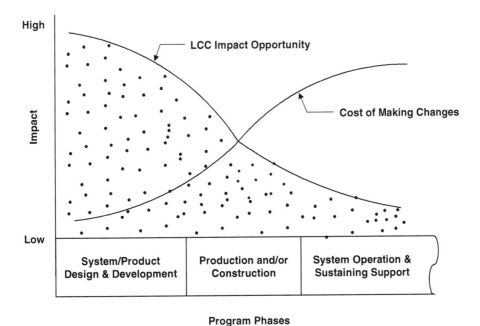

FIGURE 12.5   Activities affecting life-cycle cost.

"System engineering" constitutes an interdisciplinary and integrated approach for bringing a system into being. In essence, system engineering is "good engineering" with special areas of emphasis: (*i*) a top-down approach that views the system as a whole, versus a bottom-up-only process characteristic of many of the more traditional engineering functions; (*ii*) a life-cycle orientation that addresses all of the phases identified in Figures 12.1, 12.3, and 12.4; (*iii*) the establishment of a good comprehensive system requirement baseline from the beginning; and (*iv*) the implementation of an interdisciplinary and integrated (or team) approach throughout the system design and development process to ensure that all design objectives are addressed in an effective and efficient manner. The system engineering process is iterative and applies across all phases of the life cycle.[†]

Referring to Figure 12.4, application of the concepts and principles of systems engineering is particularly important throughout the early stages of system design and development (reflected in the first block), with special emphasis on the establishment of system-level requirements. It is during the conceptual design phase that the basic requirements for logistics and system support are first established, one way or another. It is at this stage during the initial determination of system-level requirements that the design criteria (i.e., "design-to" requirements) for the "logistics support infrastructure" are developed, and that the greatest impact on the downstream activities and LCC can be realized. It is at this early stage that logistics engineering activities should be initiated and inherent within implementation of the system engineering process. Referring to Figure 12.4, a few key system engineering activities, including the development of logistics requirements, are described through the following steps.

1. Problem ("need") identification and feasibility analysis

The system engineering process commences with the identification of a want or desire for something and is based on a real (or perceived) deficiency. For example, the current system capability is not adequate in terms of meeting certain performance goals, is not available when needed, cannot be logistically supported, or is too costly in terms of operation. As a result, a new system requirement is defined along with its priority for introduction, the date when the new system capability is required by the customer (user), and the anticipated resources necessary for acquiring the new system. Through a needs analysis, the basic functions that the system must perform are identified (i.e., primary and secondary), along with the geographical location(s) where these functions are to be performed and the anticipated period of performance. In essence, one must define the "what" requirements (versus the "how"). A complete description of need, expressed in quantitative performance and effectiveness parameters where possible, is essential.

A feasibility analysis is then accomplished with the objective of evaluating the different technological approaches that may be considered in responding to the specified need (i.e., correcting the deficiency). For instance, in the design of a communication system, should one incorporate a fiber-optic, cellular, wireless, or conventional hard-wired approach? In designing an aircraft, to what extent should one incorporate composite material? In designing a new transportation capability, to what degree should the operation of the various passenger vehicles be automated or accomplished through the use of human operators? In the development of new equipment, should packaging considerations favor "logistics transport" by air, by waterway, or by ground vehicle? At this point, it is necessary to (*i*) identify the various design approaches that can be pursued to meet the requirements; (*ii*) evaluate the most likely candidates in terms of performance, effectiveness, logistics requirements, and life-cycle

---

[*] Logistics and the design for supportability (serviceability), implemented as an integral part of the systems engineering process, are discussed in detail in: Blanchard, B.S. and W.J. Fabrycky, *Systems Engineering and Analysis*, 4th Edition, Pearson Prentice Hall, Upper Saddle River, NJ, 2006.

[†] There are different definitions and approaches to "system engineering" being implemented today, depending on one's background and experience. However, there is a common top-down, life-cycle oriented, interdisciplinary, and iterative theme throughout. A good source for definitions and activities in the field is the *International Council On Systems Engineering* (*INCOSE*), 2150 N. 107th St., Suite 205, Seattle, WA, 98133 (Web site: http://www.incose.org).

economic criteria; and (*iii*) recommend a preferred approach for application. The objective here is to select an overall technical approach, and not to select specific hardware, software, and related system components.

It is at this early point of program inception (reflected by block 1, Fig. 12.4) when logistics engineering involvement in the design process must commence. The questions are: (*i*) What type of a logistics support infrastructure is envisioned? (*ii*) Have the logistics requirements been identified and justified through the appropriate system-level trade-off analysis? (*iii*) Is the approach feasible? The objective is to determine the top-level system goals, approach, and general plan for acquisition, and the logistics support infrastructure constitutes a major element of the system in question.

2.   *System operational requirements and the maintenance concept*

Once a system need and a technical approach have been identified, it is necessary to develop the anticipated operational requirements further in order to proceed with system design as planned. At this point, the following questions should be asked: What specific mission and associated operational scenarios must the system perform? Where (geographically) and when are these scenarios to be accomplished and for how long? What are the anticipated quantities of equipment, software, people, facilities, etc., required and where are they to be located? How is the system to be utilized in terms of on–off cycles, hours of operation per designated time period, etc.? What are the expected effectiveness goals for the system (e.g., availability, reliability, design-to-LCC)? What are the expected environmental, ecological, social, cultural, and related conditions to which the system will be subjected throughout its operational life?

The establishment of a comprehensive description of operational requirements from the beginning is necessary to provide a good foundation, or baseline, from which all subsequent system requirements evolve. If one is to design and develop a system to meet a given customer (user) requirement, it is important that the various responsible members of the design team know the mission objectives and just how the system will be utilized to meet these objectives. Of particular interest are the anticipated geographical deployment and the type of operational scenarios to be accomplished. While one certainly cannot be expected to cover all future areas of operation, some initial assumptions pertaining to operational scenarios, anticipated utilization, the stresses that the system is expected to experience, etc., must be made. The question is, how can one accomplish design without having a pretty fair idea as to just how the system will be utilized? This question is particularly relevant when determining the design requirements for reliability, maintainability, supportability (serviceability) and for the logistics support infrastructure. Thus, it is appropriate to develop a few of the more rigorous operational profiles and to design with these in mind. Figure 12.6 provides a partial visualization of what might be included in defining operational requirements.

While all of this may appear to be rather obvious, it is not uncommon for the design community to identify a few of the more easily defined operational requirements, proceed with the design, modify such requirements later on, redesign to meet a changing set of requirements, and so on, which (in turn) can often result in a rather costly process with much time and many resources wasted. The objective here is to initiate a more thorough and comprehensive approach from the beginning, to provide increased visibility early and identify potential problem areas, to allow for completion of the appropriate trade-offs facilitating an effective and efficient system capability output, and to reduce the risks often inherent throughout the design process. The logistics support infrastructure must be an inherent consideration in this early establishment of system-level requirements.

The system maintenance concept, developed during the conceptual design phase, constitutes a "before-the-fact" series of illustrations and statements pertaining to the anticipated requirements for system support throughout the life cycle. The objective is to address the following questions: What logistics and maintenance support requirements are anticipated for the system throughout its life cycle? Where (geographically) and when must these support activities for the system be accomplished? To

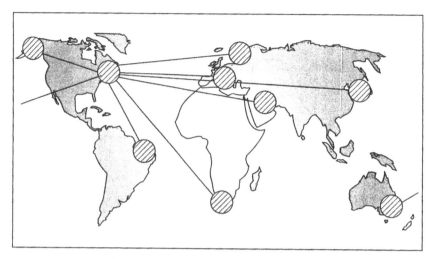

Number of Units in Operational Use per Year

Geographical Operational Areas	Year Number										Total Units
	1	2	3	4	5	6	7	8	9	10	
1. North & South America	–	–	10	20	40	60	60	60	35	25	310
2. Europe (2)	–	–	12	24	24	24	24	24	24	24	180
3. Middle East	–	–	12	12	12	24	24	24	24	24	156
4. South Africa	–	–	12	24	24	24	24	24	24	24	180
5. Pacific Rim 1	–	–	12	12	12	24	24	24	12	12	132
6. Pacific Rim 2	–	–	12	12	12	12	12	12	12	12	96
Total	–	–	70	104	124	168	168	168	131	121	1,054

Average Utilization: 4 Hours per Day, 365 Days per Year

FIGURE 12.6 System operational requirements (overall profile).

what depth (in the design of the system and its hierarchical structure) should maintenance and support be accomplished? To what level(s) should maintenance and support be accomplished (organizational, intermediate, depot, manufacturer, supplier, third-party, etc.)? Who (what organizations) will be responsible for maintenance and support at each level? What are the "design-to" effectiveness requirements for the logistics support infrastructure (e.g., availability, logistics response time, material processing time, reliability of transportation, total logistics cost)? What are the expected environmental conditions to which the system will be subjected during the performance of logistics and maintenance support functions?

Referring to Figure 12.2, the objective is to address all of the major logistics and maintenance support activities associated with both the forward and reverse flows as illustrated. These activities need to be projected further and in the context of the operational requirements for the system in question. Figure 12.7 is included as an extension to the operational requirements illustrated in Figure 12.6.

Whereas in the past these activities were primarily considered after-the-fact and further downstream in the life cycle, the objective here is to attempt to respond to the above questions at an early stage, promote life-cycle thinking early as indicated in Figure 12.5, identify potential high-risk areas that may require special attention, and build the logistics support infrastructure into the system design process

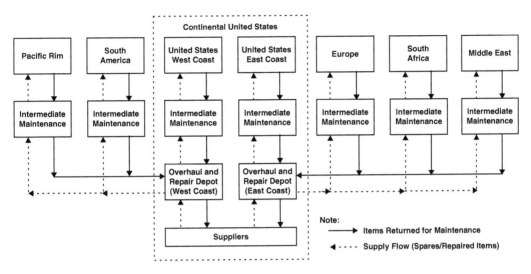

FIGURE 12.7   Top-level system maintenance and support infrastructure.

in a timely manner. The objective is to foster early front-end "visibility," even though it may be difficult (if not impossible) to define all of the basic requirements at this time.

3.   System TPMs

Evolving from the definition of system, operational requirements, and the maintenance concept are the identification and prioritization of key quantitative performance ("outcome") factors. The objective is to establish some specific "design-to" quantitative requirements as an input to the design and development process, as opposed to waiting to see how well the system will perform after the basic design has been completed. Historically, such requirements for specific equipment items, software packages, etc., have been covered partially through the specification of selected performance factors such as speed, throughput, range, weight, size, power output, accuracy and frequency. However, in most cases, the higher-level performance requirements for the system overall have not been specified. For example, to what level of operational availability should the system be designed? To what level of effectiveness must the logistics support infrastructure be designed in order to meet the required availability requirement(s) for the system? To what level of LCC should the system be designed?

By addressing only lower-level requirements for any given system, there often is the tendency to optimize design at the element level in a given system hierarchy, while at the same time suboptimizing the requirements for the system overall. Thus, it is imperative that commencing with the definition of requirements at the system level, considering the various applicable mission scenarios, constitutes a critical early step in accomplishing the activities shown in the first block of Figure 12.4. Further, these early requirements for the system form the basis for establishing lower-level requirements for design of the logistics support infrastructure.

4.   System functional analysis and requirements allocation

The functional analysis constitutes a complete description of the system in "functional" terms. This includes an expansion of all of the activities and processes accomplished through the forward and reverse flows illustrated in Figure 12.1. A function refers to a specific or discrete action (or series of actions) that is necessary to achieve a given objective, that is, an operation that the system must perform to accomplish its mission, a logistics activity that is required for the transportation of material, or a maintenance action that is necessary to restore the system for operational use. Such actions will ultimately be accomplished through the use of equipment, software, people, facilities, data, or various combinations thereof. However, at this point, the objective is to specify the "whats" and not the "hows," that is, what needs to be accomplished versus how it is to be done. The functional analysis is an iterative process commencing with the initial identification of a consumer need and breaking requirements down from the system level, to the subsystem, and as far down the hierarchical structure as necessary to identify input design criteria and/or constraints for the various elements of the system.*

Referring to Figure 12.4, the functional analysis may be initiated in the early stages of conceptual design as part of the problem (need) identification and feasibility analysis task, and can be expanded as required in the preliminary system design phase. Through the development of system operational requirements, operational functions are identified and expanded as shown in Figure 12.8. These operating functions lead

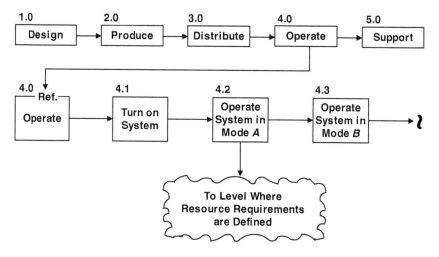

Functional Flow Block Diagram (Partial)

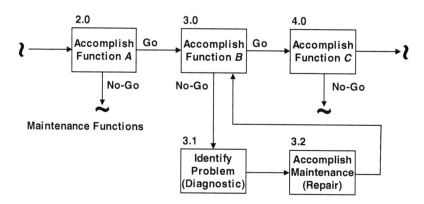

Transition from Operational Functions to Maintenance Functions

FIGURE 12.8 Functional flow diagrams (example).

* In applying the principles of system engineering, one should not identify or initiate the purchase of one piece of equipment, or module of software, or data item, or element of logistics support without first having justified the need for such through the functional analysis. On many projects, items are often purchased early based on what is perceived as a "requirement" but later determined as not being necessary. This practice can turn out to be quite costly.

to the identification of maintenance and support functions as illustrated at the bottom of the figure. The identified maintenance and support functions also constitute an expansion of the established maintenance concept. Development of the functional analysis can best be facilitated through the use of functional flow block diagrams (FFBDs), as illustrated through the expanded integrated flow presented in Figure 12.9.

Referring to Figure 12.1, logistics requirements can initially be identified by describing the specific functions to be accomplished in progressing from block 3 to block 4, from block 4 to blocks 5 and 7, and from block 7 backward to blocks 8, 6, 4, and 3, respectively. This may include a procurement function, material processing function, packaging and handling function, transportation function, warehouse storage function, maintenance function, communication function, data transmission function, and so on. The objective is to identify all of the basic functions that must be accomplished by the logistics support infrastructure for the system being addressed. Accomplishing such at this point in the life cycle enables early "visibility" which will allow for the incorporation of any necessary design changes easily and economically.

Given a good comprehensive functional description of the system, the next step is to commence with the identification of the specific requirements for hardware, software, people, facilities, data, and/or various combinations thereof. The process is to analyze each of the major blocks in the appropriate FFBD to determine the resource requirements necessary for the performance of the function in question. There are input factors, expected output requirements, controls and/or constraints, and mechanisms which must

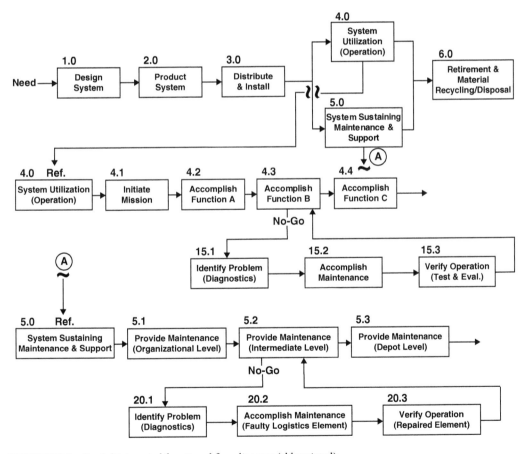

FIGURE 12.9   Partial integrated functional flow diagram (abbreviated).

---

* Refer to Blanchard, B.S., *Logistics Engineering and Management*. 6th Edition, Pearson Prentice Hall, Upper Saddle River, NJ, 2006, Chapter 4, pp. 150–172.

be determined. Through the accomplishment of design trade-offs, the best mix of resource requirements (e.g., hardware, software, people) for each function can be determined. These resources can then be combined, integrated, and lead to the identification of the various lower-level elements of the system, as illustrated in Figure 12.10.

The process of breaking the system down into its elements is accomplished by partitioning. Common functions are grouped, or combined, so as to provide a system packaging scheme with the following objectives in mind: (*i*) system elements may be grouped by geographical location, a common environment, or by similar types of equipment or software; (*ii*) individual system "packages" should be as independent as possible, with a minimum of "interaction effects" with other packages; and (*iii*) in breaking down a system into subsystems, select a configuration in which the "communication" (i.e., negative interaction effects) between the different subsystems is minimized. An overall design objective is to divide the system into elements such that only a very few (if any) critical events can influence or change the inner workings of the various packages that make up the overall system architecture. This leads to an open-architecture approach to design which, in turn, should facilitate the incorporation of system changes, technology insertions, and future improvements later on in the life cycle without causing a major configuration redesign.*

Referring to Figure 12.10, the question is, given the requirements for the system (stated in quantitative terms), what specific design-to requirements should be specified for Unit A, Unit B, logistics support infrastructure, transportation and distribution, facilities, and so on? For instance, if there is an Operational Availability (Ao) requirement of 0.90 for the system as an entity, what should be specified for the logistics support infrastructure in order to meet the system-level requirement? If, on the other hand, the system availability requirement is 0.998, then the requirements for the logistics support infrastructure may be different.

With regard to the logistics support infrastructure, the objective is to establish some specific design-to goals early (before the fact) and develop a balanced configuration that will best respond to the overall system-level requirements, rather than wait until the design is relatively "fixed" and then have to live with the results. One key performance measure of concern is the overall availability of the logistics support

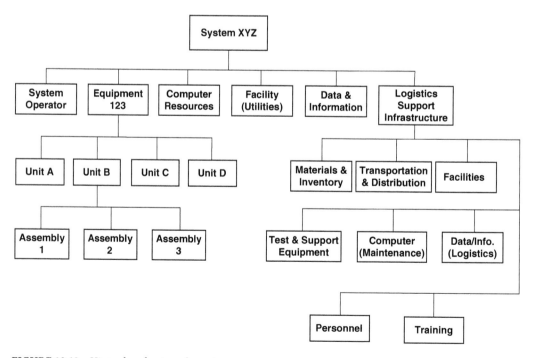

FIGURE 12.10 Hierarchy of system elements.

capability, another is logistics response time, a third is total logistics cost (TLC) or the cost per logistics support action, and so on. Top-level requirements must then be allocated (or apportioned) downward to the level necessary for providing a good and meaningful input for the design. An example of a few design-to goals are noted here:

1. The response time for the logistics support infrastructure shall not exceed four hours.
2. The procurement lead time for the acquisition of any given component shall not exceed 48 hours.
3. The reliability of the overall transportation capability shall be 0.995, or greater.
4. The transportation time between the location where on-site (organizational) maintenance is accomplished and the intermediate-level maintenance shop shall not exceed eight hours.
5. The probability of spares availability at the organizational level of maintenance shall be at least 95%.
6. The warehouse utilization rate shall be at least 75%.
7. The mean time between maintenance (MTBM) for the logistics support infrastructure shall be 1,000 or greater.
8. The time for processing logistics information shall not exceed 10 min.
9. The processing time for removing an obsolete item from the operational inventory shall not exceed 12 hours, and the cost per item processed shall be less than "x" dollars.
10. The TLC for the logistics support infrastructure shall not exceed "y" dollars per support action.

Referring to Figure 12.11, one can visualize the traceability of requirements from the top-down in order to meet such for the overall system as an entity, and performing this function at an early stage in the life cycle will facilitate the accomplishment of the necessary trade-offs and analyses, hopefully leading to an effective and efficient logistics support infrastructure capability. The specific quantitative "design-to" requirements must, of course, be tailored to the overall system-level requirements.

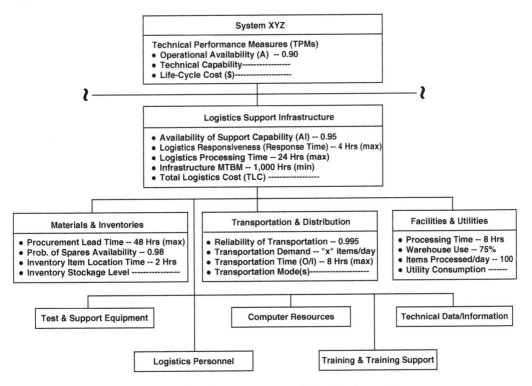

FIGURE 12.11   Allocation of technical performance measures for logistics (example).

5. System synthesis, analysis, and evaluation

Referring to Figure 12.4, given a set of input requirements from the beginning, there is an iterative and continuous process of synthesis, analysis, and evaluation, which ultimately leads to the development of an effective and efficient logistics support infrastructure configuration. For instance, at this point decisions are made pertaining to specific procurement policies, outsourcing requirements, material handling methods, selection of packaging and transportation modes, determination of inventory levels and warehousing locations, establishment of SCs, application of automation techniques, development of information processing and database requirements, determining maintenance levels of repair, and so on. Accomplishing these design-related analyses is facilitated through the selective application of the many and various operations research (OR) models or tools discussed throughout the other chapters of this handbook and in the literature.

6. System design integration

System design begins with the identification of a customer (consumer) need and extends through a series of steps as noted in Figure 12.4. Design is an evolutionary top-down process leading to the definition of a functional entity that can be produced, or constructed, with the ultimate objective of delivering a system that responds to a customer requirement in an effective and efficient manner. Inherent within this process is the integration of many different design disciplines, as well as the proper application of various design methods, tools, and technologies. Figure 12.12 provides an example showing many of the different design characteristics that must be considered and properly integrated in order to meet the specified requirements at the system level.

Effective design can best be realized through implementation of the system engineering process. Logistics engineering must be an integral part of this process, along with other design disciplines as applicable (e.g., electrical engineering, industrial engineering, mechanical engineering, reliability

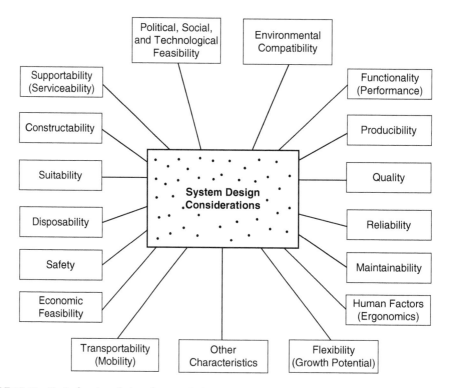

FIGURE 12.12 Typical system design characteristics.

engineering). The role of logistics engineering is twofold: (*i*) to ensure that the prime mission-related elements of the system are supportable (serviceable) through the incorporation of the proper design characteristics or attributes; and (*ii*) to design the logistics support infrastructure to provide the life-cycle support required. In this capacity, logistics engineering can serve as a design integration function across the broad spectrum of the system and throughout its development.

12.3.2   Logistics in the Production and/or Construction Phase

Referring to the four life cycles in Figure 12.3, system-level design requirements (including those for the logistics support infrastructure) evolve from the first life cycle which, in turn, provides an input for the three lower-level life cycles. The more "traditional" logistics requirements and associated SCs, particularly those in the commercial sector, have evolved primarily around the second life cycle. There are logistics engineering functions associated with the design and evaluation of the production or construction process, the development of supplier requirements and supply chains, the development of distribution and warehousing requirements, and so on. However, these requirements must be properly integrated within the context of the whole, that is, the entire spectrum of activity illustrated in Figure 12.1 and the four life cycles shown in Figure 12.3.*

12.3.3   Logistics in the System Operation and Sustaining Support Phase

Throughout the system operational or utilization phase (refer to Fig. 12.4), logistics functions will include providing the necessary support in response to:

1. Changes in system-level requirements and/or when new technologies are inserted for the purposes of enhancement. Each time a new requirement evolves, the system engineering process is implemented as appropriate; that is, there will be some redesign effort, synthesis and analysis, test and evaluation, etc. Such system-level changes will usually result in changes involving not only the prime mission-related elements of the system but the logistics support infrastructure as well.
2. Scheduled and unscheduled maintenance activities for the system and its elements as required. This will involve procurement functions, material handling tasks, transportation and distribution activities, maintenance personnel and facilities, etc. Logistics activities in this area are reflected by the reverse flow in Figure 12.1.

   While logistics activities for system life-cycle support are often not properly addressed from the beginning, the need for such is indeed essential if the system is to ultimately accomplish its planned mission, both effectively and efficiently.

12.3.4   Logistics in the System Retirement and Material Recycling/Disposal Phase

Referring to the fourth life cycle in Figure 12.3, logistics requirements for this phase pertain to the:

1. Retirement and phase-out of system components from the inventory throughout the system operational phase, and the subsequent recycling of these items for other uses and/or for disposal. This function is supplemental to those activities presented in Section 12.3.3.

---

* Logistics in the more "traditional" sense refers to the wide spectrum of activities described in the literature and taught primarily in "business-oriented" programs in the academic community. Such coverage is also described in the other chapters of this handbook. The emphasis herein is to integrate these activities from a system's perspective and within the context of its entire life cycle.

2. Support required when the system (and all of its elements) is no longer needed and is ultimately retired from the operational inventory. This function relates to the recycling and/or disposal of components, the refurbishment of land and facilities for other uses, related data and documentation, and so on.

While this phase of the life cycle is often ignored altogether, the logistics requirements can be rather extensive here, particularly if new facilities are required (for the purposes of material decomposition), new ground handling equipment is needed, special environmental controls are necessary, and so on. Again, the anticipated logistics requirements here must be addressed from the beginning, that is, in conceptual design along with the many other requirements pertaining to the system overall.

12.4 Summary and Conclusions

As a prerequisite to determining the specific logistics requirements for any given system, a good understanding of the overall environment is necessary, that is, the geographical location where the system is likely to be deployed and utilized, nature and culture of the operating agency or organization (the "user"), availability of appropriate technologies and associated resources, system procurement and acquisition processes, political structure, and so on. Additionally, it should be recognized that systems today are operating in a highly "dynamic" world and the need for agility and flexibility is predominant.

While individual perceptions on today's challenges will differ depending on personal experiences and observations, there are a number of trends that appear to be significant. For example, there is more emphasis today on total systems versus the components of systems; the requirements for systems are constantly changing; systems are becoming more complex with the continuous introduction of new technologies; the life cycles of many current systems are being extended for one reason or another while, at the same time, the life cycles of most technologies are becoming relatively shorter (due to obsolescence); there is a greater degree of outsourcing than practiced in the past; and there is more globalization and greater international competition today.

In response to some of these challenges, one needs to view logistics and the various elements of the supply chain (SC) in a much broader context than in the past. More specifically:

1. A total top-down systems approach must be assumed, with the "logistics support infrastructure" included as a major subsystem and oriented to a specific set of mission objectives. Viewing the components of such on an individual-by-individual basis is no longer feasible.
2. A total life-cycle approach to logistics must be implemented. There are logistics requirements and activities in each and every phase of the system life cycle, and these requirements must be treated as an integrated entity since the activities in any one phase could have a significant impact on those in the other phases. If one is to minimize the technical and management risks in the day-to-day decision-making process, then such decisions must be made in the context of the whole.
3. The ultimate logistics support infrastructure configuration must be agile and highly flexible, incorporating an open-architecture approach in design. System-level requirements are constantly changing, and the integration of these requirements (both horizontally and vertically) with other systems are becoming more complex. A new approach to design is necessary to facilitate the incorporation of future changes at minimum total life-cycle cost.
4. Logistics requirements must be established early in the life cycle and in conjunction with the development of system-level requirements from the beginning during the conceptual design phase. This is essential if one is to influence and "optimize" the design for maximum supportability and economic feasibility.
5. The accomplishment of logistics objectives for any type of system can best be realized through implementation of the system engineering process. "Logistics engineering" must be an inherent and active part of this process from inception.

To summarize, the nature of logistics is life-cycle oriented and involves the integration of many different elements, both internally and externally. The elements of logistics must be properly integrated within (as illustrated in Fig. 12.2), integrated with the prime mission-related elements of the system in question, and integrated externally with comparable components of other systems operating in an overall higher-level hierarchy. Thus, one might consider logistics as an integrating system's function.

12.5   Case Study—Life-Cycle Cost Analysis

One of the key TPMs for a system is its projected LCC, which is an indicator of the overall economic value of the system in question. Past experience is replete with instances where a large percentage of the total cost of a given system can be attributed to downstream activities pertaining to logistics and system maintenance and support, that is, the logistics support infrastructure as described throughout this section of the handbook. Further, as illustrated in Figure 12.5, the LCC for a system is highly dependent on design and management decisions made early in the life cycle, and the greatest opportunity for influencing LCC occurs early in the conceptual and preliminary system design phases. Thus, it is at this early stage in the system life cycle that it is essential that the logistics support infrastructure be introduced and addressed within the context of the overall systems design and development process. Further, it is at this early stage that the implementation of life-cycle cost analysis (LCCA) methods can be applied to properly assess various potential system design alternatives and their impact on logistics and system support. Given the significance of LCC as a measure of system economic value and, in particular, logistics support, it was decided to include an abbreviated LCCA case study in this section of the handbook.

In accomplishing an LCCA, there are certain steps that the analyst should perform to acquire the desired result. For the purposes of illustration, the following represents a generic approach:

1. *Define system requirements.* Define system operational requirements and the maintenance concept. Identify applicable TPMs and describe the system in functional terms, utilizing the functional analysis at the system level as required (refer to Figs. 12.6 through 12.11).

2. *Describe the system life cycle.* Establish a baseline for the development of a cost breakdown structure (CBS) and for the estimation of costs for each year of the projected life cycle. Show all phases of the system life cycle and identify the major activities in each phase (refer to Figs. 12.1, 12.3, and 12.4).

3. *Develop a CBS.* Provide a top-down and/or bottom-up cost structure to include all cost categories for the initial allocation of costs (top-down) and the subsequent collection and summary of costs (bottom-up). Develop the appropriate cost-estimating relationships (CERs), estimate the costs for each activity in the life cycle and for each category in the CBS, develop a typical cost profile, and summarize the costs through the CBS network.

4. *Select a cost model for analysis and evaluation.* Select (or develop) a mathematical or computer-based model to facilitate the life-cycle costing process. The model, developed around the applicable CBS, must be valid for and sensitive to the specific system configuration being evaluated. Accomplish a sensitivity analysis by evaluating input–output data relationships and to verify the model application.

5. *Evaluate the applicable baseline system design configuration being considered.* Apply the computerized model in evaluating the baseline design configuration being considered for adoption. Develop a cost profile and a CBS summary, identify the high-cost contributors, establish the critical cause-and-effect relationships, highlight those system elements that should be investigated for possible opportunities leading to design improvement and potential cost reduction, and recommend design changes as feasible. It is at this stage in the LCCA process that the analyst can pinpoint the costs associated with the proposed logistics support infrastructure, its elements, and their respective percent contribution to the total.

6. *Identify feasible design alternatives and select a preferred approach.* After accomplishing an LCC evaluation for the given baseline configuration, it is then appropriate to extend the LCCA to cover

the evaluation of multiple design alternatives (as applicable). Develop a cost profile and CBS summary for each feasible design alternative, compare the alternatives equivalently, perform a break-even analysis, and select a preferred design approach.

When accomplishing a complete LCCA for a large system, the detailed steps and the data requirements can be rather extensive and beyond the limits of coverage in this handbook. However, through the information presented herein, derived from an actual case study of a large communications system, it is hoped that the process and results are complete enough to demonstrate the importance of a life-cycle costing application to logistics.

12.5.1   Description of the Problem

A large metropolitan area has a need for a new communication system network capability (i.e., identified as System XYZ herein) that will enable day-to-day active communication between each and all of the following nodes: (*i*) a centralized city operational terminal located in the city center; (*ii*) three remote ground district operational facilities located in the city's suburban areas; (*iii*) 50 ground vehicles patrolling the city and within a 30-mile range; (*iv*) five helicopters flying at low altitude and within a 50-mile range; (*v*) three low-flying aircraft within a 200-mile range; and (*vi*) a centralized maintenance facility located in the city's outskirts. The proposed network needs to enable live two-way voice and data communication, 24 hours per day, throughout all of its branches and to any one of the stated nodes as required.

In response to this new system requirement, a need and feasibility analysis was accomplished, a solicitation for proposal was distributed to all known qualified potential sources of supply, and two prospective suppliers responded, each with a different design approach. The objective at this point is to evaluate each of the two supplier proposals, on the basis of system life-cycle cost, and to select a preferred approach, that is, Configuration A or Configuration B.

1.   System operational requirements and the maintenance concept

Referring to Section 12.3.1, the first major step in accomplishing a LCCA is to establish a good "baseline" description of system operational requirements, maintenance concept, primary operational TPM requirements, and top-level system functional analysis. Replicating the material presented in Section 12.3.1, paragraphs 1, 2, 3, and 4, and in Figures 12.6 and 12.7, for the proposed new communication system, network capability is required. While the specific requirements may change, establishing a good initial foundation, upon which to build the LCCA is essential. The level of detail will, of course, vary with the goals and depth of required analysis.

2.   The system life cycle

Having described the basic operational and maintenance support requirements for System XYZ, the next step is to present these requirements in the context of a proposed life-cycle framework. The objective is to identify the applicable phases of the life cycle and all of the activities within each phase. Figure 12.13, which constitutes a simplified abstraction taken from Figure 12.3, provides an illustration of the framework for the LCCA. This, in turn, forms the basis for collecting and categorizing costs for the analysis, that is, research and development cost, production and/or construction cost, operation and maintenance cost, and system retirement cost.

3.   The CBS

Given the planned program phases and the anticipated activities in each phase (shown in Fig. 12.13), the next step is to develop a CBS, or a top-down and/or bottom-up structure for the purposes of cost estimation and the collection of costs by category. The proposed CBS for System XYZ is presented in Figure 12.14 and must include all of the costs pertaining to the system, that is, direct and indirect costs, contrac-

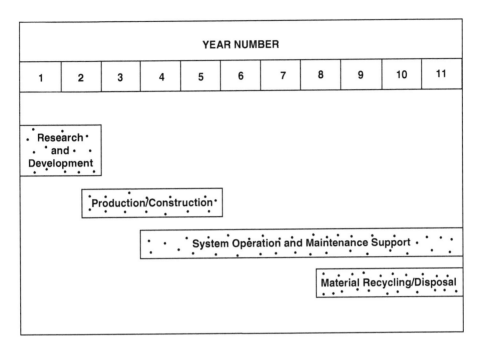

FIGURE 12.13 System XYZ life-cycle plan.

tor and supplier costs, customer (user) costs, design and development costs, production costs, hardware costs, software costs, data costs, logistics costs, and so on.

Referring to Figure 12.14, the objective is to estimate the applicable costs for each of the categories indicated. In estimating LCC, this becomes a bottom-up effort, employing the application of various CERs and activity-based costing (ABC) methods, and utilizing the appropriate analytical models and/ or tools to help facilitate the process. In developing a CBS, the analyst needs to know what is included (or left out), and how the various costs are developed. While a detailed description of what is included in each category of a CBS is required to provide the visibility desired, the summary structure in Figure 12.14 is considered to be sufficient for the purposes herein.

4. *Cost estimation and the development of cost profiles for the proposed design configurations being evaluated—Configuration A and Configuration B*

Within the context of the System XYZ life-cycle plan (Fig. 12.13) and the CBS (Fig. 12.14), the costs for each of the two proposed design configurations being evaluated were determined and are presented as shown in Figure 12.15. The costs for each of the four major categories (i.e., research and development cost, production and/or construction cost, etc.) were determined for each configuration, utilizing bottom-up estimating methods, and are summarized in Figure 12.15.

Referring to Figure 12.15, the costs are summarized in terms of the estimated inflated budgetary costs for the planned 11-year life cycle, which is reflected by the top profile, or identified as the total cost, that is, $7,978,451 for Configuration A and $8,396,999 for Configuration B. A second summary profile is included in terms of present value (PV) cost, required for the evaluation of comparable alternatives on the basis of economic equivalence. A 6% cost of capital was assumed for this LCCA effort.

5. *Evaluation of alternative design configurations and the selection of a preferred approach*

On the basis of the results shown in Figure 12.15, it appears that Configuration A is the preferred approach, because the present value (PV) cost of $5,927,885 is less than that for the other configuration. The question is, how much better is Configuration A, and at what point in time does this configuration assume a

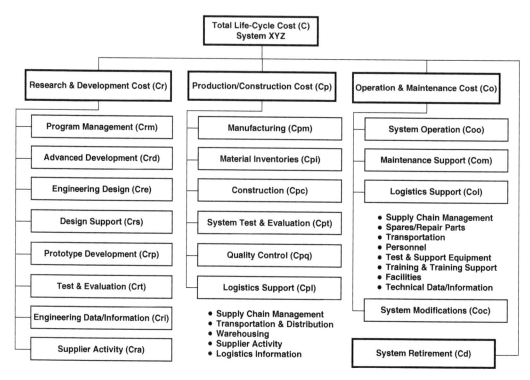

FIGURE 12.14 Cost breakdown structure for system XYZ.

position of preference? It should be noted that on the basis of acquisition costs only (i.e., Categories Cr and Cp), it appears as though Configuration B would be preferred ($4,417,404 for B and $4,509,271 for A). However, based on the overall LCC, Configuration A is preferred. Relative to the time of preference (i.e., when A assumes the point of preference), the analyst conducted a break-even analysis as illustrated in Figure 12.16. From the figure, it can be seen that Configuration A assumes a favorable position at about the 7-year, 7-month point in the projected life cycle. It was decided in this instance that this was early enough for the selection of Configuration A.

Cost Category	Life Cycle Year 1	2	3	4	5	6	7	8	9	10	11	Total ($)
Configuration A												
Research & Development (Cr)	615,725	621,112										1,236,837
Production/Construction (Cp)		364,871	935,441	985,911	986,211							3,272,434
Operation & Maintenance (Co)				179,203	207,098	448,248	465,660	483,945	503,122	523,297	544,466	3,355,039
System Retirement (Cd)									27,121	41,234	45,786	114,141
Total Cost ($)	615,725	985,983	935,441	1,165,114	1,193,309	448,248	465,660	483,945	530,243	564,531	590,252	7,978,451
Present Value Cost – 6% ($)	580,875	877,525	785,396	922,887	891,760	316,015	309,770	303,627	313,851	315,234	310,945	5,927,885
Configuration B												
Research & Development (Cr)	545,040	561,223										1,106,263
Production/Construction (Cp)		379,119	961,226	982,817	987,979							3,311,141
Operation & Maintenance (Co)				192,199	225,268	456,648	472,236	592,717	613,005	625,428	650,342	3,827,843
System Retirement (Cd)								20,145	35,336	45,455	50,816	151,752
Total Cost ($)	545,040	940,342	961,226	1,175,016	1,213,247	456,648	472,236	612,862	648,341	670,883	701,158	8,396,999
Present Value Cost – 6% ($)	514,191	836,904	807,045	930,730	906,659	321,937	314,089	384,510	383,753	374,621	369,370	6,143,809

FIGURE 12.15 Life-cycle cost profile for system XYZ.

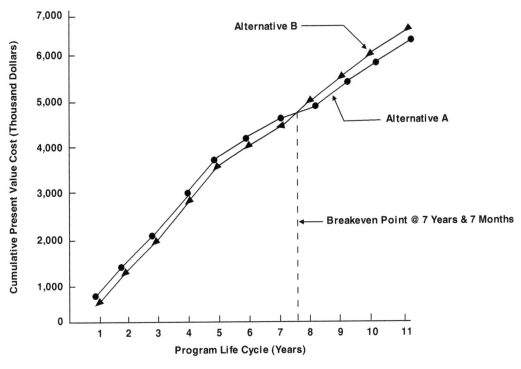

FIGURE 12.16 Break-even analysis for system XYZ.

6. *Further analysis and enhancement of the selected configuration*

Having initially selected Configuration A as being preferred over the alternative, the next steps are to further evaluate the costs that make up the $7,978,451 for this configuration, identify the high-cost contributors, determine cause-and-effect relationships, and re-evaluate System XYZ design to determine whether improvements can be implemented which will result in an overall reduction in LCC. A breakout of the costs for this configuration is presented in Figure 12.17.

Referring to Figure 12.17, for example, it should be noted that the costs associated with logistics activities (i.e., *Cpl* and *Col*) make up about 21.38% of the total. Within this spectrum, the categories of spares and/or repair parts and transportation represent high-cost contributors (4.57% and 3.73%, respectively) under Category *Col*. Additionally, transportation and distribution costs within Category *Cpl* are also relatively high (2.75%). Through a re-evaluation of the basic design configuration, the extensive requirements for spares and/or repair parts could perhaps be reduced through some form of reliability improvement, particularly for critical items with relatively high failure rates. For transportation, it may be possible to repackage elements of the system such that internal transportability attributes in the design can be improved, or to select alternative modes of transportation that will still meet the TPM requirements for the system overall, but at a lesser overall cost.

Through implementation of this process on an iterative basis, experience has indicated that significant system design improvements can often be realized. It should be noted that by improving one area of concern, the result could lead to an improvement in another area. For example, if improvement can be made in the spares and/or repair parts area (within Category *Col*), this may result in a reduction of the maintenance support cost (Category *Com*) as well. There are numerous interactions that could occur throughout the analysis process, and care must be exercised to ensure that improvement in any given area will not result in a significant degradation in another.

Configuration A

Cost Category	Cost ($) (Undiscounted)	Percent (%)
1. Research & Development (Cr)	1,236,660	15.50
(a) Program Management (Crm)	79,785	1.00
(b) Advanced Development (Crd)	99,731	1.25
(c) Engineering Design (Cre)	276,852	3.47
(d) Design Support (Crs)	193,876	2.43
(e) Prototype Development (Crp)	89,359	1.12
(f) Test & Evaluation (Crt)	116,485	1.46
(g) Engineering Data/Information (Cri)	75,795	0.95
(h) Supplier Activity (Cra)	304,777	3.82
2. Production/Construction (Cp)	3,272,762	41.02
(a) Manufacturing (Cpm)	1,716,166	21.51
(b) Material Inventories (Cpi)	453,176	5.68
(c) Construction (Cpc)	95,741	1.20
(d) System Test & Evaluation (Cpt)	228,184	2.86
(e) Quality Control (Cpq)	76,593	0.96
(f) Logistics Support (Cpl)	702,902	8.81
(1) Supply Chain Management	39,892	0.50
(2) Transportation & Distribution	219,408	2.75
(3) Warehousing	168,345	2.11
(4) Supplier Activity	263,289	3.30
(5) Logistics Information	11,968	0.15
3. Operation & Maintenance (Co)	3,354,939	42.05
(a) System Operation (Coo)	1,458,461	18.28
(b) Maintenance Support (Com)	768,325	9.63
(c) Logistics Support (Col)	1,002,891	12.57
(1) Supply Chain Management	79,785	1.00
(2) Spares/Repair Parts	364,615	4.57
(3) Transportation	297,596	3.73
(4) Personnel	153,984	1.93
(5) Test & Support Equipment	46,275	0.58
(6) Training & Training Support	24,733	0.31
(7) Facilities	20,744	0.26
(8) Technical Data/Information	15,159	0.19
(d) System Modifications (Coc)	125,262	1.57
4. System Retirement (Cd)	114,092	1.43
GRAND TOTAL	7,978,451	100.00

FIGURE 12.17   Cost breakdown structure summary.

12.5.2   Summary

The implementation of the LCCA process, particularly during the early stages of system design and development, can provide numerous benefits, to include: (*i*) influencing the overall system design for maximum effectiveness and efficiency from a total life-cycle perspective; (*ii*) facilitating the design of the logistics support infrastructure capability from the beginning, when the incorporation of any required changes can be accomplished easily and at minimum cost; and (*iii*) providing early front-end visibility by identifying potential high-cost areas and the risks associated with such. Additionally, LCCA can be applied at any stage in the system life cycle for the purposes of assessment, and for the identification of high-cost areas and the major contributors for such. This case-study approach addresses the steps and process for accomplishing a good LCCA effort.

Index

A

AAR. *See* Association of American Railroads (AAR)
Ad valorem taxes, **2**–3
Added value, **4**–8
Additive demand-price models, **6**–25
Additive demand-price relationship, **6**–16 to **6**–18
Adjacency graph property, **5**–44 to **5**–45
Administration costs, **2**–2, **2**–6 to **2**–7
AEIOUX representation, **5**–12
AGV. *See* Automated guided vehicles (AGV)
AHP. *See* Analytical hierarchy process (AHP)
AI. *See* Artificial intelligence (AI)
Air cargo, **2**–6
   decisions, **10**–3
   growth, **10**–11
   trucking and, **10**–8
ALDEP, **5**–8, **5**–38
Alexander the Great, **1**–3
Algorithms. *See specific name*, e.g., Heuristic
American Trucking Association, **2**–5
Analytical hierarchy process (AHP), **11**–9, **11**–10, **11**–16
Annealing techniques, **3**–15 to **3**–16
Annual cost of maintenance by operator, **4**–11
Approximate reasoning algorithms, **3**–19, **3**–25, **3**–26, **3**–27
Approximate reasoning model, **3**–24
Artificial intelligence (AI), **3**–1 to **3**–2
Assembly tree, **5**–36
Asset(s), **1**–13
   return on, **4**–11
   turnover, **4**–14
   utilization, **4**–14
Assigning vehicles, **3**–26, **3**–27 to **3**–28
Assisted design, **5**–47 to **5**–48
Association of American Railroads (AAR), **2**–5
AT & T run rules, **4**–6
ATA. *See* American Trucking Association
Attribute data, **4**–3, **4**–5, **4**–7, **4**–16
AutoCad, **5**–41
Automated guided vehicles (AGV), **7**–11, **7**–13
Automated storage retrieval systems, **7**–12, **7**–14, **7**–19 to **7**–26
Average case analysis, **3**–14
Average daily demand forecasts, **5**–30
Average processor, **5**–31

B

Backorder models, **6**–18 to **6**–**19**
Balanced Scorecard approach, **4**–2
Bar code labeling, **4**–12
Between-the facility logistics, **11**–10 to **11**–14

Bin shelving, **8**–9 to **8**–10, **8**–11
   area layout, **8**–9
   operations analysis, **8**–10
   storage medium, **8**–8
   time windows, **8**–9
B.J.'s Warehouse Club, **7**–18
Blackhawk helicopter, **1**–10
Block layout, **5**–6
   adjacency graph property, **5**–44, **5**–45
   continuous space, **5**–40
   defined, **5**–5
   editing, **5**–5
   optimal, **5**–44
   planar adjacency graph property, **5**–44
   transposition, **5**–5
   travel network and, **5**–42
Blocked aisles, **7**–5, **7**–6
Boundary condition, **6**–9, **6**–10, **6**–20, **6**–22
British Air Ministry, **3**–2
Brute force logistics, **1**–11 to **1**–12
Bunching, **11**–4
Business logistics, **2**–1
   costs, **2**–2 to **2**–7
      administrative, **2**–6 to **2**–7
      carrying, **2**–3 to **2**–4
      total, **2**–3
      transportation, **2**–4 to **2**–6

C

CAD. *See* Computer-aided drawing and design software (CAD)
Calculating dispatchers' preference, **3**–25 to **3**–26
Carrying costs, **2**–3 to **2**–4
   inventory, **2**–2, **2**–7, **4**–11
Carton flow area layout, **8**–11
Carton flow rack, **8**–10 to **8**–11, **8**–19, **8**–22
Carton flow storage medium, **8**–7
Carton pick operations, **8**–8, **8**–12
Case study
   automated storage retrieval systems, **7**–19 to **7**–26
   distribution system design, **9**–16 to **9**–17
   electronic manufacturing logistics, **4**–9 to **4**–19
   Gulf War, **1**–9 to **1**–14
   inventory control theory, **6**–24
   LCC, **12**–20 to **12**–25
   logistics metrics, **4**–9 to **4**–18
   material handling system, **7**–19 to **7**–25
   performance metrics, **4**–15 to **4**–19
   transportation system, **10**–12 to **10**–13
CATIA, **5**–41
Cause and effect diagram, **4**–5, **4**–20

CBS. *See* Cost breakdown structure (CBS)
Center for Engineering Logistics and Distribution, 4–7 to 4–8
Centerline, calculation of, 4–16
CER. *See* Cost-estimating relationship (CER)
Charlemagne, 1–4
Check sheet, 4–5
China, 2–1
Classical heuristic algorithms, 3–8 to 3–9
Classical theory of sets, 3–18, 3–30
Clinical laboratory delivery process, 11–5, 11–6
CLM. *See* Council of Logistics Management (CLM)
Clustering by Sweep algorithm, 3–13
Clustering-routing approach, 3–13
COI. *See* Cube per order index (COI)
Combined delivery, 11–8, 11–9
Computer-aided drawing and design software (CAD), 5–41
Concave costs, 6–10
Consolidation, 8–15
Construction heuristic, 5–38
Construction phase, 12–18
Containers, 4–9, 4–14
Contaminated materials, 7–19
Control charts, 4–5 to 4–7
    AT & T run rules, 4–6
    for attribute data, 4–7
    construction of, 4–7, 4–8
    error types, 4–6
    exponentially weighted moving average, 4–7
    fraction nonconforming type, 4–9
    in logistics area, 4–5
    types of, 4–6 to 4–7
    for variable type data, 4–7, 4–8
Converse, Paul, 1–8
Conveyor system, 5–5, 7–2, 7–9 to 7–10
CORELAP, 5–8, 5–38, 5–58
Cost(s), 5–2, 5–47, 5–49
    activities affecting, 12–8
    actualized marginal, 5–52
    administration, 2–2, 2–6 to 2–7, 6–1
    allowed moving, 5–27
    analysis, 8–21 to 8–22
    assignment, 5–39, 5–40
    average, 6–4, 6–5
    carrying, 2–2, 2–3 to 2–4, 2–7
    closing, 5–49
    components, 5–52
    concave, 6–10, 6–10 to 6–11
    contraction, 5–51, 5–52
    data, 8–22
    dynamic center relocation, 5–52
    echelon holding, 6–3, 6–4, 6–6
    effectiveness, 12–7
    emergency shipment, 6–14
    engineering, 5–59
    estimation, 5–17
    expected, 6–12
    factory, 7–1
    fixed, 5–17, 5–28, 5–50
    holding, 6–2, 6–3, 6–4, 6–5, 6–6, 6–7, 7–5
    implementation, 5–28, 5–39, 5–50, 5–52
    insurance, 2–3
    interaction, 5–54
    inventory, 2–7, 7–5
        carrying, 2–2, 2–7, 4–11
        risk, 2–3
        service, 2–3
    inventory *vs.* unit load, 7–5
    life cycle, 7–8, 12–7, 12–8
    long-term, 7–4
    maintenance, 4–11, 7–8
    marginal, 5–52, 6–13
    material movement, 7–1
    measurable, 7–8
    MHD, 7–15, 7–17
    moving, 5–26, 5–27, 5–28, 5–50, 5–52
    negligible, 5–28
    operating, 5–17, 5–28, 7–8, 7–17
    ordering, 6–9
    overhead, 7–18
    per operation, 4–11
    per piece, 4–11
    per transaction, 4–11
    per unit of throughput, 4–11
    per unit variable, 6–2
    positive setup, 6–22
    rate, 6–3
    replenishment, 6–2, 6–3, 6–9
    set-up, 6–2, 6–5, 6–6, 6–7, 6–14, 6–15
    shortage, 6–12, 6–19
    structures, 5–37
    total, 6–5, 6–12
    total logistics, 12–11, 12–16
    transformational, 5–28, 5–31
    transition, 5–52
    transportation, 2–4 to 2–6, 5–57, 7–5, 7–15, 7–17
    trucking, 2–4
    of trucks, 7–11
    unit, 5–24
    unit overage, 6–13
    unitary, 5–54, 5–55
    variance, 4–11
    warehousing, 8–19 to 8–22
Cost breakdown structure (CBS), 12–20, 12–23, 12–25
Cost-estimating relationship (CER), 12–20
Cost-minimization, 5–2
Costco, 7–18
Council of Logistics Management (CLM), 1–1, 3–28, 4–2
Council of Supply Chain Management Professionals (CSCMP), 1–1, 4–1
Courier delivery, 11–6 to 11–9, 11–10
Cradle to grave perspective, 1–8

CRAFT, **5**–8, **5**–38, **5**–39
Cranes, **7**–12
Credit term, **4**–11
Crossover, **3**–16, **3**–17
CSCMP. *See* Council of Supply Chain Management Professionals (CSCMP)
Cube per order index (COI), **7**–5, **7**–6, **7**–7
Customer(s)
    distribution design, **9**–6
    metric, **4**–2
    orders, **7**–18, **7**–19, **7**–22
    service, **11**–3, **11**–4, **11**–10, **11**–14
        facility, **7**–18
Cut trees, **5**–45 to **5**–47, **5**–59
Cycle sub-time
    distribution/filling, **4**–13
    planning/design, **4**–13
    reverse logistics, **4**–13
    sourcing, **4**–13
    transportation, **4**–13
Cycle time, **4**–9, **4**–10, **4**–13

D

Data entry accuracy, **4**–3, **4**–12
Decomposition based heuristic algorithms, **3**–12 to **3**–13
Defect concentration diagram, **4**–5
Defense Logistics Agency (DLA), **1**–12
Defuzzification, **3**–21
Defuzzifier, **3**–19
Deliveries, **4**–15, **4**–16
    nonconforming, **4**–16
    on-time, **4**–12
    point of use, **4**–13
    supplier direct, **4**–13
Delivery zones, **4**–15
Demand forecast, **5**–30, **5**–31
    average daily, **5**–30
    multi-year demand, **5**–30
    novelty T-shirts, **6**–24
    transportation system, **10**–11 to **10**–12
Department area summary, **8**–19
Department of Defense. *See* U. S. Department of Defense (DOD)
Deregulation transportation, **1**–8, **2**–7
Design aggregation, **5**–3 to **5**–7
Design methodologies
    assisted, **5**–47 to **5**–48
    evolution, **5**–37
    global optimization, **5**–49 to **5**–50
    heuristic, **5**–38 to **5**–39
    holistic metaheuristics, **5**–48 to **5**–49
    interactive, **5**–41
    interactive optimization-based, **5**–44 to **5**–47
    manual, **5**–47 to **5**–48
    mathematical programming based, **5**–39 to **5**–41
    metaheuristic, **5**–42 to **5**–44
    space available, **5**–27
Design space available, **5**–27
Deterministic models, **6**–1 to **6**–11. *See also* Economic order quantity (EOQ) models
Disassembly network, **5**–36
Disassembly tree, **5**–36
Dispatchers' preferences, **3**–25 to **3**–27
Dispatching in truckload trucking
    fuzzy logic approach, **3**–21 to **3**–28
        numerical example, **3**–28
        propose solution to, **3**–23 to **3**–27
        statement of problems, **3**–22 to **3**–23
Distribution design case, **9**–16 to **9**–17
Distribution system design, **9**–1 to **9**–19
    case study, **9**–16 to **9**–17
    data analysis and synthesis, **9**–5 to **9**–7
    engineering design principles, **9**–3 to **9**–4
    Geoffrion and Graves model, **9**–13 to **9**–15
    heterogeneous data, **9**–3 to **9**–4
    logistic data components, **9**–5 to **9**–6
    models, **9**–8 to **9**–15
    schematic, **9**–2
    sensitivity and risk analysis, **9**–15 to **9**–16
DLA, Defense Logistics Agency (DLA)
Document accuracy, **4**–12
DOD. *See* U. S. Department of Defense (DOD)
Drucker, Peter, **1**–8
Dutch National Mobility Model (DNMM), **10**–13
Dutch railway system, **10**–12 to **10**–13
Dynamic probabilistic discrete model, **5**–50 to **5**–55

E

Echelon holding costs, **6**–3, **6**–4, **6**–6
    redistributing, **6**–7 to **6**–8
Economic(s), **2**–1 to **2**–9
Economic order quantity (EOQ) models, **6**–1 to **6**–11
    constructing, **6**–6 to **6**–8
    echelon holding costs, **6**–3, **6**–4, **6**–6
    inter-setup intervals, **6**–8
    multi-product assembly system, **6**–5 to **6**–8
    two facilities, **6**–3 to **6**–5
Economic value added, **4**–11
ECT. *See* European Combined Terminals (ECT)
EDI. *See* Electronic data interchange (EDI)
Eisenhower, Dwight, **1**–10
Electronic data interchange (EDI), **12**–3
Electronics manufacturing logistics, **4**–13
Elemental process, **5**–24
Empty miles, **4**–3, **4**–14
Empty trailers/containers, **4**–14
Encoded values of variable x, **3**–17
Engine, inference, **3**–19
Engineered storage area (ESA), **4**–10, **4**–12, **4**–15
Engineering design principles, **9**–4
Engineering logistics, **1**–2 to **1**–3

Engineering tool chest, 3–1 to 3–30
    algorithms complexity, 3–13 to 3–14
    fuzzy logic approach to dispatching, 3–21 to 3–28
        numerical example, 3–28
        propose solution to, 3–23 to 3–27
        statement of problems, 3–22 to 3–23
    heuristic algorithms, 3–7 to 3–14
    mathematical programming, 3–4 to 3–7
    operation research, 3–2 to 3–4
    randomization optimization techniques, 3–15 to 3–18
Enterprise resource planning (ERP), 1–9
EOQ. *See* Economic order quantity (EOQ) models
Ergonomic material handling, 7–4
ERP. *See* Enterprise resource planning (ERP)
Error types, 4–6
ESA. *See* Engineered storage area (ESA)
Euclidean distance, 5–6, 5–7, 5–10
European Combined Terminals (ECT), 7–19
Exchange heuristic algorithms, 3–10 to 3–12
Expected product demand, routings and, 5–16
Expected traffic layout, 5–16
Expedite ratio, 4–13
Expenditures, 2–1 to 2–2
    categories, 2–2 to 2–7
    freight forwarder, 2–4
    before interest and taxes, 4–11
    operating, 4–11
    trucking costs, 2–4
Exponentially weighted moving average, 4–6, 4–7

F

Facilities location and layout design, 5–1 to 5–61, 8–18
    design aggregation and granularity levels, 5–3 to 5–7
    dynamic evolution, 5–28 to 5–32
    existing design, 5–26 to 5–28
    exploiting processing and spatial flexibility, 5–19 to 5–24
    flow and traffic, 5–12 to 5–17
    illustrative layout design, 5–17 to 5–19
    integrated location and layout design optimization, 5–50 to 5–55
    layout hierarchical illustration, 5–4
    network and facility organization, 5–32 to 5–36
    network deployment, 5–4
    qualitative proximity relationships, 5–9 to 5–12
    space representation, 5–7 to 5–9
    uncertainty, 5–25 to 5–26
Fast movers, 8–3, 8–6, 8–11, 8–12, 8–19
Feasibility analysis, 12–6, 12–7, 12–21
    problem identification and, 12–9 to 12–10
Federal Highway Administration (FHWA), 2–7
FedEx, 10–8 to 10–13
FFBD. *See* Functional flow block diagrams (FFBD)
FHWA. *See* Federal Highway Administration (FHWA)

Fill rate, 4–13
Financial metrics, 4–2, 4–8, 4–10, 4–11 to 4–12
Flow(s)
    based design skeleton, 5–18
    circulation and, 8–19
    estimation, 5–20
    estimation matrix, 5–13
    traffic design and, 5–12 to 5–17
Flying blind logistics, 1–12
Focused logistics, 1–8
Forecasting
    accuracy, 4–12
    demand, 5–30, 5–31
    freight demand, 10–11 to 10–12
    multi-year demand, 5–30
    transportation systems, 10–11 to 10–12
Forklift truck, 8–3, 8–4, 8–5, 8–8, 8–22
Forwarders, 2–4, 2–6
Free flow distance, 5–6, 5–7, 5–10
Freight demand forecasting, 10–11
Freight forwarder, 2–4, 2–6
Functional flow block diagrams (FFBD), 12–14
Functional flow map of operations, 8–4
Fuzzifier, 3–19
Fuzzy logic
    basic elements of sets and systems, 3–18 to 3–21
    dispatching in truckload trucking using, 3–21 to 3–28
        numerical example, 3–28
        propose solution to, 3–23 to 3–27
        statement of problems, 3–22 to 3–23
Fuzzy sets, 3–18 to 3–21, 3–24, 3–25, 3–30

G

Gantry Crane and Hoist, 7–13 to 7–14
GDP. *See* Gross domestic product (GDP)
General building description, 8–17 to 8–18
Genetic algorithms, 3–16 to 3–18
Geoffrion and Graves model, 9–13 to 9–15
Geographical information systems (GIS), 5–41
Geographical locations, 9–5
GIS. *See* Geographical information systems (GIS)
Global transportation infrastructure, 10–10 to 10–11
Global Transportation Network, 1–13
Globalization, 2–1
Goods, 7–18
Google Earth, 5–41
Granularity levels, 5–3, 5–6, 5–7
Gravity roller conveyor, 7–3
Gross domestic product (GDP), 2–1 to 2–2, 2–7, 2–8
    worldwide export volume *vs.*, 2–2
Gross profit margin, 4–11
Gulf War, 1–9 to 1–14
    background, 1–9 to 1–11
    lessons learned, 1–11 to 1–12
        applying, 1–12 to 1–14

H

Handheld barcode scanner, **8**–4, **8**–5
Hannibal, **1**–9
Hazardous materials, **7**–19
Helicopter, **1**–10
Heuristic algorithms, **3**–7 to **3**–14
    2-OPT, **3**–11 to **3**–12
    assigning vehicles, **3**–27 to **3**–28
    based on random choice, **3**–9
    classical, **3**–8 to **3**–9
    complexity, **3**–13 to **3**–14
    decomposition based, **3**–12 to **3**–13
    exchange, **3**–10 to **3**–12
    greedy, **3**–9 to **3**–10
    NN, **3**–10
    transportation request, **3**–27 to **3**–28
Hill-climbing, **3**–9
Histogram, for variable type data, **4**–5, **4**–18
Ho Chi Minh trail, **1**–7
Hoists, **7**–12
Holistic metaheuristics, **5**–48 to **5**–49
HoloPro, **5**–48 to **5**–49
Honeycombing loss, **7**–5, **7**–6
Hospital(s)
    delivery system components, **11**–4 to **11**–10
    floor layouts, **11**–5
    hierarchy structure, **11**–10
    service trends, **11**–11
Human courier(s)
    process, **11**–6
    robot couriers, **11**–4 to **11**–10
    velocity, **11**–8

I

Idleness, **4**–14
Illustrative layout design, **5**–17 to **5**–19
Illustrative set, **5**–36
Inbound staging, **8**–4, **8**–17, **8**–19
Increase in profit adjusted revenues per CWT, **4**–11
Incremental discount model, **6**–10
India, **2**–1
Inference engine, **3**–19
Information technology (IT), **1**–2, **9**–3, **9**–8, **12**–3
Infrastructure, **5**–22
    fixed, **5**–17
    logistics, **3**–2, **5**–11, **12**–2, **12**–5, **12**–6, **12**–7, **12**–8
    shared, **5**–9
    support, **12**–2, **12**–5, **12**–6, **12**–7, **12**–8
    transformation, **5**–2
    transportation, **10**–8 to **10**–10
Insurance costs, **2**–3
Integer programming model, **3**–6 to **3**–7
    formulation, **3**–7
Integrated delivery network, **11**–11, **11**–15

Integrated location and layout design optimization, **5**–50 to **5**–55
Integrating system function, **12**–1 to **12**–25
    life cycle cost analysis, **12**–20 to **12**–25
    system life cycle, **12**–4 to **12**–18
    total system approach, **12**–2 to **12**–3
Inter-setup intervals, **6**–8
Interactive design methodologies, **5**–41
    optimization-based, **5**–44 to **5**–47
Interests, **2**–3, **2**–4
    operating expenses before, **4**–11
International Business Machines (IBM), **11**–2
International markets, **2**–9
International Society of Logistics, **1**–2
Inventory, **7**–19
    accuracy, **4**–12
    control theory, **6**–1 to **6**–26
        case study, **6**–24
        deterministic models, **6**–1 to **6**–11
        stochastic models, **6**–11 to **6**–24
    costs, **2**–7
        carrying, **2**–2, **2**–7, **4**–11
        risk, **2**–3
        service, **2**–3
        unit load *vs.*, **7**–5
    days in, **4**–13
    on hand, **4**–11
    JIT management model, **2**–4
    model, **6**–9 to **6**–11
    shrinkage, **4**–11
    turns, **4**–14
IT. *See* Information technology (IT)

J

Japan, **1**–6, **1**–7 to **1**–8
Jibs, **7**–9, **7**–12
JIT. *See* Just-in-time (JIT)
Just-in-time (JIT), **1**–8, **4**–9
    inventory management model, **2**–4

K

Key performance indicators (KPI), **4**–2
King Khalid Military City (KKMC), **1**–10
KKMC. *See* King Khalid Military City (KKMC)
Korean War, **1**–6 to **1**–7
KPI. *See* Key performance indicators (KPI)

L

Labor utilization, **4**–14
Layout
    evaluation, **5**–22
    myopically generated plan, **5**–29
    representation for design purposes, **5**–6
LCC. *See* Life cycle cost (LCC)
LCL. *See* Lower control limit (LCL)

Lean logistics, **1**–8
Life cycle cost (LCC), **12**–7, **12**–16
    activities affecting, **12**–8
    analysis, **12**–20 to **12**–25
    case study, **12**–20 to **12**–26
    minimum total, **12**–19
Linear programming, **3**–4 to **3**–6
Lippert, Keith, **1**–12
Load factor, **4**–14
Load types, **8**–2
Loading, on-time, **4**–12
Location-allocation, **9**–9, **9**–10 to **9**–12
Logistician professional associations, **1**–1 to **1**–2
Logistics
    brute force, **1**–11 to **1**–12
    business, **2**–1 (*See also* Business logistics)
    cradle to grave perspective, **1**–8
    data, **4**–3 to **4**–4
    defined, **1**–1 to **1**–2, **3**–1
    economic impact, **2**–1 to **2**–9
    emergence as science, **1**–8 to **1**–9
    engineering tool chest, **3**–1 to **3**–30 (*See also*
        Engineering tool chest)
    engineering *vs.* business, **1**–2 to **1**–3
    etymology, **1**–1
    expenditures, **2**–1
    flying blind, **1**–12
    focused, **1**–8
    historical perspectives, **1**–1 to **1**–14
    lean, **1**–8
    military, **1**–3 to **1**–8
    operating expenses, **4**–11
    performance-based, **12**–6
    performance metrics, **4**–8 to **4**–9
    precision-guided, **1**–11
    productivity, **2**–7 to **2**–9
    support infrastructure, **12**–2, **12**–5, **12**–6, **12**–7,
        12–8
Logistics management, **1**–2
    boundaries and relationships, **1**–2
Lost sales models, **6**–16 to **6**–18
    additive demand-price relationship, **6**–16 to **6**–18
    multiplicative demand-price relationship, **6**–18
Lower control limit (LCL), **4**–5

M

Maintenance, **12**–10 to **12**–12, **12**–21
    cost, **4**–11
    mean time between, **12**–16
Management problem, **3**–4, **3**–5
Management science, **3**–1
Manual design, **5**–47 to **5**–48
Manufacturing
    electronic, **4**–9 to **4**–19
    JIT, **4**–9

    pressboard, **4**–3
    service industries, **11**–14
MAPINFO, **5**–41
Marius, **1**–3
Market location, **2**–7
Marshalling, on-time, **4**–12
Material(s)
    burden, **4**–14
    contaminated, **7**–18
    recycling/disposal, **12**–1, **12**–2, **12**–5, **12**–6, **12**–18
        to **12**–19
Material handling
    automation, **7**–8
    case study, **7**–19
    environmental, **7**–8
    equation, **7**–15
    equipment, **7**–9 to **7**–15
    ergonomics, **7**–4
    life cycle, **7**–8
    planning, **7**–2
    principles, **7**–2 to **7**–8
    rate, **4**–11
    space utilization, **7**–5 to **7**–6
    standardization, **7**–2
    system, **7**–6
    unit load, **7**–4 to **7**–5
    warehouse functions, **7**–17, **7**–18
    work, **7**–2
Material handling device (MHD), **7**–9
Material Handling Research Center (MHRC), **4**–7
Material handling system, **7**–1 to **7**–25
    case study, **7**–19 to **7**–25
Mathematical formulation, **3**–2, **3**–3, **3**–8
Mathematical model, **3**–2 to **3**–5
Mathematical programming, **3**–4 to **3**–7
    integer, **3**–6 to **3**–7
    linear, **3**–4 to **3**–6
Mean time between maintenance (MTBM), **12**–16
Measurement system development, **4**–2
Metric(s)
    cycle time, **4**–9, **4**–13
    logistics, **4**–9 to **4**–18
    performance, **4**–1
        categories, **4**–2
        delivery time, **4**–15 to **4**–17
        establishing and monitoring, **4**–2
        exceptions to standard packaging, **4**–18
        framework, **4**–3, **4**–10
        groups, **4**–8, **4**–10 to **4**–15
            cycle time, **4**–8, **4**–10, **4**–13
            financial, **4**–8, **4**–10, **4**–11 to **4**–12
            quality, **4**–8, **4**–10, **4**–12
            resource metrics, **4**–8, **4**–10, **4**–14
        nature of, **4**–2
        in pressboard manufacturing, **4**–3
        scrutinizing for added value, **4**–8
        shortages, **4**–17 to **4**–18

SMART, **4**–2
resource, **4**–14
MHD. *See* Material handling device (MHD)
Miles, empty, **4**–3, **4**–14
Military logistics, historical examples, **1**–3 to **1**–8
Mobile robot delivery systems, **5**–4
Modified work place, **7**–4
Moving costs, **5**–50, **5**–52
allowed, **5**–27
estimated, **5**–28
extreme, **5**–27
impact of, **5**–28
important, **5**–27
negligible, **5**–26
nonnegligible, **5**–26
significant, **5**–27
Moving people *vs.* moving goods
design challenges, **10**–4
differences and similarities, **10**–3 to **10**–5
performances measures, **10**–4 to **10**–5
shared systems, **10**–5
transportation systems, **10**–2 to **10**–4
MTBM. *See* Mean time between maintenance (MTBM)
Multi-facility product oriented organization, **5**–35
Multi-period inventory model, **6**–9 to **6**–11
Multi-product assembly system, **6**–3, **6**–5 to **6**–8
Multi-year demand forecast, **5**–30
Multi-year expected average processor, **5**–31
Multi-year two-sigma robust processor, **5**–31
Multi-year uncertainty of average daily demand
forecast, **5**–30
Multiplicative demand-price relationship, **6**–18
Mutation, **3**–16
Muther's AEIOUX representation, **5**–12, **5**–39
Myopically generated dynamic layout plan, **5**–29

N

NAICS. *See* National American Industry Classification
System (NAICS)
Napoleon, **1**–4 to **1**–5
National American Industry Classification System
(NAICS), **11**–1
National Council of Physical Distribution
Management, **1**–1
National Science Foundation Industry/University
Cooperative Research Center Program, **4**–7
Nebraska Medical Center (NMC), **11**–12, **11**–13
Need identification, **12**–7, **12**–9 to **12**–10, **12**–13
Net profit margin, **4**–11
Network efficiency, **4**–14
NMC. *See* Nebraska Medical Center (NMC)

O

O-notation, **3**–14
Objectives, **3**–3, **3**–4
Obsolescence, **2**–3, **2**–6

Off-line shipments, **4**–13
Oil pipeline transportation, **2**–3, **2**–4, **2**–6
On-time delivery, **4**–12
On-time entry into system, **4**–12
On-time loading, **4**–12
On-time marshaling, **4**–12
On-time pick up, **4**–12
On-time put away, **4**–12
Operation Desert Farewell, **1**–10 to **1**–11
Operation Desert Shield, **1**–10 to **1**–11
Operation Desert Storm, **1**–10
Operation Enduring Freedom, **1**–12 to **1**–14
Operational functions, **12**–13
Operations research, **3**–2 to **3**–4
defined, **3**–2
history, **3**–2
problem solving steps, **3**–3 to **3**–4
Optimal value function, **6**–10
Optimization techniques, **3**–15 to **3**–18
Order characteristics, **8**–4
Order picking, **8**–16
Phoenix Pharmaceuticals, **7**–23 to **7**–25
trucks, **7**–11
Order processing, **8**–16

P

P-chart attributes, **4**–16, **4**–17, **4**–18, **4**–19
Pack rate, **4**–14
Packing, **8**–15
area, **8**–5
operations, **8**–11, **8**–16, **8**–18
Pagonis, William G. (Gus), **1**–10
Pallet floor stacking area layout, **8**–14
Pallet floor storage, **8**–12
Pallet handling, **8**–13
Pallet pick operations, **8**–8
Pallet rack, **8**–11 to **8**–12
area, **8**–5
area layout, **8**–12
lower level of, **8**–13
storage medium, **8**–7
Pallet reserve storage area, **8**–4, **8**–6
Palletizers, **7**–11
Pareto chart, **4**–5, **4**–11, **4**–18, **4**–19
Payables outstanding past credit term, **4**–11
Perfect Order Index (POI), **4**–2
Performance
analysis, **8**–21
measures, **10**–4 to **10**–5 (*See also* Performance
metrics)
moving people *vs.* moving goods, **10**–4 to **10**–5
warehousing, **8**–19 to **8**–21
Performance-based logistics, **12**–6
Performance metrics (PM), **4**–1
case study, **4**–15 to **4**–19
categories, **4**–2

delivery time, **4**–15 to **4**–17
establishing and monitoring, **4**–2
exceptions to standard packaging, **4**–18
framework, **4**–10
groups, **4**–8, **4**–10 to **4**–15
    cycle time, **4**–8, **4**–10, **4**–13
    financial, **4**–8, **4**–10, **4**–11 to **4**–12
    quality, **4**–8, **4**–10, **4**–12
    resource metrics, **4**–8, **4**–10, **4**–14
nature of, **4**–2
in pressboard manufacturing, **4**–3
scrutinizing for added value, **4**–8
shortages, **4**–17 to **4**–18
SMART, **4**–2
Pharmaceutical courier delivery, **11**–6
Pharmacy delivery process, **11**–6
Pharmacy time demand process, **11**–8
Philip, **1**–3
Phoenix Pharmaceuticals, **7**–19 to **7**–25
    order picking, **7**–23 to **7**–25
Pick up, on-time, **4**–12
Piece packing, **8**–15
    area, **8**–5
    operations, **8**–11
Piece pick operations, **8**–6 to **8**–7, **8**–12, **8**–16, **8**–19 to
    8–20
Planar adjacency graph property, **5**–44
Plant Design and Optimization Suite, **5**–47
Point-of-use deliveries, **4**–13
Point-of-Use/Pull System, **4**–15
Positive setup costs, **6**–22
Precision-guided logistics, **1**–11
Pressboard manufacturing, **4**–3
Printed wiring board (PWB), **4**–12
Problem identification and feasibilty analysis, **12**–9 to
    12–10
Process monitoring, statistical methods, **4**–4 to **4**–7
Process orientation, types of, **5**–34
Processor layouts, **5**–5, **5**–6, **5**–19, **5**–23, **5**–47
Product(s)
    conversion factors, **8**–3
    demand, **5**–16
    dimensions, **8**–3
    distribution system design, **9**–5 to **9**–6
    routings, **5**–16
    storage requirements, **8**–3
Productivity
    analysis, **8**–20 to **8**–21
    past twenty-five years, **2**–7 to **2**–8
    on road, **4**–14
Professional associations for logisticians, **1**–1 to **1**–2
Profit
    adjusted revenues per CWT, **4**–11
    per square foot, **4**–14
Programming
    integer model, **3**–6 to **3**–7
    linear, **3**–4 to **3**–6
    mathematical, **3**–4 to **3**–7
Put away, on time, **4**–12
PWB. *See* Printed wiring board (PWB)
Pyxis HelpMate robotics courier, **11**–5

Q

Quadratic assignment problem, **5**–40 to **5**–41, **5**–57
Qualitative proximity relationships, **5**–2, **5**–9 to **5**–12,
    5–17 to **5**–18, **5**–37, **5**–39, **5**–45
Quality metrics, **4**–10, **4**–12

R

Radio frequency identification (RFID), **7**–7
    ROI, **1**–13 to **1**–14
    Wal-Mart, **1**–13 to **1**–14
Radio frequency transmitters, **8**–14
Rail transportation, **2**–4, **2**–5, **2**–7
Random choice algorithms, **3**–9
Randomization optimization techniques, **3**–15 to **3**–18
    engineering tool chest, **3**–15 to **3**–18
    genetic algorithms, **3**–16 to **3**–18
    simulated annealing, **3**–15 to **3**–16
Ratio of inbound to outbound, **4**–14
Receiving and stowing, **8**–4 to **8**–6
Receiving rates, **4**–14
Record accuracy, **4**–12
Rectilinear distance, **5**–6, **5**–7, **5**–10, **8**–8
    computation of, **5**–55
Recursive formula, **6**–9, **6**–11
Rental car industry, **11**–3
    revenue management problem, **3**–4
        optimal solution, **3**–6
        solution space, **3**–5
Reproduction, **3**–16, **3**–17, **3**–18
Resource metrics, **4**–14
Resource Optimization and Innovation, **11**–13
Responsibility-based center typology, **5**–33
Retrieval systems, **7**–12
Return on assets, **4**–11
Return on investment (ROI), **1**–13 to **1**–14, **4**–11
Returns processing, **8**–9
Revenue(s)
    growth percentage, **4**–11
    logistics, cycle sub-time, **4**–13
    management
        optimal solution, **3**–6
        rent-a-car problem, **3**–5 to **3**–6
        solution space, **3**–5
    per associate, **4**–14
    per square foot, **4**–14
Rewarehousing, **8**–9
RFID. *See* Radio frequency identification (RFID)
Robot(s), **5**–4, **7**–11
    couriers, **11**–4 to **11**–10
    performance measures, **11**–8
    speed modeling, **11**–8

storage, 7–26
Robust processor, **5**–31
ROI. *See* Return on investment (ROI)
Romans, **1**–3 to **1**–4
Roulette wheel selection, 3–17
Rules, **3**–20
Run rules, 4–6
Russia, **1**–4 to **1**–5
Russian Trade Federation, 2–1

S

Sam's Club, 7–18
Scatter diagram, 4–5
Schwarzkopf, Norman, 1–**10**, 1–**11**
SCM. *See* Supply chain management (SCM)
Service industries
    consumer behavior, **11**–11
    customers, **11**–2
    logistics, **11**–1 to **11**–16
    manufacturing, **11**–14
Service sector firms, **11**–2
Sets theory, 3–18, 3–30
Shipments
    off-line, 4–13
    rates, 4–14
Shipping operations, **8**–8, **8**–16
Shortages, 4–17 to 4–18
Shrinkage, inventory, 4–11
Simulated annealing techniques, 3–15 to 3–16
Single-point crossover operator, 3–17
Site layout, hierarchical illustration, 5–4
Slow movers, **8**–11
Small companies, 2–8
Society of Logistics Engineers (SOLE), **1**–2
SOLE. *See* Society of Logistics Engineers (SOLE)
SolidWorks, 5–41
Solution space, **3**–5, **3**–8, **3**–15
Sort-while-pick cart, **8**–10
Sorting, **8**–15
    operations, **8**–8, **8**–16
South Korea, 2–1
Space
    continuous *vs.* discrete, 5–8
    cushions, **11**–8
    facilities location and layout design, 5–7 to 5–8
    requirement, 5–17, 5–18
    utilization, 7–5 to 7–6
Specialized processor, continuous *vs.* discrete, 5–24
SSUs. *See* Strategic Service Units (SSUs)
Staggers Act, 2–7
Staging operations, **8**–8, **8**–15, **8**–16
Standard packaging exceptions, 4–18
Statistical methods for process monitoring, 4–4 to 4–7
Statistical process control (SPC), 4–4 to 4–7
    control charts, 4–5 to 4–7
        AT & T run rules, 4–6

        for attribute data, 4–7
        construction of, 4–7, 4–8
        error types, 4–6
        exponentially weighted moving average, 4–7
        fraction nonconforming type, 4–9
        in logistics area, 4–5
        types of, 4–6 to 4–7
        for variable type data, 4–7, 4–8
    seven tools, 4–4 to 4–5
Stochastic models, **6**–11 to **6**–24
    joint pricing and inventory control, 6–15 to 6–19
    lemma, **6**–22 to **6**–24
    multiple period, 6–19 to 6–22
    newsvendor problems, 6–11 to 6–15
    positive set-up cost, **6**–22
    scenarios, 5–25
Stochastic scenarios, 5–25
Stock keeping unit (SKU), **4**–15
Stock-to-non-stock ratio, 4–13
Storage capacity, **8**–19
Storage robots, 7–26
Stowing, **8**–3 to **8**–4
Strategic Service Units (SSUs), **11**–12
Superimposed qualitative relationships, layout with,
        5–19
Supplier, direct delivery, 4–13
Suppliers, distribution system design, **9**–6
Supply chain management (SCM), **1**–2, **4**–1
    boundaries and relationships, 1–2
        defined, 1–2
Supply chain, warehousing role, **8**–2
Support infrastructure, logistics, **12**–2, **12**–5, **12**–6,
        12–7, **12**–8
Support operations, **8**–9
Sustaining support phase, **12**–18 to **12**–19
Sweep algorithm, **3**–12, **3**–13
System analysis, **12**–16
System design
    characteristics, **12**–18
    development phase and, **12**–6 to **12**–7
    integration, **12**–16 to **12**–17
System element hierarchy, **12**–15
System engineering, **12**–7, **12**–9
System evaluation, **12**–16
System functional analysis and requirement allocation,
        12–13 to **12**–16
System life cycles, **12**–5, **12**–21 to **12**–22
    concurrent approach, **12**–6
    integrating function of, **12**–4 to **12**–18
    system engineering, **12**–7
System of rules, disjunctive, **3**–20, **3**–21
System operation, **12**–18 to **12**–19
    logistics support activities, **12**–3
    requirements, **12**–10 to **12**–12, **12**–21
System retirement, **12**–19
System synthesis, **12**–16
System TPM, **12**–13

T

T-shirts, **6**–24
Target (retail store), **2**–7
Taxes
   ad valorem, **2**–3
   operating expenses before, **4**–11
Technical performance measures (TPM), **12**–6, **12**–16
   allocation, **12**–17
   system, **12**–13
Temporary storage, **7**–18, **11**–18
The Logistics Institute (TLI), **4**–7
Tiger International, **10**–11
Time periods, **9**–5
TLC. *See* Total logistics cost (TLC)
Tool chest, **3**–1 to **3**–30
   algorithms complexity, **3**–13 to **3**–14
   fuzzy logic approach to dispatching, **3**–21 to **3**–28
      numerical example, **3**–28
      propose solution to, **3**–23 to **3**–27
      statement of problems, **3**–22 to **3**–23
   heuristic algorithms, **3**–7 to **3**–14
   mathematical programming, **3**–4 to **3**–7
   operation research, **3**–2 to **3**–4
   randomization optimization techniques, **3**–15 to
      3–18
Total logistics cost (TLC), **2**–3, **12**–15
Total Quality Management (TQM), **1**–8
Total system's approach, **12**–2 to **12**–3
TPM. *See* Technical performance measures (TPM)
TQM. *See* Total Quality Management (TQM)
Tracking accuracy, **4**–12
Traffic facilities location and layout design, **9**–12 to
   9–15
Trailer/tractor ratio, **4**–14
Trailer turns, **4**–14
Trailers, **4**–14
TRANSCOM, **1**–12
Transformation channels, **9**–7
Transformation facilities, **9**–7
Transportation
   case study, **10**–12
   characteristics, **10**–5 to **10**–6
   consolidation *vs.* operational frequency, **10**–9 to
      10–10
   costs, **2**–4 to **2**–6
      per unit, **4**–12
   deregulation, **1**–8, **2**–7
   dispatchers, **3**–23 to **3**–25
   domestic and international infrastructure, **10**–10
   freight demand forecasting, **10**–11
   global example, **10**–10 to **10**–11
   infrastructure, **10**–8 to **10**–10
   modes, **10**–5 to **10**–7
   moving people *vs.* moving goods, **10**–2 to **10**–4
   multi-mode, **10**–8
   oil pipeline, **2**–3, **2**–4, **2**–6

   rail, **2**–5
   request assigning vehicles, **3**–26
   selection, **10**–6 to **10**–8
   solutions, **10**–9
   systems, **10**–1 to **10**–13
Transportation Command (TRANSCOM), **1**–12
Travel
   aisle, **5**–10
   empty, **5**–12
   evaluation, **5**–15
   flow, **5**–12
   loaded, **5**–12
   network, **5**–10
   trip-based, **5**–17
Trends, **4**–12
Triangularity, **3**–9
Trip-based travel, **5**–17
Truck(s)
   cost, **5**–17, **7**–9, **7**–11, **8**–22, **10**–4, **11**–4
   dispatching, **3**–21 to **3**–28
   forklift, **7**–15, **8**–3, **8**–22, **9**–6
   intercity, **2**–3
   local, **2**–3
   miles traveled, **2**–7, **4**–14
   order-picking, **7**–12
   revenues, **2**–4
   in war, **1**–6
Trucking costs, **2**–4
Truckload trucking
   basic elements, **3**–18 to **3**–21
   fuzzy logic approach to dispatching in, **3**–21 to
      3–28
   numerical example, **3**–28
   problem statement, **3**–22 to **3**–23
   propose solution to, **3**–23 to **3**–27
Turnover analysis, **8**–20
Two-optimal tour heuristic algorithms, **3**–11 to **3**–12
Two-optimal tour(2-OPT) heuristic algorithms, **3**–11
   to **3**–12

U

U. S. Department of Commerce
   service sector, **11**–1
U. S. Department of Defense, **1**–8, **1**–13
U. S. Transportation Command, **1**–12
Unit load *vs.* inventory costs, **7**–5
Upper control limit (UCL), **4**–5, **4**–6

V

Value added economics, **4**–11
Value added services, **7**–18
Vehicle routing, **3**–4
Vehicle routing problem (VRP), **3**–12
Vietnam War, **1**–7
Visio, **5**–41

W

Wal-Mart, 2–7
  growth, 2–8
  RFID, 1–13 to 1–14
Warehouse Education and Research Council (WERC), 4–2
Warehouse functions, 7–18
Warehouse location problem, 9–12 to 9–13
Warehouse management system (WMS), 8–12, 8–14
Warehouse material-handling devices, 7–12
Warehousing, 8–1 to 8–22
  facility layout and flows, 8–16 to 8–18
  functional departments and flows, 8–3 to 8–8
  management, 8–14 to 8–15
  material handling, 7–17
  performance and cost analyses, 8–19 to 8–21
  product and order descriptions, 8–2 to 8–3
  sorting, packing, consolidation and staging, 8–13
  storage department descriptions and operations, 8–9 to 8–12

supply chain role, 8–2
Washington, George, 1–9 to 1–10
Water transportation, 2–5 to 2–6
WebLayout, 5–47 to 5–48
Window characteristics, 8–10
Wiring board, 4–12
With-in-the facility logistics, 11–3
WMS. *See* Warehouse management system (WMS)
Work space modified, 7–4
World trade
  growth, 2–1
World Trade Organization, 2–1, 2–2
World War I, 1–5
World War II, 1–5 to 1–6
Worldwide export volume, gross domestic product *vs.*, 2–2
Worse case analysis, 3–14

Z

Zero-inventory ordering policy, 6–2

Milton Keynes UK
Ingram Content Group UK Ltd.
UKHW051948071024
449327UK00026B/2214